D1288598

THE PINEAL GLAND

Editor

Russel J. Reiter, Ph.D.

SERIES OUTLINE

Volume I

Anatomy and Biochemistry

A survey of advances in pineal research. Structural and functional relationships in the nonmammalian pineal organ. Fluorescence histochemistry of the pineal gland. Functional morphology of sympathetic nerve fibers in the pineal gland of mammals. Ultrastructure of the mammalian pinealocyte. Pineal biochemistry: comparative aspects and circadian rhythms. General biochemistry of the pineal gland of mammals. Indole metabolism in the mammalian pineal gland. Pharmacology of the rat pineal gland. Hormone effects on the pineal gland. Methods for measuring pineal hormones.

Volume II

Reproductive Effects

The pineal and reproduction in fish, amphibians, and reptiles. The pineal and reproduction in birds. Reproductive effects of the pineal gland and pineal indoles in the Syrian hamster and the Albino rat. Pineal involvement in the photoperiodic control of reproduction and other functions in the Djungarian hamster *Phodopus Sungorus*. Melatonin levels as they relate to reproductive physiology. Arginine vasotocin and vertebrate reproduction. Other pineal peptides and related substances — physiological implications for reproductive biology. Potential sites of action of pineal hormones within the neuroendocrine-reproductive axis.

Volume III

Extra-Reproductive Effects

Circadian organization in fish, reptiles, and amphibians: role of the pineal gland. Circadian organization in birds and mammals: role of the pineal gland. Behavioral aspects of the pineal clock. Pineal as a biological clock. Pineal in thermoregulation. Pineal-thyroid and pineal-adrenal interactions. Consequences of the pineal and pinealectomy in the deer. Pineal and tumors: experimental and clinical aspects. Pineal methoxyindoles in the human. Psychiatric aspects of the pineal gland.

The Pineal Gland

Volume II
Reproductive Effects

Editor

Russel J. Reiter, Ph.D.

Professor
Department of Anatomy
The University of Texas
Health Science Center
San Antonio, Texas

CRC Press, Inc.
Boca Raton, Florida

Library of Congress Cataloging in Publication Data

Main entry under title:

The Pineal gland.

Bibliography: p.
Includes index.
CONTENTS: v. 1. Anatomy and biochemistry.
—v. 2. Reproductive effects.
1. Pineal body. I. Reiter, Russel J.
QP188.P55P553 v.2 599.01'42 80-15927
ISBN 0-8493-5713-6 (Series)
ISBN 0-8493-5716-0 (v.2)

Direct all inquiries to CRC Press, Inc., 2000 N.W. 24th Street, Boca Raton, Florida 33431.

International Standard Book Number 0-8493-5713-6

Library of Congress Card Number 80-15927
Printed in the United States

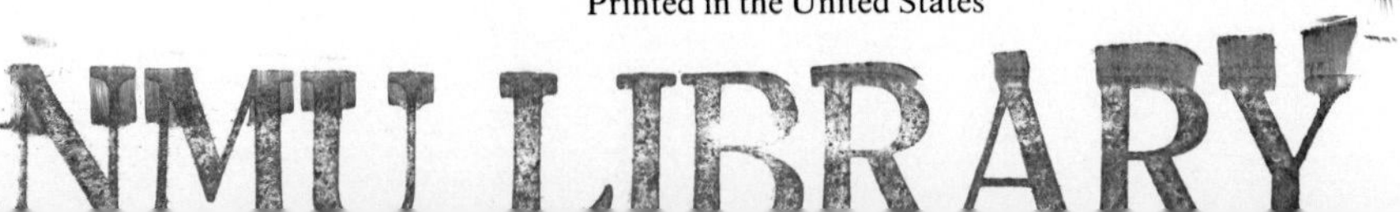

SERIES PREFACE

Until about 15 years ago the pineal gland, or epiphysis cerebri, was one of the few organs in the body to which no legitimate function had been assigned. Indeed, virtually every experimentalist regarded this organ as an embryological vestige of little functional import. Over the years many scientists had undertaken investigations into the physiology of this then enigmatic organ. Almost without exception, however, they rapidly became discouraged because of the negative, equivocal, or contradictory data that were obtained. These failures were primarily the result of a lack of knowledge concerning the regulatory processes of the gland. A pivotal finding was made about two decades ago when a potential pineal hormonal envoy was isolated and identified. This observation, coupled with the acquisition of knowledge that light and darkness are important in controlling pineal activity and the demonstration of a dramatic effect of the pineal on the reproductive physiology of a mammal, laid the foundation for subsequent productive investigations.

It is these successes, primarily garnered within the last 15 years, to which the present series entitled *The Pineal Gland* addresses itself. The series will include three volumes, the first being devoted to the anatomy and biochemistry of the gland, the second to the reproductive effects and the third to the extra-reproductive consequences of the pineal. The authors have been given great latitude in their approach to the topics they were asked to discuss. It was, however, requested that they be comprehensive in their presentation and objective in their appraisal of the published findings.

It is my hope that the three volumes in this series will usher in a new era of pineal research by providing the necessary background information to anyone entering the field and by serving as an up-to-date and comprehensive repository of information for all individuals already in the field. The pineal has now been linked, either directly or indirectly, to virtually every organ system in the body. For this reason the books also should be of widespread interest to individuals in a variety of research disciplines.

The editor is grateful to CRC Press for undertaking the publication of this new and important series. The educational benefit acquired by the editor already has made it a worthwhile venture for him.

Russel J. Reiter

THE EDITOR

Dr. Russel J. Reiter is Professor of Anatomy at The University of Texas Health Science Center at San Antonio, San Antonio, Texas. Dr. Reiter received his B.A. from St. John's University, Collegeville, Minnesota and his M.S. and Ph.D. from Bowman Gray School of Medicine at Wake Forest College, Winston-Salem, North Carolina. Dr. Reiter is a member of 18 professional and/or scientific organizations and has been the recipient of a United States Public Health Service Career Development Award. Dr. Reiter has published more than 230 research papers including 40 reviews and monographs. He has also written four books on the subject of the pineal gland and has edited three others. He serves on the Editorial Board of the following journals: *Neuroendocrinology, Endocrine Research Communications, Journal of Neural Transmission, Progress in Reproductive Biology, Endocrinology, Substance and Alcohol Actions/Misuse,* and *Proceedings of the Society of Experimental Biology and Medicine.* His current research interest include the biochemistry and the physiology of the mammalian pineal gland in terms of its effects on the reproductive system.

CONTRIBUTORS

Bryant Benson, Ph.D.
Professor and Head
Department of Anatomy
University of Arizona
Tucson, Arizona

David E. Blask, Ph.D., M.D.
Assistant Professor
Department of Anatomy
University of Arizona College of Medicine
Tucson, Arizona

Victor L. de Vlaming, Ph.D.
Visiting Assistant Professor
Department of Animal Physiology
University of California
Davis, California

Ietskina Ebels, D.Sc.
Senior Research Fellow
Department of Organic Chemistry
State University of Utrecht
Utrecht, The Netherlands

Klaus Hoffmann, Ph.D.
Research Fellow
Max-Planck-Institut fur Verhaltensphysiologie
Andechs
Federal Republic of Germany

Harry Lynch, Ph.D.
Lecturer
Laboratory of Neuroendocrine Regulation
Department of Nutrition & Food Science
Massachusetts Institute of Technology
Cambridge, Massachusetts

James M. Olcese, Ph.D.
Assistant Professor
Department of Biology
Southwestern at Memphis
Memphis, Tennessee

Charles L. Ralph, Ph.D.
Professor and Chairman
Department of Zoology and Entomology
Colorado State University
Fort Collins, Colorado

Mary K. Vaughan, Ph.D.
Associate Professor
The University of Texas Health Science Center
San Antonio, Texas

Richard J. Wurtman, M.D.
Professor and Director
Laboratory of Neuroendocrine Regulation
Department of Nutrition & Food Science
Massachusetts Institute of Technology
Cambridge, Massachusetts

TABLE OF CONTENTS

Chapter 1

THE PINEAL AND REPRODUCTION IN FISH, AMPHIBIANS, AND REPTILES

Victor de Vlaming and James Olcese

TABLE OF CONTENTS

I. INTRODUCTION

Reproduction of most temperate latitude ectothermic vertebrates is cyclic. Annual reproductive rhythms of many ectotherms are controlled or synchronized by seasonal day-length variations.[1-3] That the pineal complexes of most ectothermic vertebrates are located on the dorsal surface of the diencephalon and contain elements resembling retinal cones imply that these organs receive and perhaps convey photic information to other central nervous system (CNS) centers. Indeed, neural tracts do project from ectotherm pineal complexes and impulse activity in these tracts is affected by illumination (or its absence). As will be discussed below, there is relatively good evidence that ectotherm pineal organs also possess endocrine functions which are affected by light-dark (LD) cycles. These characteristics of the ectotherm pineal complex set the stage for suspecting that in some species it is involved in mediating photoperiod effects on reproductive processes.

The varigated morphology of pineal complexes among the vertebrates leads one to speculate that diverse functions were perhaps acquired through evolution. Little beyond preliminary morphological descriptions in a few species is known concerning the pineal complexes of most ectothermic groups. Comparative studies may provide a better understanding of the functions and evolution of this organ.

As Ralph[4] pointed out, deciphering the role of pineal organs among the vertebrates has been biased until recently by experiments designed to test for functions ascribed to the pineal glands of a limited number of mammalian species. Although a pineal organ-reproduction relationship appears well established in some mammalian species, such a relationship has not been thoroughly investigated among other vertebrate groups.

For optimal functioning, an organism's physiological processes must be temporally organized (daily and seasonally). This need is met by circadian and annual physiological rhythms. Most physiological and behavioral circadian rhythms can be entrained to light-dark cycles. Compelling evidence from studies with mammals and birds (endotherms) establishes the pineal gland as a link in one of the pathways by which photoperiod synchronizes circadian rhythms.[5-13] Functioning as a neuroendocrine transducer, the endotherm pineal gland converts light-dark cycle information into an output which contributes to the temporal organization of these physiological and behavioral rhythms. The output of the pineal gland may be integrated (centrally or peripherally) with output from other entraining pathways. With the possible exception of some avian species, the pineal organ does not appear to serve as the major self-sustained, driving oscillator (e.g., biological clock). Essential to note is that pineal complex output among ectotherms could be neural and/or hormonal since it is a sensory organ and probably a neuroendocrine organ (see below).

A pineal organ-annual reproductive rhythm relationship could well be related to or based upon the role of pineal organs in circadian organization and entrainment to light-dark cycles. Specifically, pineal organs perhaps convert photic environment information into an output which modulates the temporal organization of endocrine processes, including those involved in phasing reproductive events. Links between reproduction and circadian rhythms are being established. For example, some evidence has been gathered which indicates that annual physiological cycles depend upon circadian rhythms of blood hormone titers or changes in phase relationships of these oscillations.[14-16] Indeed, seasonal and/or photoperiod-dependent shifts of 24-hour physiological rhythms are documented among ectotherms.[17-24] Furthermore, photosexual responses, as well as other aspects of reproductive function in birds and mammals, have CNS circadian rhythms as a basis.[25-30]

II. AGNATHA

The pineal complex of cyclostomes is positioned on the dorsal diencephalon and consists of a parapineal organ situated slightly ventral to the epiphysis.[31-34] Both organs are saccular, the lumen being continuous with the third ventricle. Photoreceptor elements, resembling retinal cones, occur in both the parapineal and pineal organs; the outer segments of these sensory cells project into the lumen. Terminals of the sensory cells are associated with ganglion cells. The nerve tract from the epiphysis (pinealofugal) projects towards the posterior commissure, whereas the tract from the parapineal organ projects towards the habenular commissure.

Two different types of electrophysiological responses have been observed in pineal sense organs: (1) achromatic responses consisting of an inhibition of impulse activity in response to light stimuli of all wavelengths, and (2) chromatic responses consisting of an impulse frequency increase upon stimulation with medium and long wavelengths (excitatory component) and a decrease of impulse frequency upon stimulation with short wavelengths, including ultraviolet radiation (inhibitory component).[33-36] Similar neurophysiological characteristics, with minor variations depending on group and/or species, typify the pineal complexes of elasmobranchs, teleosts, anuran amphibians, and lacertilian reptiles. Therefore, electrophysiological data will generally not be presented in the following sections. One is prompted, nonetheless, to ask — Do these electrophysiological similarities imply a common pineal organ photosensory function among the ectotherms? To what uses, one wonders, is this sensory information put?

Electron microscope studies[31,33] indicate that the sensory cells of the lamprey pineal organ may also serve a secretory function. Melatonin precursors were identified by a fluorescence technique in the pineal and parapineal organs of *Lampetra planeri.*[37]

The activity of hydroxyindole-*O*-methyltransferase (HIOMT), the enzyme which converts *N*-acetylserotonin to melatonin in the pineal complex of larval *Geotria australis,* shows a 24-hr rhythm, peak activity during the dark phase.[38] There is an indication that the epiphyseal complex of *L. planeri* is involved with a circadian chronometric mechanism.[39-41] Larvae and adults of this species undergo a circadian color change which persists in intact animals maintained in darkness and in blinded animals exposed to a light-dark cycle. Removal of the pineal complex abolishes the rhythmic color response in larval *L. planeri;* both blinding and pineal removal are necessary to abolish the rhythm in adults. Administration of melatonin promotes color changes similar to those mediated by external lighting in intact *Lampetra* larvae.[40] According to Joss,[41] the pineal organ may not play a similar role among other lamprey genera.

Lampreys breed only once, generally in spring or early summer, and then die.[42] Presumably, temperature plays the predominant role in regulating gonadal growth among lampreys, with photoperiod having little significance.[43] Nonetheless, the work of Eddy[44] indicates that the pineal complex of *L. planeri* may be involved in mediating photoperiod effects; specifically, pinealectomy prevents metamorphosis, a process critically linked to season. Oocyte growth and development of secondary sex characteristics are slightly delayed (assessment during April) in *L. fluviatilis* pinealectomized at the outset of the autumn upstream spawning migration.[45] Also, spermatogenesis is delayed in pinealectomized males. Thus, within the genus *Lampetra,* available data suggest that the pineal complex is progonadal. Although neither season nor environmental conditions are mentioned, labeled melatonin is concentrated by the pineal complex and pituitary gland of larval *Geotria australis,* another lamprey, to a much greater extent than by other tissues.[38] The meaning of these data remains to be elucidated, but an effect on endocrine function is implied.

III. CHONDRICHTHYES

Among the few elasmobranchs examined, an epiphyseal stalk projects from the dorsal diencephalon to a skull cartilage concavity where it expands into a saccular end vesicle.[34,46,47] The pineal organ parenchyma consists of receptor cells, supporting (glial) cells, and ganglion cells. The outer segments of the photoreceptors project into the end-vesicle lumen. Ganglion cells, which are associated with the bases of sensory cells, may constitute the epiphyseal stalk (pineal nerve tract) which projects to the posterior commissure with a few fibers diverging to the habenular commissure. In the dogfish shark, dense-cored vesicles originate near the Golgi zone of pineal photoreceptor cells.[47] These vesicles accumulate in the basal process, connoting a secretory function for these sensory cells.

Annual reproductive cycles occur in the few viviparous species studied,[48,49] whereas oviparous species seem to have an extended egg-laying period.[50] We are aware of only one study of environmental influences on elasmobranch reproduction. In that study, Dobson and Dodd[51] note that temperature is the primary factor affecting the annual testicular cycle of *Scyliorhinus canicula.* Nonetheless, preliminary experiments suggest that pinealectomy of *S. canicula* produces a rise in pituitary ventral-lobe gonadotrophin content; injection of pineal extracts into pinealectomized fish reduces ventral lobe gonadotrophin content to nondetectable levels.[52] That the pineal organ of *S. canicula* affects peripheral functions is supported by the report of Wilson and Dodd.[53] Blinding abolishes the contraction of melanophores which occurs when the sharks are placed over a light background. Pinealectomy does not alter the background response, but obliterates the paling response evoked by exposure to darkness.

IV. OSTEICHTHYES

Sarcopterygii — The pineal complex of the coelancanth, *Latimeria chalumnae* (Crossopterygii) consists of two vesicles representing parapineal and pineal organs; both occur intracranially on the dorsal diencephalon and show a photoreceptor morphology.[54] A parapineal may be absent among adult dipnoans.[55]

Actinopterygii — Hill[56] examined morphologically the pineal complex of *Amia calva* (Holostei). We are not aware of any physiological studies dealing with the pineal organ of chondrostean fishes. The teleostei are by far the most extensively studied group of fishes.

Seasonal day length changes control, synchronize, or modulate, the reproductive cycles of many temperate latitude fishes.[1,57,58] Available data indicate that the photosexual response has a circadian rhythm basis.[59-62] Given that the pineal organ of teleosts is a photosensory organ possessing a circadian metabolic pattern which is affected by photoperiod (see below), a pineal organ-reproduction relationship is not unlikely.

Among adult teleostean fishes the parapineal is retained by a few species only. Thorough morphological description of teleostean pineal organs are available.[1,32,34,63] The pineal organ generally consists of an expanded, saccular, and well-vascularized end vesicle located closely beneath the parietal bones (sometimes occupying and attached to a bone concavity) and the hollow epiphyseal stalk connecting the pineal body with the posterio-dorsal diencephalon (Figures 1 and 2). The end vesicle is largely composed of three cell types: photosensory cells similar to the cones of the retina, supporting cells of an ependymal glial type, and sensory nerve (ganglion) cells (Figures 2 and 3). Outer segments of the sensory cells protrude into the end-vesicle lumen (Figures 2 and 3), whereas the basal segments of these cells apparently synapse with the sensory neurons. The pineal nerve tract consists in part or totally of the sensory neuron axons.

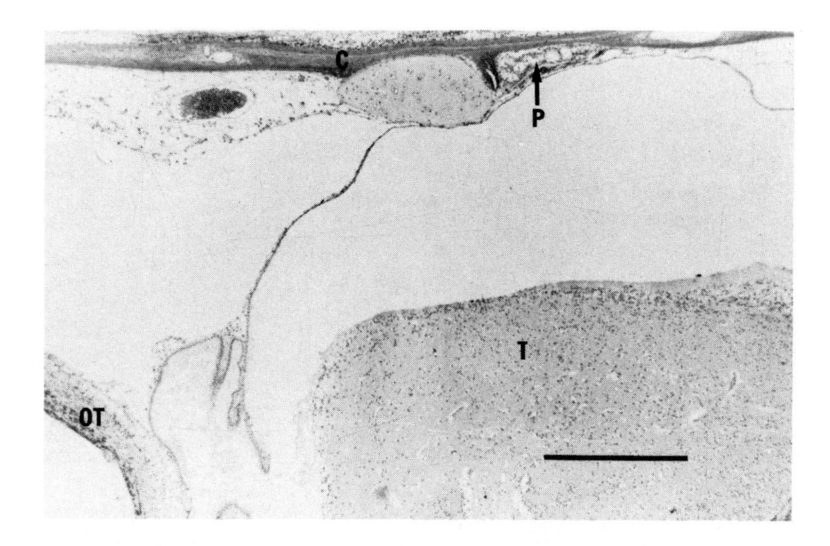

FIGURE 1. Sagittal section through the dorsal diencephalon region of the goldfish, *Carassius auratus*, brain showing the close apposition of the pineal organ (arrow) to the overlying cranium. OT-optic tectum; T-telencephalon. Bar indicates 0.5 mm. Courtesy of John McNulty.

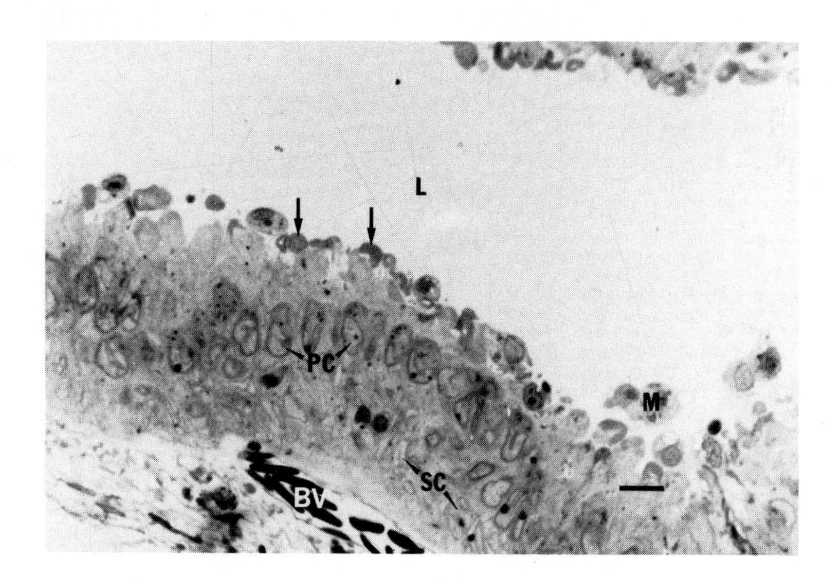

FIGURE 2. Light micrograph of *Carassius auratus* pineal organ showing the prominent central lumen (L) lined by outer segments of photoreceptor cells (arrows). Occasional macrophages (M) are seen within the lumen. Supportive-cell (SC) nuclei and photoreceptor-cell (PC) nuclei are easily identified. Bar indicates 10 μm. Courtesy of John McNulty.

Among this diverse vertebrate group, pineal organ morphology is rather variable. Most species studied have saccular, photoreceptor-rich pineal organs, whereas the epiphyses of some species are compact and photoreceptor poor. The relationship between age and pineal morphology has not been thoroughly studied. Above the end vesicle, the brain case is frequently depigmented or has less pigmentation than other skull areas.

Ultrastructure studies connote secretory characteristics of both sensory and support-

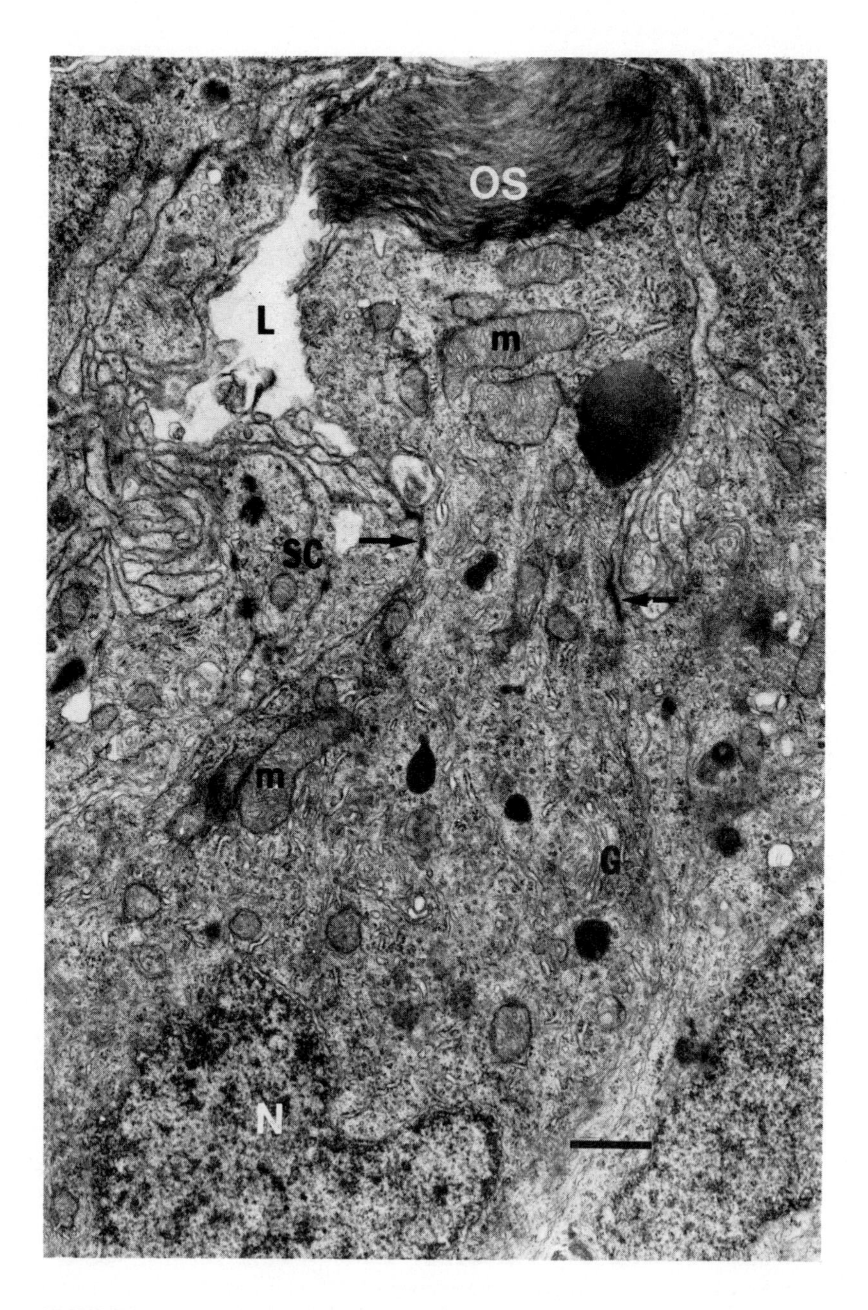

FIGURE 3. Low magnification micrograph of photoreceptor cell of *Carassius auratus* pineal organ with its basal nucleus (N), constricted neck region, and expanded inner segment which projects into the central lumen (L). Part of the outer segment (OS) of the cell is also seen. Supportive cells (SC) border the photoreceptor cell forming junctional complexes at the neck region (arrows). M-mitochondria; G-Golgi. Bar indicates 1 μm. Courtesy of John McNulty.

ing cells (Figure 4).[64-66] A light microscopy investigation of the *Symphodus melops* pineal organ reveals an accumulation of aldehyde fuchsin granules during the day with a depletion of these granules during the dark phase.[67] Although Chèze[67] interprets these data as reflecting a diurnal fluctuation of secretion, we believe that the variation was in glycogen content. Urasaki[68] proposes that pineal organ supporting cells are much more prevalent in *Oryzias latipes* exposed to a LD 10:14 than to a LD 14:10 regime.

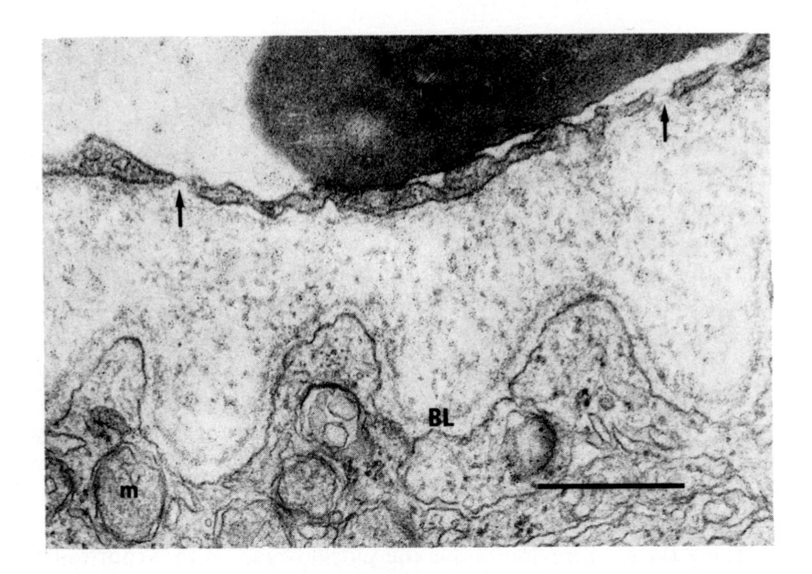

FIGURE 4. Part of a blood vessel bordering the pineal organ of *Carassius auratus*. The capillary endothelial cell is fenestrated (arrows), lending support to the proposed secretory role of the teleostean pineal organ. Bar indicates 0.5 μm. Courtesy of John McNulty.

Hafeez et al.[69] examined the effects of environmental illumination differences on nuclear morphometrics of pineal organ receptor and supporting cells in larval rainbow trout, *Salmo gairdneri.* Receptor and supporting cell activity (using nuclear and nucleolar diameters as an index) appear greatest in fish exposed to continuous darkness; the morphological data implies that supporting cell activity is lowest in animals maintained under continuous lighting. Blinding and pineal masking indicate that receptor cells are affected by incident light on the pineal organ skull region, but not by photic influences via the lateral eyes. Conversely, supporting cells are influenced by information received via the lateral eyes only. That blinding alters pineal organ activity in *S. gairdneri* is indicated also by the data of Smith and Weber (see below).[70] Hafeez et al.[69] suggest that the trout pineal organ could serve as a comparator, since it is affected by photic information received directly and via the lateral eyes.

Receptor cells of the teleostean pineal organ consistently reveal a normal cyclic breakdown of the outer segments. Hafeez and Zerihun[71] hypothesize that release of bioactive metabolites into the pineal lumen, and hence cerebrospinal fluid of the third ventricle, could accompany this process. Photoperiod conditions and season are seldom reported by investigators of pineal organ morphology. Ultrastructural examination of potential seasonal, diurnal, and photoperiod-dependent variations of pineal organ morphology could prove informative.

Understanding the pineal organ photosensory function could be enhanced by locating the CNS termination sites of epiphyseal sensory neurons. The central projections and terminations of the pineal nerve in *S. gairdneri* were examined by Hafeez and Zerihun[72] using cobalt chloride iontophoresis. Only pinealofugal fibers were detected, these projecting over an extensive area of the di- and mesencephalon. The possibility of direct innervation of the hypothalamic preoptic nucleus was suggested. This[72] and other studies provide ample morphological evidence for communications and interactions between the pineal organ and other brain regions.

Pinealopetal innervation and/or the existence of interneurons within the teleostean end vesicle are implied by some studies.[65,73-76] The occurrence of pinealopetal nerve fibers within the epiphyseal stalk of *S. gairdneri* was denied by Hafeez and Zerihun.[72] Innervation of the end-vesicle cells could come, however, via another route.

Although undoubtedly a sensory organ, an unequivocal ultrastructural definition of pineal organ secretory activity has yet to be achieved (see above). Neuroendocrine functions, nonetheless, have been attributed to the teleostean pineal organ based upon biochemical and fluorescence histochemical evidence of an indoleamine metabolism in its parenchymal cells. Both ependymal supporting and receptor cells contain and/or actively incorporate 5-hydroxytryptophan and serotonin;[71,77-79] the metabolic derivative of serotonin, melatonin, also occurs in the teleostean pineal organ.[80,81] HIOMT activity has been demonstrated in the pineal organ of some teleosts.[82] A diurnal fluctuation of pineal HIOMT activity in juvenile steelhead trout, *S. gairdneri,* maintained outdoors in Oregon during June and July, was reported.[83] Peak HIOMT activity occurs 6 to 7 hr after onset of darkness, the nadir about 1 hr before the onset of darkness. With steelhead trout parr, the diurnal variation of pineal HIOMT activity is of greater amplitude in fish exposed to a LD 8:16 regime than in animals maintained on a LD 16:8 regime.[70] Additionally, HIOMT activity of pineals collected during the dark phase of the photoregimes is much greater in trout exposed to a LD 8:16 as compared to a LD 16:8 regime. These data clearly illustrate a photoperiod effect on pineal organ activity. Capping the skull region above the pineal organ of the trout fails to abolish the diurnal fluctuation of pineal HIOMT activity.[70] Bilateral enucleation or exposure of the trout to constant darkness abolishes the pineal HIOMT activity fluctuation. Unfortunately, adequate controls were not included in the blinding experiment nor was the photoregime reported. Furthermore, pineal HIOMT activity in the blinded fish was equivalent to light phase activity levels. Background color of the trout's environment affects both the diurnal fluctuation and absolute level of pineal HIOMT activity.[84]

Plasma melatonin levels (using a radioimmunoassay) during the dark phase are double those during the light phase in *Salmo gairdneri* maintained on a LD 12:12 photoperiod.[85] Scotophase, but not photophase melatonin levels are lowered in pinealectomized trout although the nocturnal elevation of plasma melatonin still occurs.[86] Sectioning trout optic nerves does not extinguish the day-night variation of plasma melatonin titers.[86] These results reveal that the pineal organ is not the sole source of melatonin. Indeed, organ cultured trout retinae synthesize melatonin from serotonin.[87] Gern's findings lend credence to the theory that the teleostean pineal organ functions in a neuroendocrine capacity. Of note also is that both radioimmunoassay and bioassay argue for the presence of arginine vasotocin in the teleostean pineal organ.[88]

According to our contention, a pineal-reproduction relationship could rely upon an involvement of this organ in temporal organization of neuroendocrine and/or hormonal rhythms. Indeed, in *S. fontinalis* the entrainment of circadian activity rhythms to light-dark cycles involves the pineal organ.[89] Furthermore, pinealectomy disrupts (i.e., abolishes or phase shifts) diurnal variations of plasma and liver metabolites as well as some hypothalamic and hormonal cycles.[23,90-93] That the pineal organ contributes significantly to controlling circadian organization and rhythmicity of physiological and behavioral events among teleosts has been convincingly demonstrated by the investigations of Kavaliers.[94-100] His findings indicate that the pineal organ is not the sole source of circadian organization, but it may serve as a synchronizer or coupler of endogenous, self-sustained oscillators.

Pineal-produced melatonin could be a mediator of this organ's actions upon circadian rhythms. For example, melatonin administration decreases locomotor activity of juvenile sockeye salmon, *Oncorhynchus nerka,* only during the light phase of a LD 12:12 regime.[101] Furthermore, melatonin is extremely effective in inducing melanosome aggregation in both dermal and epidermal melanophores of the siluroid catfish, *Parasilurus asotus.*[102] Precursors to melatonin are without effect, but some of its possible metabolites show a moderate aggregating action. The thorough study of Fujii and

Miyashita,[102] as well as other investigators (see bibliography of their paper), provide substantial evidence that melatonin possesses peripheral actions in several teleostean fishes.

Appearance of secondary sexual characteristics is delayed in guppies, *Lebistes reticulatus,* fed dessicated bull pineals.[103] Pinealectomy of males of the same species leads to a slight acceleration of testicular development.[104] A pineal organ-reproductive system relationship was not demonstrated in studies of a few other teleostean species.[105-107]

In the medaka, *Oryzias latipes,* reproductive effects of long photoperiod (LD 14:10) exposure vary seasonally; ovarian enlargement is stimulated by long photoperiods most dramatically during May and June.[108] During these 2 months, blinding retards the galvanizing effects of a long photoperiod on ovarian size. Medaka pinealectomized in late February and exposed to continuous light or natural day lengths in Japan until early May have smaller ovaries than intact fish maintained under the same regimes.[109] Therefore, the medaka pineal organ appears to be a component in a pathway by which long photoperiods promote gonadal recrudescence.[109-112] Pinealectomy of ripe female medaka, subsequently maintained on a LD 14:10, 26°C regime, does not inhibit ovulation or spawning when assessed up to 2 weeks postoperative.[113] This observation argues for a role of the pineal organ prior to oocyte maturation and ovulation.

Urasaki[108] contends that the medaka pineal organ is an essential component in the transmission of photic stimuli from the eyes to the gonads. However, blinded-pinealectomized fish have smaller ovaries than blinded fish exposed to a LD 14:10 photoperiod during May.[108] These data suggest to us that at certain times of the year, photic information received via the eyes and pineal organ may stimulate outputs which are additive (using gonadal size as an index) at some CNS level. Yet another possibility is that the medaka pineal organ serves as a comparator of photic information received by different receptors. Urasaki's[68,108,110-112] investigations indicate that the pineal organ makes a slightly greater contribution than the eyes with regard to the gonadal stimulatory effects of long photoperiods.

Medaka obtained during mid-October were blinded, pinealectomized, sham-pinealectomized, or blinded-pinealectomized and exposed to continuous darkness or the natural declining day length in Japan.[110] Under both lighting regimes, mean ovarian gonosomatic index (GSI) in the groups had the following rank: sham-pinealectomized < blinded < pinealectomized < blinded-pinealectomized. Thus, both the pineal organ and eyes seem to be involved in suppressing gonadal activity in medaka maintained on a short photoperiod or in continuous darkness. Moreover, the greater the deprivation of photoreceptors which could mediate the effects of prolonged dark phases, the less the gonadal inhibition. A working interpretation of these data is that both retinae and pineal organ secrete melatonin under short photoperiod and continuous-darkness conditions. If melatonin is inhibitory to gonadal activity, such an interpretation would seem tenable.

Melatonin administration inhibits oocyte development and alters pituitary gonadotroph cytology in medaka exposed to a LD 14:10, 26°C regime during March.[114] During winter, ovarian size is larger in controls than in pinealectomized fish treated with melatonin and exposed to a LD 16:8 regime, but not a LD 8:16 regime.[111] Such results could be obtained if short photoperiods evoke pineal organ and retinal melatonin production. In vitro experiments demonstrate that melatonin at relatively low doses (0.1 μg/mℓ) slightly represses, but did not totally inhibit, medaka oocyte maturation.[113]

During the period of December to April with medaka exposed to photoregimes of 8 to 14 hr of light per day, ovarian weights of blinded fish were greater than in intacts.[68,108,111] These results connote an inhibitory role for the eyes similar to that noted in autumn, but being photoperiod independent. Even though we find these results difficult to interpret, they do confirm further that the mechanisms involved in mediating and integrating photic environment information are rather complex. Of particular note is that melatonin treatment reverses the stimulatory effects of blinding on ovarian GSI in fish exposed to a LD 8:16 regime during winter.[111] Some of the seemingly variable results obtained throughout Urasaki's studies could be related to the lack of consistency with regards to temperature conditions.

In the cyprinid teleost, *Notemigonus crysoleucas,* the effects of pinealectomy vary depending on the phase of the natural reproductive cycle when the organ is removed, as well as with the photoperiod-temperature regime under which the experimental animals are maintained.[115] Long photoperiod-warm temperature regimes accelerate gonadal maturation in this species, whereas short photoperiod-warm temperature regimes induce gonadal regression. Pinealectomy blocks the stimulatory effects of a LD 15.5:8.5, 25°C regime on *Notemigonus* ovarian activity during the preparatory (January), prespawning (March to April) and early spawning (May) periods; similar results are obtained with males, except that epiphysectomy during May has no obvious effects on testicular activity. During the prespawning and spawning seasons, pinealectomy retards the gonadal regressive effects of a LD 9:15, 25°C regime in both male and female *Notemigonus.* Pinealectomy does not ameliorate the repressive effects of this regime during the preparatory period. No pronounced effects of pinealectomy on gonadal activity are noted in fish exposed to LD 15.5:8.5 or LD 9:15 regimes at 12° to 15°C between January to June.

Data presented by de Vlaming and Vodicnik[91] imply that in *Notemigonus,* the pineal organ exerts its effects on reproduction, at least in part, via the hypothalamus and pituitary. An experiment conducted during the fall with fish exposed to a LD 15.5:8.5 or a LD 9:15 photoperiod at 15° or 25°C reveals that pinealectomy alters pituitary gonadotrophin cells under all four regimes. Pinealectomy, in this experiment, also repressed ovarian activity in fish maintained on the long photoperiod regimes. Although their assay is somewhat suspect, de Vlaming and Vodicnik[91] report that epiphysectomy results in changes of pituitary gonadotrophin content, especially in photoperiod groups of *Notemigonus* maintained at 15°C during February. The diurnal variation of pituitary gonadotrophin content observed in sham operated fish exposed to a LD 15.5:8.5, 15°C regime during early April is extinguished by pinealectomy. Hypothalami taken from pinealectomized fish exposed to a LD 15.5:8.5, 25°C regime during late April presumably contain less gonadotrophin-releasing activity than control tissues.[91] On the other hand, pinealectomy results in an elevation of hypothalamic gonadotrophin releasing activity in fish exposed to a LD 9:15, 25°C regime. Apparently the pineal organ of *N. crysoleucas* is a component in the pathway(s) by which photoperiod affects reproductive mechanisms. Removal of this organ can be stimulatory or inhibitory to gonadal activity depending upon photoperiod-temperature conditions.

Goldfish, *Carassius auratus,* taken from winter conditions of light (i.e., LD 8:16) and exposed to various photoregimes at 20°C during different times of the year show gonadal responses to photoperiods that are seasonally variable.[116] Gonadal enlargement is stimulated by longer photoperiods only during spring. Pinealectomized goldfish exposed to an LD 8:16, 13°C regime between 9 January and 3 May possess larger gonads than controls, indicating an antigonadal effect of this organ during the prespawning period.[116] Gonadal size of pinealectomized and sham-operated fish do not differ in fish exposed to short photoperiod conditions at 20°C during June to August, October to December and April to May. The repressive effects of the goldfish pineal organ under short photoperiod conditions during spring has been confirmed by de Vlaming and Vodicnik.[117] Specifically, in goldfish exposed to an LD 8:16 regime at 20 to 24°C during April or May, pinealectomy reverses the inhibitory effects of the photoregime on gonadal activity. Pituitary, but not plasma, gonadotrophin levels (sampled

4 hr after the onset of the light phase) as assayed by RIA are significantly greater in pinealectomized than sham-operated goldfish exposed for 21 days to a LD 8:16, 20°C regime during March.[92]

According to Hontela and Peter[93] ovarian GSI does not differ among blinded, pinealectomized, blinded-pinealectomized, and sham-operated goldfish held for a short term on a LD 8:16, 21°C regime during mid-March. However, goldfish undergoing ovarian recrudescence and sexually mature females exhibit diurnal fluctuations of serum gonadotrophin levels.[118] Since these fluctuations are altered by photoperiod, Hontela and Peter[93] examined the effects of pinealectomy and blinding on serum gonadotrophin levels (assayed by RIA) in goldfish at two different times during the day, as well as at various times during the year and under different photoperiod conditions. Although differences were not noted in serum gonadotrophin levels at the two sample times in sham operated and blinded goldfish exposed to the LD 8:16, 21°C regime during mid-March, differences were recorded in the pinealectomized and blinded-pinealectomized groups. These data imply, but do not prove, that pinealectomy under these conditions results in the establishment of a diurnal serum gonadotrophin fluctuation absent from the controls. One can also glean from the results presented by Hontela and Peter[93] that an experimental treatment could alter the circadian rhythm of a given hormone without causing a significant difference (compared to a control group) in the level of the hormone at a single sample time.

Pinealectomy seemingly has no effect on gonadal activity in goldfish maintained on a LD 8:16, 20°C regime during June, February, or March.[117] Neither pinealectomy, blinding, or exposure to continuous darkness has a pronounced effect on gonadal size during the October to December period when goldfish are held under seminatural decreasing or short photoperiod conditions at 6 to 13°C.[119,120] Vodicnik et al.[92] found that pinealectomy has no effect on pituitary and serum gonadotrophin levels throughout a 24-hr period or on ovarian GSI in goldfish maintained on a LD 15.5:8.5, 24°C regime during November. These findings have been confirmed with goldfish exposed to a LD 16:8, 21°C regime during November; a diurnal fluctuation of serum gonadotrophin levels, however, occurs in blinded fish.[93] Serum estradiol-17β levels are lowered in blinded, but not pinealectomized goldfish exposed to a seminatural decreasing photoperiod at 13°C during October to November.[120]

According to Fenwick,[116] pinealectomy has no obvious effects on gonadal size in goldfish maintained on long photoperiods at 20°C during June to August, October to December, and April to May. On the contrary, de Vlaming and Vodicnik[117] submit that pinealectomy of goldfish during February, May, or June blocks the stimulatory effects on the gonads of a LD 16:8 regime (20 to 24°C). Ovaries of epiphysectomized fish are smaller than those of fish maintained under continuous darkness in June. Vodicnik et al.[119] reported that exposure of goldfish to continuous darkness results in a greater ovarian inhibition than does pinealectomy of fish exposed to a seminatural increasing photoperiod during May (experiments conducted simultaneously at 20°C). These results lead one to consider the possibility that the eyes and pineal organ have an additive effect with regard to the stimulatory effects of long photoperiods on gonadal activity. Pertinent here is the observation that the phototactic response of the goldfish depends upon the presence of both the pineal organ and the eyes.[121] Pinealectomy and blinding are equally effective, however, at inducing ovarian regression in goldfish exposed to a seminatural increasing photoperiod during March; gonadal condition in the blinded and pinealectomized groups is equivalent to that of fish maintained under continuous darkness.[119]

Blinding alone has no effects on ovarian GSI in experiments conducted during November, mid-March, and late-April with goldfish held on LD 16:8, 21°C regimes.[93] In mid-March, ovarian GSI is smaller in pinealectomized and blinded-pinealectomized animals than in sham-operated fish. Hontela and Peter's[93] data indicate that pinealectomy, and possibly blinding, repress or phase shift the serum gonadotrophin daily cycle recorded in control fish exposed to a LD 16:8, 21°C regime during mid-March and late-April. Vodicnik et al.[92] claimed that pituitary gonadotrophin levels of pinealectomized goldfish are greater than in controls when samples are taken at 10 hr, but not at 4 hr, after the onset of the light phase of a LD 15.5:8.5, 22°C regime (March experiment); plasma gonadotrophin levels, however, are lower in pinealectomized fish than in controls only at the earlier sampling time. Perhaps the pineal organ influences gonadal development through, at least in part, alteration of the daily cycles of serum gonadotrophin levels.

Ovarian size is smaller in both epiphysectomized and blinded as compared to control goldfish exposed to a seminatural increasing photoregime at 20°C during March; serum estrogen levels, however, are lowered only in blinded fish.[120] Delahunty et al.[120] contended that pinealectomy has no effect on ovarian GSI or serum estradiol-17β levels in goldfish maintained under a seminatural increasing photoperiod (18°C) during April to May or under a LD 16:8, 12°C regime during June. These data agree with those of Fenwick,[116] but not with other reports from de Vlaming's laboratory. The potential role of the goldfish pineal organ under long photoperiod conditions remains to be elucidated, yet we ascribe the results presented by Delahunty et al.[120] to a more careful pinealectomy procedure.

Fenwick[80] held goldfish on a LD 16:8, 20°C regime during the spring, one group receiving 20-μg melatonin injections for 50 days. Melatonin retarded the increase in gonadal size which occurred in the controls, but does not induce gonadal regression. Histological examination confirmed that all spermatogenic and oogenic stages were present in the gonads of melatonin treated fish, implying a reduced rate of development. Examination of pituitary cytology indicated that melatonin reduced gonadotrophin secretion.

Although the results summarized above are not always consistent, they do intimate that the goldfish pineal organ is a component in the pathway(s) through which photoperiod affects gonadal activity. Output from the pineal organ may modulate gonadotrophin secretion rhythms. Evidence points to a pro- and antigonadal role of the goldfish pineal organ, the effects of pinealectomy depending on time of year and temperature-photoperiod conditions. Regardless of the photoperiod-temperature regime, most of the effects of pinealectomy are observed in spring concomitant with the rapid phase of gonadal recrudescence. A hypothesis worth considering is that one role of the pineal of this species may be to reduce the rate of gonadal recrudescence during winter when day length is short, but nonetheless, increasing. That the small size of goldfish gonads during late summer and autumn relates to repressive effects of the pineal organ is not supported by existing data.

Hypothalamic serotonin, as well as other monoamines, are implicated in the control of endocrine rhythms.[122,123] A hypothalamic serotonergic mechanism apparently is a component of photosexual responses among some birds and mammals.[124] In the ferret, a hypothalamic serotonin circadian rhythm is concerned with transmitting day-length information.[125] Furthermore, this hypothalamic serotonin rhythm may be driven by melatonin.[125] Hypothalamic monoamine oxidase (MAO) is a component in the serotonergic control of the goldfish pituitary.[126] Activity and the daily variation of hypothalamic MAO in the goldfish is regulated, in part, by photoperiod.[23,127] Studies by Olcese and de Vlaming[23] illustrated that pinealectomy alters hypothalamic MAO activity of goldfish exposed to a short, but not long photoperiod. Photoperiod also affects the day-night fluctuation of hypothalamic serotonin activity (turnover) in the goldfish.[128] Unpublished data from our laboratory (Olcese and de Vlaming) show that the hypothalamic serotonin activity day-night fluctuation is disturbed by pinealectomy, and that administration of melatonin to pinealectomized and intact fish held on a LD 16:8, 20°C regime evoked a 180° shift in the serotonin activity day-night variation. Reproductive effects of the pineal organ could involve modulation of hypothalamic monoaminergic mechanisms.

In addition to the studies summarized above, data are available which illustrate that melatonin retards gonadal development or induces gonadal regression. The effects of melatonin administration (20-day treatment) 6 to 8 hr after the onset of the light phase in a LD 12:12 photoperiod were evaluated in female catfish (*Heteropneustes fossilis*) during the prespawning and spawning periods.[129] The ovarian cycle of this fish is partially synchronized by photoperiod. In the prespawning period melatonin treatment inhibits vitellogenesis, induces follicular atresia, and results in a reduction in the number of pituitary gonadotrophs. Ovarian regression and a lowering of pituitary gonadotroph number occurs in melatonin-treated catfish during the spawning season. In another Asian catfish, *Mystus tengara,* melatonin injection 2 hr after light phase onset (LD 12:12) during the spring rapid-ovarian-recrudescence phase arrests vitellogenesis and increases the frequency of atretic oocytes.[130] Seawater *Fundulus similis,* collected during January along the Texas coast, were exposed to a LD 13:11 or a LD 10:14 photoperiod; subgroups under each photoperiod regime were treated with relatively low doses of melatonin.[131] Ovarian and testicular GSIs are smaller in melatonin treated as compared to control fish on a LD 13:11 regime, but not on a LD 10:14 regime. Should melatonin occur in the plasma of these fish and the levels be higher under short photoperiod conditions, such results could be interpreted without difficulty. During May, melatonin treatment inhibits testicular enlargement in *F. similis* exposed to either a LD 13:11 or LD 10:14 regime. These results show that the effects of melatonin administration can be photoperiod dependent, and that the gonadal effects of this indoleamine at a given photoperiod can vary with time of year.

It would not seem unreasonable, in view of the studies summarized above, to suspect that in some species pineal and/or retinal-produced melatonin mediates the gonadal regressive effects of declining day length during late summer, the inhibitory effects of short day lengths during fall and winter and/or retards rapid gonadal growth in winter when day length is increasing, yet still short.

In attempting to gain an overview, one of the most striking aspects of the literature dealing with teleostean fishes is that pinealectomy and/or blinding can, in the same species, be stimulatory, inhibitory, or have no effect on gonadal activity. The response to pinealectomy or blinding depends on time of year and photoperiod-temperature conditions. With regard to the pineal, available data point to an inhibitory function on the gonads under short photoperiod conditions and a stimulatory function on the reproductive system under long photoperiod conditions; both of these responses generally can be demonstrated during late winter to early spring only. Monitoring other (nonreproductive) physiological or endocrine parameters reveals that the effects of pinealectomy or melatonin treatment among teleostean fishes are generally photoperiod-dependent.[18,90,131,134] Combined, all of these data provide compelling testimony for the pineal organ functioning as a component in the pathway(s) by which seasonal day-length variations control or synchronize annual physiological cycles.

V. AMPHIBIA

The pineal complex of anurans consists of the epiphysis cerebri, an intracranial evagination of the dorsal diencephalon, and the frontal organ, an extracranial outgrowth of the epiphysis. In urodeles and caecilians the frontal organ is lacking. The epiphysis and frontal organ possess photosensory elements that resemble retinal cone cells in the structure of the outer segments.[32,34] Hartwig and Baumann[135] demonstrated the presence of a photopigment in the epiphysis of *Rana temporaria,* whereas a different photolabile substance occurs in the frontal organ of *R. catesbiana.* Other studies reveal the presence of serotonin, or related indoleamines, in the photoreceptor cells of the amphibian epiphysis[32,138] or in the epiphysis as a whole.[137,138] Many of the outer segments in the amphibian epiphysis project horizontally into the lumen, which is continuous with the third ventricle.[34,139]

Further evidence in favor of a photoreceptor role for the amphibian pineal complex derives from studies regarding pineal innervation.[34,35] In anurans, both components of the pineal complex emit neurons to, as well as receive neurons from, the brain. Nerve tracts from the frontal organs of several ranid species can be traced into various midbrain regions.[35,140] Pinealopetal fibers from the limbic and visual centers of the diencephalon-mesencephalon areas course into the amphibian pineal complex.[139,141] In the toad, *Bufo arenarum,* thick bundles of catecholaminergic neurons connect the pineal organ with the posterior commissure.[142] Since experimental sympathectomy has no effect while transection of the pineal nerve tract causes the loss of pineal adrenergic fluorescence, it can be concluded that these neurons are pinealopetal, but probably do not arise from the superior cervical ganglia (as in mammals).

Little research has focused on the biochemistry of the amphibian pineal complex. HIOMT was first shown to be present in the amphibian brain by Quay.[143] Interestingly, retinal HIOMT activity is substantially greater than pineal HIOMT activity in amphibians, when tissues are collected in the middle of the light phase during the winter months. That retinal HIOMT activity exceeds that found in the pineal organ is confirmed in studies with *R. pipiens,* sampled in the middle of the dark phase (on a LD 12:12 regime).[144] Daily variations of both retinal and pineal HIOMT are abolished by exposure to constant light or darkness. Furthermore, the daily variations of HIOMT activity in a LD 12:12 regime are extinguished by maintaining animals at a constant temperature. This is a noteworthy finding, for it indicates that the observed HIOMT variations are primarily a function of temperature fluctuations.

Both serotonin and melatonin occur in the anuran pineal organ.[32,136,145] Recently, an investigation with *Ambystoma tigrinum* demonstrated a typical rhythm of plasma melatonin during a LD 12:12 regime (i.e., scotophase levels higher than photophase levels) which is abolished by pinealectomy.[146] However, this pinealectomy effect is due entirely to lowered scotophase levels; photophase levels are unaffected. A tenable hypothesis is that the night time elevations of plasma melatonin are pineal-dependent; since pineal ablation does not eliminate the presence of circulating melatonin, it seems reasonable to postulate a nonpineal source of melatonin, the retina being the most likely candidate.

A role for the amphibian pineal complex in rhythmic phenomena and certain photoperiodic responses seems likely. For example, it has been known for some time that larval amphibians blanch when placed in darkness. Whereas this response is abolished by pinealectomy,[147,148] neither removal of the frontal organ, blinding, or hypophysectomy interferes with the blanching response of young larvae,[149] indicating that the pineal organ acts via humoral pathways to stimulate dermal melanophore contractions. Melatonin is a powerful melanophore-contracting agent in larval amphibians. Light-driven daily variations in the mitotic rate of ventral tailfin epidermal cells are also pineal-dependent in *Xenopus* larvae.[150]

Activity rhythms of numerous amphibian species are photoperiod-entrained.[151] Adler[151,152] and others[153] offered convincing evidence that the amphibian pineal complex is an extraocular photoreceptor which perceives light cues utilized for control of circadian locomotor rhythms. However, whether it is the only extraocular photoreceptor remains to be determined. By influencing activity rhythms, the pineal complex could affect the thermal conditions of the animal. In view of the importance of temperature in the amphibian reproductive cycle,[2] such an influence of the pineal complex on activity rhythms could have important consequences for gonadal development.

The amphibian pineal complex has the prerequisite neurophysiological characteristics to serve as an illumination "dosimeter" for the regulation of body temperature. Since exposure to sunlight typically results in a greater heat absorption by the animal, the monitoring of illumination intensities and duration by the pineal complex may play an important thermoregulatory function. This view was initially proposed by Steyn[154] and is vigorously advocated by Ralph.[155,156] Recently, further support for a thermoregulatory role of the amphibian pineal complex was provided.[157] Administration of melatonin (4 mg/kg body weight, given at midphotophase of a LD 12:12 regime) to *Necturus maculosus* results in the abolition of the daily behavioral thermoregulatory cycle. Furthermore, melatonin significantly decreases the selected temperatures of salamanders in a linear thermal gradient.

That melatonin and the pineal complex can affect various behaviors implies a central site of action. Removal of the frontal organ (with or without the pineal organ) presumably increases synthesis of neurosecretory material in the preoptic nucleus, regardless of photoperiod conditions.[158] Since the preoptic nucleus is an important gonadotrophic control area, such evidence implies a functional relationship between the pineal complex and the pituitary-gonadal axis.

Both photoperiod and temperature appear to influence gonadal function in amphibians. According to Mazzi[159] the temperature influence is channeled through the hypothalamus. Photoperiod effects on the spermatogenic cycle of *Plethodon* (a urodele)[160] and *Rana* (an anuran)[161] have been reported, although it seems probable that photoperiod acts only to facilitate the action of temperatures.

The number of studies concerned directly with the elucidation of pineal-gonadal interactions in amphibians is limited. Pinealectomy is without effect on ovarian maturation in larval *Alytes obstetricians.*[162] In *Bufo melanostictus,* pinealectomy and blinding both act to decrease spermatogenesis in a manner similar to constant darkness exposure.[163] However, steroidogenic activity in the Leydig cells (based on nuclear morphology) presumably is increased by pinealectomy, while being inhibited by removal of the lateral eyes. Melatonin treatment (1 μg/g body weight for 5 days) reverses the steroidogenic effects of pinealectomy.[163] Unfortunately, in many of the above studies, exact photoperiod and temperature conditions, as well as other variables (e.g., time of year, injection time for melatonin, etc.) were not specified. Nonetheless, it appears that photoperiod effects, at least in *Bufo,* may be mediated through both the eyes and the pineal complex. In *Rana esculenta,* spermatogenesis is stimulated by injections of human chorionic gonadotrophin and this stimulation can be blocked by injections of bovine pineal homogenates (environmental conditions not given), but not by melatonin.[164,165] These studies however did not consider the issue of melatonin injection time (relative to the LD cycle), an important variable for demonstrating melatonin effects in fish, birds, and mammals.[90,166,167]

Pinealectomized and blinded male *Rana esculenta* show equally impaired testicular responsiveness to stimulating photoperiod and temperature conditions (LD 12:12 or LD 24:0 at 17°C in winter).[161] Furthermore, pituitary gonadotrophs in pinealectomized and in blinded frogs fail to respond to such stimulating conditions. De Vlaming and co-workers[131] demonstrated significant inhibitory effects of melatonin (4 μg per animal, given 10 hr after light onset for 28 days in December — LD 16:8, 30°C) on gonadal weight, ovarian vitellogenesis and seminiferous tubule diameter in the frog *Hyla cinearea.* Melatonin results in a significant reduction of in vitro ovulation of *Rana pipiens* oocytes.[168]

VI. REPTILIA

The pineal complex among reptiles represents the full spectrum of morphological possibilities. In crocodilians, the pineal complex is absent altogether,[169] in many lacertilians (lizards) it includes both an epiphysis and an extracranial parietal eye (analogous to the amphibian frontal organ), while in ophidians (snakes) and chelonians (turtles) it has lost nearly all photosensory elements to become a secretory structure somewhat akin to the mammalian pineal organ.[31,34]

The parietal eye of lizards (when present) is located beneath a transparent scale at the midline of the skull. It is the most highly developed extraocular photoreceptor, possessing a cornea, a lens, cone-like photosensory cells, and a nerve tract to the epiphysis and into the brain.[35,170] A relationship between parietal eye occurrence and the geographical distribution of lizards has been demonstrated.[171] In general, the parietal eye is absent among low-latitude lizards, implying that the role of this structure may be to facilitate survival and reproductive success at higher latitudes where there is greater day-length variation.

Electrophysiological responses to illumination have been recorded from both the parietal eye and epiphysis of lizards.[35] The pineal organ and the parietal eye are linked physically and functionally so that the photosensitivity of the latter is modulated by neuronal feedback from the pineal organ during the day.[172] During scotophase, however, the parietal eye transmits "off" responses to the pineal organ, which is very sensitive to certain neurotransmitters at this time (in contrast to photophase) even though its feedback effects on the parietal eye are diminished. This temporal interaction between the parietal eye and the pineal organ could be significant in the context of circadian rhythms.

The lacertilian epiphysis has received considerable attention from the phylogenetic standpoint. Collin[31] argues that lizards represent a turning point in the design of the vertebrate pineal organ. Whereas the photoreceptor cell outer segments of the lizard parietal eye are quite numerous and complete,[173] the pineal organ generally displays rudimentary photosensory capabilities.[174-176] Most of the "sensory" cells in the lizard pineal organ are generally independent of nerve cells, having assumed a secretory role in place of their previous sensory role. Various lines of evidence support this idea.[31] Furthermore, these rudimentary photoreceptor cells show selective uptake of labeled melatonin precursors.[177] Accompanying the transformation of the epiphysis into a predominantly secretory organ is the presence of pinealopetal autonomic innervation.[178,179] The epiphysis of turtles and snakes show a continuation of this trend toward a secretory function; there is little evidence on which to propose a photosensory role.[34,180,181] In turtles and snakes, the epiphysis is quite similar to the mammalian pineal organ, being secretory in appearance and receiving sympathetic innervation.[34,182] Although pinealofugal fibers have been reported in *Natrix,* their actual origin, destination, and function remain uncertain.[183]

Serotonin is present in the epiphysis of lizards, snakes, and turtles,[143,183-186] but not in the lizard's parietal eye.[187] Pineal serotonin content in *Lacerta muralis* undergoes both circadian and circannual rhythms during exposure to natural environmental conditions.[184] Temperature and photoperiod interactions are at least partly involved in the regulation of these rhythms. HIOMT is also present in the pineal complexes of lizards, snakes, and turtles.[143] In *Lampropholes guichenoti,* pineal HIOMT activity undergoes daily fluctuations, with peak levels occurring at the onset and end of scotophase.[188] However, in *Sceloporus occidentalis,* the absolute levels of pineal HIOMT activity do not depend on a light-dark cycle, since they are unchanged in constant light or darkness.[189] Daily variations of plasma melatonin have been demonstrated in scincid lizards[190,191] with peak levels occurring during the scotophase. These observed rhythms of plasma melatonin are seasonally dependent in their amplitude, which may relate to photoperiod differences prior to capture. In support of this idea, it was shown that exposure of *Trachydosaurus rugosus* to constant light or darkness abolished the plasma melatonin rhythm.[191] This would seem to support the view that the rhythmic release of melatonin into the reptilian bloodstream may serve to inform the animal's physiology about the prevailing photoperiod conditions. Interestingly, serum melatonin has been detected in mature alligators, which lack a pineal complex, although there is no daily rhythm.[192] This suggests that nonpineal (retinal?) melatonin may be secreted at a steady, basal rate.

Pineal serotonin and melatonin content undergo circadian variations, which are 180° out-of-phase with each other in *Testudo hermanni.*[184,186] Of special interest is the demonstration that these circadian rhythms disappear during winter "hibernation" concomitant with decreased absolute levels of these indoleamines.[185,186,193] Elevated levels and rhythms of the indoleamines in the pineal organ of *Testudo* appear to be closely related to reproductive functions since they are most evident during the summer breeding season. Recent data of Owens and colleagues[194] seem to support this hypothesis; specifically, adult female sea turtles (*Chelonia mydas*) captured in the field during mating or nesting seasons, showed the lowest plasma melatonin levels of any sampled group. Seasonal variations in the pineal content of arginine vasotocin were reported for *Testudo.*[88]

Environmental cues are utilized by reptiles for maintaining a number of rhythmic phenomena, the most well described of which is the circadian rhythm of locomotor activity.[30,195,196] In lacertilians, locomotor rhythms can be entrained by light-dark cycles[195] even after the lateral eyes, the parietal eye, and the pineal organ are removed. However, under conditions where the locomotor activity rhythm of *Sceloporus olivaceus* is free running in constant dim light, removal of the parietal eye is without effect whereas pinealectomy causes marked changes in the period (τ) of the circadian rhythm.[197] This indicates that the pineal organ plays a role in the circadian organization of lizards. Melatonin administration by means of subcutaneous silastic capsules (10 μg released per day) to lizards has significant effects on free-running locomotor rhythms during exposure to constant light or darkness,[198] again implicating the pineal organ in circadian organization.

Two valuable reviews on the role that the ectothermic pineal complex may play in both circadian and circannual thermoregulation appeared recently.[155,156] Since both temperature and light are important variables for reproduction in reptiles[3,199] and since exposure to natural sunlight typically results in heat absorption, a thermoregulatory role for the pineal complex does not seem unreasonable, and may well influence reproduction by this indirect means. The parietal eye of lizards in particular has been studied from this perspective. In both *Xantusia vigilis* and *Sceloporus occidentalis,* removal of the parietal eye leads to a greater frequency of exposure to light as compared to control animals.[200,201] Furthermore, reproductive activity is accelerated and thyroid activity is stimulated in these parietalectomized lizards. Roth and Ralph[202] present evidence to argue that such "phototaxic" effects in lizards are in fact not a behavioral response to light, but rather to the thermal nature of most light sources, including sunlight. In this context, the early data of Clausen and Poris[203] demonstrating a stimulatory effect of parietalectomy on testicular recrudescence in *Anolis carolinensis* can be compared with the negative findings of Licht and Pearson.[204] Not inconceivable is that these contradictory results relate to the thermal qualities of the light sources which were used.

Few investigations have dealt with the effect of pinealectomy on the reptilian hypothalamic-hypophyseal-gonadal axis. Augmentation of neurosecretory activity (using morphological criteria) in the paraventricular nucleus of the turtle *Emys leprosa* following pinealectomy has been claimed.[205] Such treatment also produces hyperactive thyroids (determined cytologically) and represses testicular development in this species.[206] In female *Anolis carolinensis* (maintained on a LD 6:18, 31°C regime), pinealectomy is effective in stimulating ovarian follicular development only during the winter.[207] As the breeding season approaches, the ovarian response to pineal removal declines and eventually disappears. Melatonin (10 μg/day administered at midphotophase for 12 days) blocks the stimulatory effect of pinealectomy in winter, but has no effect at other times of the year.[207] Testicular regression is produced by administration of melatonin (5 μg/day for 20 days) to adult *Callisaurus draconoides* captured in the reproductive season and maintained outdoors.[208] Male *Calotes versicolor* maintained on natural photoperiod-temperature conditions are reproductively stimulated (using histological criteria) by the removal of the pineal organ.[209] When these pinealectomized lizards are exposed to LD 6:18, 31°C regimes in May, they are seen to be less regressed than control animals. Furthermore, when pinealectomized *Calotes* are placed on a LD 15:9, 30°C regime in November, testicular development exceeds that found in respective controls.[210] Thus, it appears that the pineal organ of *Calotes,* in contrast to *Anolis,* is inhibitory to reproduction throughout most of the year. It should be noted, however, that postoperative acclimation for *Anolis* was only 2 weeks,[207] whereas in the experiments with *Calotes*[210] it varied from 6 to more than 9 weeks. Possibly, pinealectomy effects depend on experimental duration as well as season, and environmental conditions. Vivien[211] reported that hCG injections to sexually quiescent snakes, *Tropidonotus natrix,* provokes cytological activation of the pinealocytes. Such findings give credence to the view that the reptilian pineal complex may be involved in reproductive phenomenon, although much remains to be clarified.

VII. CONCLUSIONS

One must, in concluding, first address the question, "Among the ectotherms is there unequivocal evidence of a pineal organ-reproduction relationship?" Considering data available at this time we contend that the pineal organ is not essential for reproductive processes. Substantial evidence, coming especially from investigations with teleostean fishes, compels us to the supposition that the pineal organ does modulate or "fine tune" reproductive cycles in some ectothermic vertebrates. Perhaps the epiphysis of some of these vertebrates plays an additive (with other photosensors and/or neuroendocrine centers) or subsidiary role in photoperiod control or modulation of the reproductive system.

In those species where there is evidence that the pineal organ affects reproduction, can it be concluded that the action is due solely to a photoreceptor role? Most studies testify to a pineal organ-reproduction relationship being photoperiod dependent. The pineal organ may be, therefore, a component in the pathway or pathways by which light-dark cycles influence reproduction. Few data point to the pineal organ as the sole receptor of the photic information which affects reproductive control centers. Furthermore, a pineal organ-reproduction relationship could be related to the fact that the epiphysis is a neuroendocrine transducer whose output is regulated by the photic environment. Thus at our present stage of understanding we cannot decipher whether the pineal organ-reproduction relationship is a consequence of the sensory and/or neuroendocrine nature of this organ.

Given that there appears to be a relationship between ectotherm pineal organs and reproduction, one could pursue the question of whether the actions are direct or indirect (these, of course, are not mutually exclusive). Our definition of a direct action would be one which affects any component of the hypophysiotrophic centers-pituitary-gonad axis. Considering all of the pineal organ research among ectotherms, there are

Table 1
EFFECTS OF PINEALECTOMY OR MELATONIN TREATMENT ON REPRODUCTION AMONG ECTOTHERMS[a]

Pinealectomy	Melatonin
Agnatha	
Delays oocyte growth in *Lampetra*	—
Delays SPG[b] in *Lampetra*	—
Chondrichthyes	
↑ GtH[c] content in pituitary of *Scyliorhinus*	—
Osteichthyes	
↑ Testicular development in *Lebistes*	↓ Oocyte development in *Oryzias*
↓ Or ↑ GSI[d] in ♀ *Oryzias*	↓ In vitro oocyte maturation in *Oryzias*
No effect on spawning in *Oryzias*	Reversed blinding effects on GSI in ♀ *Oryzias*
↑ Or ↓ gonadal maturation in *Notemigonus*	Alters pituitary GtH cells in *Oryzias*
Alters pituitary GtH content and rhythm in *Notemigonus*	↓ GSI in *Carassius*
↓ Or ↑ GtH releasing activity in *Notemigonus* hypothalami	↓ GtH secretion in *Carassius*
↑, ↓ Or no effect on GSI in *Carassius*	↓ Vitellogenesis in *Heteropneustes*
↑ Or no effect on serum and pituitary GtH in *Carassius*	↑ Ovarian regression in *Heteropneustes*
Alters serum GtH rhythm in *Carassius*	↓ Pituitary gonadotrophs in *Heteropneustes*
No effect on serum estrogen in *Carassius*	↑ Ovarian regression in *Mystus*
	↓ GSI of *Fundulus*
Amphibians	
No effect in ♀ *Alytes*	Reverses PX effects on StG[e] in *Bufo*
↓ SPG in *Bufo*	No effect on induced SPG in *Rana*
↑ StG in *Bufo*	↓ GSI in *Hyla*
↓ Testicular response to stimulatory L/D in *Rana*	↓ Vitellogenesis in *Hyla*
↓ Pituitary response to stimulatory L/D in *Rana*	↓ Seminiferous tubule diameters in *Hyla*
	↓ In vitro oocyte maturation in *Rana*
Reptiles[f]	
↓ Testicular development in *Emys*	Blocks PX effects in *Anolis*
↑ Or no effect on ovarian development in *Anolis*	Produces testicular regression in *Callisaurus*
↑ Testicular development in *Calotes*	

[a] See text for further details
[b] SPG = spermatogenesis
[c] GtH = gonadotrophin
[d] GSI = gonosomatic index
[e] StG = gonadal steroidogenesis
[f] Excluding parietalectomy

indications that melatonin influences the gonads and pituitary directly. Data obtained with studies on teleostean fishes suggest that the pineal organ, via either neural and/or endocrine output, participates in the regulation of gonadotrophin secretion and hypothalamic function. The pineal organ could influence reproduction indirectly if its output modulates the physiological status of organisms. Possibilities here are numerous since the pineal organ has been linked to CNS function, circadian organization, metabolism, electrolyte balance, growth, and thermoregulation among the ectotherms.

Taking the meager data presently available, one would tentatively conclude that the pineal organ is antigonadal among lizards and elasmobranchs, yet progonadal among cyclostomes and anuran amphibians (Table 1). According to some investigators, a paradox exists in some teleostean fishes where pinealectomy can be either pro- or antigonadal. The effects of pineal organ removal on the reproductive system in these fishes depends on season and photoperiod-temperature conditions. It is not surprising to us, if, as we suggested above, the pineal organ is a component in the pathway or pathways (sensory and/or neuroendocrine) by which photoperiod modulates CNS and physiological function, that pinealectomy would have variable effects in animals exposed to different photoperiods. Nonetheless, there are several characteristics of the pineal organ of most ectotherms which result in assessment of their functions being somewhat complex. The pineal sense organs of ectothermic vertebrates apparently transfer photic information via both neural and humoral channels to CNS centers and neuroendocrine effector centers. Peripheral actions of a hormonal output are also possible. There is evidence also among many ectotherms that pineal cells are innervated. Such pinealopetal innervation could alter the sensitivity and/or activity of the epiphysis. It is becoming more evident that temperature can influence pineal organ activity. Furthermore, pineal organs represent only one component of photosensory and photoneuroendocrine systems which feed, perhaps, into complex pathways of photic information integration.

Rhythms in the CNS (including neuroendocrine centers) organize the physiological functions of the whole organism. These CNS rhythms can be entrained to light-dark shifts and are altered by day-length variations. Pineal organs of ectothermic vertebrates (including all vertebrate classes discussed herein) are most consistently linked with chronophysiological events and photoperiod-affected parameters. Therefore, the probable significance of the pineal organ is as a component in the pathway or pathways by which light-dark cycles modulate physiological processes, particularly circadian and annual organization. Assessment of pineal organ function may be rendered more difficult if it is but a component whose output is channeled into a complex network. Moreover, the effects of pinealectomy and/or treatment with melatonin can give rise to misleading interpretations if the circadian structure and rhythmic phase relations within the monitored parameter are not considered.

We have discovered with goldfish that the pineal end vesicle or portions of it frequently remain attached to the parietal bone following pinealectomy unless the bone is actually scraped (Figure 1). This could be a common occurrence at least among fishes. Thus what has been considered pinealectomy may be a severing of the pineal stalk and removal of the saccus dorsalis only. In such a case, the sensory function of the pineal, as well as the CNS input into the epiphysis could be disrupted, but an endocrine function might remain. Should the end vesicle secrete melatonin when the pineal tract is severed, one might erroneously conclude that the presumably removed pineal promotes gonadal development. This possibility is particularly pertinent to the theory that the pineal organ of some teleostean fishes is progonadal under long photoperiod conditions during the spring. Future investigators will have to present a more rigorous demonstration that pinealectomy has actually been completed. Associatively, severing of the epiphyseal tract is highly likely when performing sham pinealectomies.

Studies have shown that melatonin treatment can retard gonadal development or induce gonadal regression in reptiles, amphibians, and teleosts. Since this indoleamine occurs in the bloodstream of several ectotherms, it may be a mediator of some pineal actions. However, melatonin is also produced by other tissues, notably the retinae. High affinity binding of melatonin in crude membrane preparations of bovine brain tissue was recently demonstrated.[212] Similar demonstrations of melatonin binding in ectotherm tissues would better establish the endocrine role of this indoleamine.

Studies oriented towards delineating the interaction of the pineal organ and eyes on photoperiod-synchronized daily and annual rhythms under various photoperiod-temperature regimes could prove informative. A link with ectotherm reproduction could gain support if estrogen receptors could be demonstrated within the pineal organ. Notably lacking among the ectotherms are investigations which probe the effects of castration and/or sex steroid treatment on pineal organ morphology or on *N*-acetylserotonin transferase (NAT) activity. Pertinent data could be provided from electron microscope studies and NAT activity determinations on pineal organs from animals exposed to various photoperiod-temperature regimes. Does the diurnal rhythm of pineal melatonin secretion vary under different photoperiod-temperature regimes, under constant environmental conditions, or in blinded ectotherms?

The pineal organ is not enigmatic, it is our finite understanding and limited approach to the study of its function(s) which cause us to imagine it as enigmatic. Without a search for identity, the pineal organs of vertebrates continue to play their roles.

ACKNOWLEDGMENTS

We thank Dr. John McNulty for providing us with the micrographs presented herein and Dr. Martin Kavaliers for allowing us to quote manuscripts in press. VdV wishes to express gratitude to F.F. for support during this effort.

REFERENCES

1. **de Vlaming, V. L.,** Environmental and endocrine control of teleost reproduction, in *Control of Sex in Fishes,* Schreck, C. B., Ed., Virginia Polytechnic Institute and State University, Blackburg, 1974, 13.
2. **Jørgensen, C. B., Hede, K.-E., and Larsen, L. O.,** Environmental control of annual ovarian cycles in the toad, *Bufo bufo* L.: role of temperature, in *Environmental Endocrinology,* Assenmacher, I. and Farner, D. S., Eds., Springer-Verlag, New York, 1978, 28.
3. **Lofts, B.,** Reptilian reproductive cycles and environmental regulators, in *Environmental Endocrinology,* Assenmacher, I. and Farner, D. S., Eds., Springer-Verlag, New York, 1978, 37.
4. **Ralph, C. L.,** Pineal control of reproduction: nonmammalian vertebrates, in *The Pineal and Reproduction,* Reiter, R. J., Ed., S. Karger, Basel, 1978, 30.
5. **Kincl, F. A., Chang, C. C., and Zbuzova, V.,** Observations on the influence of changing photoperiod on spontaneous wheel-running activity of neonatally pinealectomized rats, *Endocrinology,* 87, 38, 1970.
6. **Quay, W. B.,** Physiological significance of the pineal during adaptations to shifts in photoperiod, *Physiol. Behav.,* 5, 353, 1970.
7. **Quay, W. B.,** Precocious entrainment and associated characteristics of activity patterns following pinealectomy and reversal of photoperiod, *Physiol. Behav.,* 5, 1281, 1970.
8. **Reiter, R. J.,** Endocrine rhythms associated with pineal gland function, in *Biological Rhythms and Endocrine Function,* Hedlund, L. W., Franz, J. M., and Kenny, A. D., Eds., Plenum Press, New York, 1975, 43.
9. **Menaker, M. and Zimmerman, N.,** Role of the pineal in the circadian system of birds, *Am. Zool.,* 16, 45, 1976.
10. **Banerji, T. K. and Quay, W. B.,** Adrenal dopamine β-hydroxylase activity: 24-hour rhythmicity and evidence for pineal control, *Experientia,* 32, 253, 1976.
11. **Banerji, T. K., Quay, W. B., and Kachi, T.,** Hypothalamic dopamine-β-hydroxylase activity: fluctuations with time of day and their modifications by intracranial surgery, adrenalectomy, and pinealectomy, *Neurochem. Res.,* 3, 281, 1978.
12. **Gwinner, E.,** Effects of pinealectomy on circadian locomotor activity rhythms in European starlings, *Sturnus vulgaris, J. Comp. Physiol.,* 126, 123, 1978.
13. **Hendel, R. C. and Turek, F. W.,** Suppression of locomotor activity in sparrows by treatment with melatonin, *Physiol. Behav.,* 21, 275, 1978.
14. **Meier, A. H.,** Chronoendocrinology of vertebrates, in *Hormonal Correlates of Behavior,* Eleftheriou, B. E. and Sprott, R. L., Eds., Plenum Press, New York, 1975, 469.
15. **Kuo, C.-M. and Watanabe, W. O.,** Circadian responses of teleostean oocytes to gonadotropins and prostaglandins determined by cyclic AMP concentration, *Ann. Biol. Anim. Biochem. Biophys.,* 18, 949, 1978.
16. **Meier, A. H., Fivizzani, A. J., Spieler, R. E., and Horseman, N. D.,** Circadian hormone basis for seasonal conditions in the gulf killifish, *Fundulus grandis,* in *Comparative Endocrinology,* Gaillard, P. J. and Boer, H. H., Eds., Elsevier/North-Holland, Amsterdam, 1978, 141.

17. **Delahunty, G., Olcese, J., Prack, M., Vodicnik, J. J., Schreck, C. B., and de Vlaming, V. L.**, Diurnal variations in the physiology of the goldfish, *Carassius auratus, J. Interdiscip. Cycle Res.*, 9, 73, 1978.
18. **Delahunty, G., Schreck, C. B., and de Vlaming, V. L.**, Effects of photoperiod on plasma corticoid levels in the goldfish, *Carassius auratus* — role of the pineal organ, *Comp. Biochem. Physiol.*, 65A, 355, 1980.
19. **Simpson, T. H.**, An interpretation of some endocrine rhythms in fish, in *Rhythmic Activity of Fishes,* Thorpe, J. E., Ed., Academic Press, New York, 1978, 55.
20. **Gillet, C., Breton, B., and Billard, R.**, Seasonal effects of exposure to temperature and photoperiod regimes on gonad growth and plasma gonadotropin in goldfish *(Carassius auratus), Ann. Biol. Anim. Biochem. Biophys.*, 18, 1045, 1978.
21. **Peter, R. E., Hontela, A., Cook, A. F., and Paulencu, C. R.**, Daily cycles in serum cortisol levels in the goldfish: effects of photoperiod, temperature and sexual condition, *Can. J. Zool.*, 56, 2443, 1978.
22. **Sauerbier, I.**, Seasonal variations in the circadian rhythm of tissue catecholamines in the frog (*Rana temporaria* L.), *Comp. Biochem. Physiol.*, 61C, 157, 1978.
23. **Olcese, J. and de Vlaming, V.**, Daily variation of, photoperiod and pinealectomy effects on, hypothalamic monoamine oxidase activity in the goldfish, *Comp. Biochem. Physiol.*, 63C, 363, 1979.
24. **Spieler, R. E.**, Diel thythms of circulating prolactin, cortisol, thyroxine, and triiodothyronine levels in fishes: a review, *Rev. Can. Biol.*, 38, 301, 1979.
25. **Quay, W. B.**, Regulation and reproductive effects of hypothalamic 24-hour neurochemical rhythms, in *Biology of Reproduction, Basic and Clinical Studies,* Velardo, J. T. and Kasprow, B. A., Eds., Pan American Congress of Anatomy, New Orleans, 1972, 99.
26. **Suzuki, Y., Homma, K., Takahashi, M., Horikoshi, H., and Lin, Y.-C.**, Participation of circadian rhythms in infradian reproductive activities, in *Biological Rhythms in Neuroendocrine Activity,* Kawakami, M., Ed., Ikaku Shoin, Tokyo, 1974, 151.
27. **Eskes, G. A. and Zucker, I.**, Photoperiodic regulation of the hamster testis: dependence on circadian rhythms, *Proc. Natl. Acad. Sci. U.S.A.*, 75, 1034, 1978.
28. **Stetson, M. H.**, Circadian organization and female reproductive cyclicity, in *Aging and Biological Rhythms,* Samis, H. V., Jr. and Campobianco, S., Eds., Plenum Press, New York, 1978, 251.
29. **Turek, F.**, Diurnal rhythms and the seasonal reproductive cycle in birds, in *Environmental Endocrinology,* Assenmacher, I. and Farner, D. S., Eds., Springer-Verlag, New York, 1978, 144.
30. **Rusak, B. and Zucker, I.**, Neural regulation of circadian rhythms, *Physiol. Rev.*, 59, 449, 1979.
31. **Collin, J. P.**, Differentiation and regression of the cells of the sensory line in the epiphysis cerebri, in *The Pineal Gland,* Wolstenholme, G. E. W. and Knight, J., Eds., Churchill Livingstone, Edinburgh, 1971, 79.
32. **Ueck, M.**, Vergleichende Betrachtungen zur neuroendokrinen Aktivitat des Pinealorgans, *Fortschr. Zool.*, 22, 167, 1974.
33. **Morita, Y.**, Direct photosensory activity of the pineal, in *Brain-Endocrine Interaction II: the Ventricular System in Neuroendocrine Mechanisms: Proceedings,* Knigge, K. M. and Kobayashi, H., Eds., S. Karger, Basel, 1975, 376.
34. **Hamasaki, D. I. and Eder, D. J.**, Adaptive radiation of the pineal system, *Handb. Sens. Physiol.*, 7, 497, 1977.
35. **Dodt, E.**, The parietal eye (pineal and parietal organs) of lower vertebrates, *Handb. Sens. Physiol.*, 7(3B), 113, 1973.
36. **Hanyu, I., Niwa, H., and Tamura, T.**, Salient features in photosensory function of teleostean pineal organ, *Comp. Biochem. Physiol.*, 61A, 49, 1978.
37. **Meiniel, A.**, Presence d'indolamines dans les organes pineal et parapineal de *Lampetra planeri, C.R. Acad. Sci. (Paris)*, 2870, 313, 1978.
38. **Joss, J. M. P.**, Hydroxyindole-*O*-methyltransferase (HIOMT) activity and the uptake of ^{3}H-melatonin in the lamprey, *Geotria australis* Gray, *Gen. Comp. Endocrinol.*, 31, 270, 1977.
39. **Young, J. Z.**, The photoreceptors of lampreys, *J. Exp. Biol.*, 12, 254, 1935.
40. **Eddy, J. M. P. and Strahan, R.**, The role of the pineal complex in the pigmentary effector system of the lampreys, *Mordacia mordax* (Richardson) and *Geotria australis* Gray, *Gen. Comp. Endocrinol.*, 11, 528, 1968.
41. **Joss, J. M. P.**, The pineal complex, melatonin and color change in the lamprey, *Lampetra, Gen. Comp. Endocrinol.*, 21, 188, 1973.
42. **Dodd, J. M.**, The hormones of sex and reproduction and their effects in fish and lower chordates: twenty years on, *Am. Zool.*, 15(1), 137, 1975.
43. **Larsen, L. O. and Rothwell, B.**, Adenohypophysis, in *The Biology of Lampreys,* Vol. 2, Hardisty, M. W. and Potter, I. C., Eds., Academic Press, New York, 1972, 1.
44. **Eddy, J. M. P.**, Metamorphosis and the pineal complex in the brook lamprey, *Lampetra planeri, J. Endocrinol.*, 44, 451, 1969.

45. **Joss, J. M. P.**, Pineal-gonad relationships in the lamprey *Lampetra fluviatilis, Gen. Comp. Endocrinol.*, 21, 118, 1973.
46. **Tilney, F. and Warren, L. F.**, The morphology and evolutional significance of the pineal body, *Am. Anat. Mem.*, Wistar Press, Philadelphia, 1919, No. 9.
47. **Rüdeberg, C.**, Light and electron microscopic studies on the pineal organ of the dogfish, *Scyliorhinus canicula* L., *Z. Zellforsch. Mikrosk. Anat.*, 96, 548, 1969.
48. **Teshima, K., Yoshimura, H., and Mizue, K.**, Studies on the sharks. II. On the reproduction of Japanese dogfish *Mustelus manazo* Blecker, *Bull. Fac. Fish. Nagasaki Univ.*, 32, 41, 1971.
49. **Ketchen, K. S.**, Size at maturity, fecundity and embryonic growth of the spiny dogfish *(Squalus acanthias)* in British Columbia waters, *J. Fish. Res. Bd. Can.*, 29, 1717, 1972.
50. **Sumpter, J. P. and Dodd, J. M.**, The annual reproductive cycle of the female lesser spotted dogfish, *Scyliorhinus canicula* L., and its endocrine control, *J. Fish Biol.*, 15, 687, 1979.
51. **Dobson, S. and Dodd, J. M.**, The roles of temperature and photoperiod in the response of the testis of the dogfish, *Scyliorhinus canicula* L. to partial hypophysectomy (ventral lobectomy), *Gen. Comp. Endocrinol.*, 32, 114, 1977.
52. **Dobson, S.**, Endocrine Control of Reproduction in the Male *Scyliorhinus canicula*, Ph.D. thesis, Univ. of Wales, Cardiff, 1975.
53. **Wilson, J. F. and Dodd, J. M.**, The role of the pineal complex and lateral eyes in the colour change response of the dogfish, *Scyliorhinus canicula* L., *J. Endocrinol.*, 58, 591, 1973.
54. **Hafeez, M. A. and Merhige, M. E.**, Light and electron microscope study on the pineal complex of the coelacanth, *Latimeria chalumnae* Smith, *Cell Tissue Res.*, 178, 249, 1977.
55. **Ueck, M.**, Ultrastrukturbesonderheiten der pinealen Sinneszellen von *Protopterus dolloi, Z. Zellforsch. Mikrosk. Anat.*, 100, 560, 1969.
56. **Hill, C.**, The epiphysis of teleosts and *Amia, J. Morphol.*, 9, 237, 1894.
57. **Poston, H. A.**, Neuroendocrine mediation of photoperiod and other environmental influences on physiological responses in salmonids: a review, *Tech. Pap. U.S. Fish Wildl. Serv.*, 96, 1, 1978.
58. **Peter, R. E. and Hontela, A.**, Annual gonadal cycles in teleosts: environmental factors and gonadotropin levels in blood, in *Environmental Endocrinology*, Assenmacher, I. and Farner, D. S., Eds., Springer-Verlag, New York, 1978, 20.
59. **Baggerman, B.**, Photoperiodic responses in the stickleback and their control by a daily rhythm of photosensitivity, *Gen. Comp. Endocrinol. Suppl.*, 3, 466, 1972.
60. **Sundararaj, B. I. and Vasal, S.**, Photoperiod and temperature control in the regulation of reproduction in the female catfish *Heteropneustes fossilis, J. Fish. Res. Board. Can.*, 33, 959, 1976.
61. **Chan, K. K.-S.**, A photosensitive daily rhythm in the female medaka, *Oryzias latipes, Can. J. Zool.*, 54, 852, 1976.
62. **Sundararaj, B. I., Nath, P., and Jeet, V.**, Role of circadian and circannual rhythms in the regulation of ovarian cycles in fishes: a catfish model, in *Comparative Endocrinology*, Gaillard, P. J. and Boer, H. H., Eds., Elsevier/North Holland, Amsterdam, 1978, 137.
63. **Fenwick, J. C.**, The pineal organ, in *Fish Physiology*, Vol. 4, Hoar, W. S. and Randall, D. J., Eds., Academic Press, New York, 1970, 91.
64. **McNulty, J. A.**, A comparative study of the pineal complex in the deep-sea fishes *Bathylagus wesethi* and *Nezumia liolepsis, Cell Tissue Res.*, 172, 205, 1976.
65. **McNulty, J. A.**, The pineal of the troglophilic fish, *Chologaster agassizi:* an ultrastructural study, *J. Neural Transm.*, 43, 47, 1978.
66. **McNulty, J. A.**, A light and electron microscopic study of the pineal in the blind goby, *Typhlogobius californiensis* (Pisces: Gobiidae), *J. Comp. Neurol.*, 181, 197, 1978.
67. **Chèze, G.**, Etude morphologique, histologique et expérimentale de l'épiphyse de *Symphodus melops* (Poisson, Labridè), *Bull. Soc. Zool. Fr.*, 94, 47, 1969.
68. **Urasaki, H.**, The function of the pineal gland in the reproduction of the medaka, *Oryzias latipes, Bull. Lib. Arts & Sci. Course, Sch. Med. Nihon Univ.*, 2, 11, 1974.
69. **Hafeez, M. A., Wagner, H. H., and Quay, W. B.**, Mediation of light-induced changes in pineal receptor and supporting cell nuclei and nucleoli in steelhead trout *(Salmo gairdneri), Photochem. Photobiol.*, 28, 213, 1978.
70. **Smith, J. R. and Weber, L. J.**, The regulation of day-night changes in hydroxyindole-*O*-methyltransferase activity in the pineal gland of steelhead trout *(Salmo gairdneri), Can. J. Zool.*, 54, 1530, 1976.
71. **Hafeez, M. A. and Zerihun, L.**, Autoradiographic localization of ^{3}H-5-HTP and ^{3}H-5-HT in the pineal organ and circumventricular areas in the rainbow trout, *Salmo gairdneri* Richardson, *Cell Tissue Res.*, 170, 61, 1976.
72. **Hafeez, M. A. and Zerihun, L.**, Studies on central projections of the pineal nerve tract in rainbow trout, *Salmo gairdneri* Richardson, using cobalt chloride iontophoresis, *Cell Tissue Res.*, 154, 485, 1974.
73. **Chèze, G.**, Innervation épiphysaire chez *Symphodus melops* (Poisson-Labridae), *Bull. Soc. Zool. Fr.*, 96, 53, 1971.

74. **Wake, K.**, Acetylcholinesterase-containing nerve cells and their distribution in the pineal organ of the goldfish, *Carassius auratus, Z. Zellforsch. Mikrosk. Anat.*, 145, 287, 1973.
75. **Korf, H.-W.**, Acetylcholinesterase-positive neurons in the pineal and parapineal organs of the rainbow trout, *Salmo gairdneri* (with special reference to the pineal tract), *Cell Tissue Res.*, 155, 475, 1974.
76. **Urasaki, H.**, Fine organization of nervous system and sensory cell in the pineal gland of the teleost, *Oryzias latipes, Bull. Lib. Arts & Sci. Course, Sch. Med. Nihon Univ.*, 4, 15, 1976.
77. **Oguri, M., Omura, Y., and Hibiya, T.**, Uptake of ^{14}C-labelled 5-hydroxytryptamine into the pineal organ of rainbow trout, *Bull. Jpn. Soc. Sci. Fish.*, 34, 687, 1968.
78. **Hafeez, M. A. and Quay, W. B.**, Histochemical and experimental studies on 5-hydroxytryptamine in pineal organs of teleosts *(Salmo gairdneri* and *Atherinopsis californiensis), Gen. Comp. Endocrinol.*, 13, 211, 1969.
79. **Owman, C. and Rüdeberg, C.**, Light, fluorescence, and electron microscope studies on the pineal organ of the pike, *Esox lucius* L. with special regard to 5-hydroxytryptamine, *Z. Zellforsch. Mikrosk. Anat.*, 107, 522, 1970.
80. **Fenwick, J. C.**, Demonstration and effect of melatonin in fish, *Gen. Comp. Endocrinol.*, 14, 86, 1970.
81. **Satake, N. and Morton, B. E.**, Scotophobin A causes dark avoidance in goldfish by elevating pineal *N*-acetylserotonin, *Pharmacol. Biochem. Behav.*, 10, 449, 1979.
82. **Hafeez, M. A. and Quay, W. B.**, Pineal acetylserotonin methyltransferase activity in the teleost fishes *Hesperoleucas symmetricus* and *Salmo gairdneri*, with evidence for each of constant light and darkness, *Comp. Gen. Pharmacol.*, 1, 257, 1970.
83. **Smith, J. R. and Weber, L. J.**, Diurnal fluctuations in acetylserotonin methyltransferase (ASMT) activity in the pineal gland of the steelhead trout *(Salmo gairdneri), Proc. Soc. Exp. Biol. Med.*, 147, 441, 1974.
84. **Smith, J. R. and Weber, L. J.**, Alterations in diurnal pineal hydroxyindole-*O*-methyltransferase (HIOMT) activity in steelhead trout *(Salmo gairdneri)* associated with changes in environmental background color, *Comp. Biochem. Physiol.*, 53C, 33, 1976.
85. **Gern, W. A., Owens, D. W., and Ralph, C. L.**, Plasma melatonin in the trout: day-night change demonstrated by radioimmunoassay, *Gen. Comp. Endocrinol.*, 34, 453, 1978.
86. **Gern, W. A., Owens, D. W., and Ralph, C. L.**, Persistence of the nychthemeral rhythm of melatonin secretion in pinealectomized or optic tractsectioned trout *(Salmo gairdneri), J. Exp. Zool.*, 205, 371, 1978.
87. **Gern, W. A. and Ralph, C. L.**, Melatonin synthesis by the retina, *Science*, 204, 183, 1979.
88. **Vivien-Roels, B., Guerne, J. M., Holder, F. C., and Schroeder, M. D.**, Comparative immunohistochemical, radioimmunological and biological attempt to identify arginine-vasotocin (AVT) in the pineal gland of reptiles and fishes, *Prog. Brain Res.*, 52, 459, 1979.
89. **Eriksson, L.-O.**, Die Jahresperiodik Augen-und Pinealorganloser bachsaiblinge *Salvelinus fontinalis* Mitchell, *Aquilo Ser Zool.*, 13, 8, 1972.
90. **Delahunty, G., Bauer, G., Prack, M., and de Vlaming, V.**, Effects of pinealectomy and melatonin treatment on liver and plasma metabolites in the goldfish, *Carassius auratus, Gen. Comp. Endocrinol.*, 35, 99, 1978.
91. **de Vlaming, V. L. and Vodicnik, M. J.**, Effects of pinealectomy on pituitary gonadotrophs, pituitary gonadotropin potency and hypothalamic gonadotropin releasing activity in *Notemigonus crysoleucas, J. Fish Biol.*, 10, 73, 1977.
92. **Vodicnik, M. J., Kral, R. E., de Vlaming, V. L., and Crim, L. W.**, The effects of pinealectomy of pituitary and plasma gonadotropin levels in *Carassius auratus* exposed to various photoperiod-temperature regimes, *J. Fish Biol.*, 12, 187, 1978.
93. **Hontela, A. and Peter, R. E.**, Effects of pinealectomy, blinding, and sexual condition on serum gonadotropin levels in the goldfish, *Gen. Comp. Endocrinol.*, 40, 168, 1980.
94. **Kavaliers, M.**, Pineal involvement in the control of circadian rhythmicity in the lake chub, *Couesius plumbeus, J. Exp. Zool.*, 209, 33, 1979.
95. **Kavaliers, M.**, The pineal organ and circadian organization in teleost fishes, *Rev. Can. Biol.*, 38, 281, 1979.
96. **Kavaliers, M.**, Circadian locomotor activity rhythms of the burbot, *Lota lota:* Seasonal differences in period length and effect of pinealectomy, *J. Comp. Physiol.*, 136, 215, 1980.
97. **Kavaliers, M.**, The pineal and circadian rhythms of fishes, in *Environmental Physiology of Fishes*, Ali, M. A., Ed., Plenum Press, New York, 1980, 631.
98. **Kavaliers, M.**, Circadian organization in white suckers, *Catostomus commersoni:* the role of the pineal organ, *Comp. Biochem. Physiol.*, 68A, 127, 1981.
99. **Kavaliers, M., Firth, B. T., and Ralph, C. L.**, Pineal control of the circadian rhythm of colour change in the killifish *(Fundulus heteroclitus), Can. J. Zool.*, 58, 456, 1980.

100. **Kavaliers, M. and Ralph, C. L.**, Pineal involvement in the control of behavioral thermoregulation of the white sucker, *Catostomus commersoni, J. Exp. Zool.*, 212, 301, 1980.
101. **Byrne, J. E.**, Locomotor activity responses in juvenile sockeye salmon, *Oncorhynchus nerka,* to melatonin and serotonin, *Can. J. Zool.*, 48, 1425, 1970.
102. **Fujii, R. and Miyashita, Y.**, Receptor mechanisms in fish chromatophores — IV. Effects of melatonin and related substances on dermal and epidermal melanophores of the siluroid, *Parasilurus asotus, Comp. Biochem. Physiol.*, 59C, 59, 1978.
103. **Krockert, G.**, Die Wirkung der Verfütterung von Schilddrüsen-und Zirbeldrüsen-substanz an *Lebistes reticulatus* (Zahnkarpfen), *Zeit. Gesamte Exp. Med.*, 98, 214, 1936.
104. **Pflugfelder, O.**, Wirkungen partieller Zerstörungen der Parietal Region von *Lebistes reticulatus, Arch. Entwicklungsmech. Org.*, 147, 42, 1954.
105. **Schonherr, J.**, Über die Abhängigkeit der Instinkthandlungen vom Vorderhirn und Zwischenhirn (Epiphyse) bei *Gasterosteus aculeatus* L., *Zool. Jahrb. Abt. Allg. Zool. Physiol.*, 65, 357, 1955.
106. **Rasquin, P.**, Studies in the control of pigment cells and light reactions in recent teleost fishes, *Bull. Am. Mus. Nat. Hist.*, 115, 1, 1958.
107. **Pang, P. K. T.**, The effect of pinealectomy on the adult killifish, *Fundulus heteroclitus, Am. Zool.*, 7, 715, 1967.
108. **Urasaki, H.**, The role of pineal and eyes in the photoperiodic effect on the gonad of the medaka, *Oryzias latipes, Chronobiologia,* 3, 228, 1976.
109. **Urasaki, H.**, Effects of pinealectomy on gonadal development in the Japanese killifish (Medaka), *Oryzias latipes, Annot. Zool. Jpn.*, 45, 10, 1972.
110. **Urasaki, H.**, Role of the pineal gland in gonadal development in the fish, *Oryzias latipes, Annot. Zool. Jpn.*, 45, 152, 1972.
111. **Urasaki, H.**, Effects of restricted photoperiod and melatonin administration on gonadal weight in the Japanese killifish, *J. Endocrinol.*, 55, 619, 1972.
112. **Urasaki, H.**, Effect of pinealectomy and photoperiod on oviposition and gonadal development in the fish, *Oryzias latipes, J. Exp. Zool.*, 185, 241, 1973.
113. **Iwamatsu, T.**, Studies on oocyte maturation of the medaka, *Oryzias latipes.* VII. Effects of pinealectomy and melatonin on oocyte maturation, *Annot. Zool. Jpn.*, 51, 198, 1978.
114. **Urasaki, H.**, Response of the hypophysial-ovarian system of the teleost, *Oryzias latipes,* to administration of melatonin, *Bull. Lib. Arts & Sci. Course, Sch. Med. Nihon Univ.*, 5, 15, 1977.
115. **de Vlaming, V.**, Effects of pinealectomy on gonadal activity in the cyprinid teleost, *Notemigonus crysoleucas, Gen. Comp. Endocrinol.*, 26, 36, 1975.
116. **Fenwick, J. C.**, The pineal organ: photoperiod and reproductive cycles in the goldfish, *Carassius auratus, J. Endocrinol.*, 46, 101, 1970.
117. **de Vlaming, V. and Vodicnik, M. J.**, Seasonal effects of pinealectomy on gonadal activity in the goldfish, *Carassius auratus, Biol. Reprod.*, 19, 57, 1978.
118. **Hontela, A. and Peter, R. E.**, Daily cycles in serum gonadotropin levels in the goldfish: effects of photoperiod, temperature, and sexual condition, *Can. J. Zool.*, 56, 2430, 1978.
119. **Vodicnik, M. J., Olcese, J., Delahunty, G., and de Vlaming, V.**, The effects of blinding, pinealectomy and exposure to constant dark conditions on gonadal activity in the female goldfish, *Carassius auratus, Environ. Biol. Fish.*, 4, 173, 1979.
120. **Delahunty, G., Schreck, C., Specker, J., Olcese, J., Vodicnik, M. J., and de Vlaming, V.**, The effects of light reception on circulating estrogen levels in female goldfish, *Carassius auratus:* importance of retinal pathways versus the pineal, *Gen. Comp. Endocrinol.*, 38, 148, 1979.
121. **Fenwick, J. C.**, Effects of pinealectomy and bilateral enucleation on the phototactic response and the conditioned response to light of the goldfish, *Carassius auratus, Can. J. Zool.*, 48, 175, 1970.
122. **Scapagnini, U., Gerendai, I., Clementi, G., Fiore, L., Marchetti, B., and Prato, A.**, Role of brain monoamines in the regulation of the circadian variations of activity of some neuroendocrine axes, in *Environmental Endocrinology,* Assenmacher, I. and Farner, D. S., Eds., Springer-Verlag, Berlin, 1978, 135.
123. **Kordon, C., Hery, M., Gogan, F., and Rotsztejn, W. H.**, Circadian pattern of secretion of hormones by the anterior pituitary gland with particular reference to involvement of serotonin in their rhythmic regulation, in *Environmental Endocrinology,* Assenmacher, I. and Farner, D. S., Eds., Springer-Verlag, Berlin, 1978, 161.
124. **El Halawani, M. E., Burke, W. H., and Ogren, L. A.**, Effects of drugs that modify brain monoamine concentrations on photoperiodically-induced testicular growth in coturnix quail *(Coturnix coturnix japonica), Biol. Reprod.*, 18, 148, 1978.
125. **Yates, C. A. and Herbert, J.**, Differential circadian rhythms in pineal and hypothalamic serotonin induced by artificial photoperiods or melatonin, *Nature (London),* 262, 1976.
126. **Olcese, J. M., Hall, T. R., Figueroa, H. R., and de Vlaming, V. L.**, Hypothalamic monoamine oxidase, a component in the serotonergic control of pituitary prolactin content in *Carassius auratus, Gen. Comp. Endocrinol.*, 38, 309, 1979.

127. **Olcese, J. M. and de Vlaming, V. L.**, Interaction of environmental photoperiod and temperature on hypothalamic monoamine oxidase activity in *Carassius auratus*, L., *Comp. Biochem. Physiol.*, 66A, 153, 1980.
128. **Olcese, J., Darr, C., DeMuri, B., Hall, T. R., and de Vlaming, V.**, Photoperiod effects on hypothalamic serotonergic activity in the goldfish, *Carassius auratus, Comp. Biochem. Physiol.*, 66A, 363, 1980.
129. **Sundararaj, B. I. and Keshavanath, P.**, Effects of melatonin and prolactin treatment on the hypophysial-ovarian system in the catfish, *Heteropneustes fossilis* (Bloch), *Gen. Comp. Endocrinol.*, 29, 84, 1976.
130. **Saxena, P. K. and Anand, K.**, A comparison of ovarian recrudescence in the catfish, *Mystus tengara* (Ham.), exposed to short photoperiods, to long photoperiods, and to melatonin, *Gen. Comp. Endocrinol.*, 33, 506, 1977.
131. **de Vlaming, V. L., Sage, M., and Charlton, C. B.**, The effects of melatonin treatment on gonosomatic index in the teleost, *Fundulus similis*, and the tree frog, *Hyla cinerea, Gen. Comp. Endocrinol.*, 22, 433, 1974.
132. **de Vlaming, V. L.**, Effects of photoperiod-temperature regimes and pinealectomy on body fat reserves in the golden shiner, *Notemigonus crysoleucas, Fishery Bull.*, 73, 766, 1975.
133. **Vodicnik, M. J. and de Vlaming, V. L.**, The effects of pinealectomy on pituitary prolactin levels in *Carassius auratus* exposed to various photoperiod-temperature regimes, *Endocrine Res. Comm.*, 5, 199, 1978.
134. **de Vlaming, V.**, Effects of pinealectomy and melatonin treatment on growth in the goldfish, *Carassius auratus, Gen. Comp. Endocrinol.*, 40, 245, 1980.
135. **Hartwig, H.-G. and Baumann, C.**, Evidence for photosensitive pigments in the pineal complex of the frog, *Vision Res.*, 14, 597, 1974.
136. **Owman, C., Rüdeberg, C., and Ueck, M.**, Fluoreszenz-mikroskopischer Nachwis biogener Monoamine in der Epiphysis von *Rana esculenta* und *Rana pipiens, Z. Zellforsch. Mikrosk. Anat.*, 111, 550, 1970.
137. **Meissl, H. and Donley, C. S.**, Free amino acids in the pineal organ of the rainbow trout, *Salmo gairdneri*, and the frog, *Rana esculenta, Gen. Comp. Endocrinol.*, 34, 76, 1978.
138. **Oksche, A. and Hartwig, H. G.**, Pineal sense organs — components of photoneuroendocrine systems, *Prog. Brain Res.*, 52, 113, 1979.
139. **Korf, H.-W.**, Histological, histochemical and electron microscopical studies on the nervous apparatus of the pineal organ in the tiger salamander, *Ambystoma tigrinum, Cell Tissue Res.*, 174, 475, 1976.
140. **Cadusseau, J., Gaillard, F., and Galand, G.**, Pineal response types in the frog's brain under white light exposure, *Exp. Brain Res.*, 36, 41, 1979.
141. **Zilles, K. and Nickeleit, V.**, Efferent connections from the brain to the frontal organ in *Rana temporaria* demonstrated by labeling with horseradish peroxidase, *Cell Tissue Res.*, 196, 189, 1979.
142. **Iturriza, F. C.**, Histochemical demonstration of biogenic monoamines in the pineal gland of the toad, *Bufo arenarum, J. Histochem. Cytochem.*, 15, 301, 1967.
143. **Quay, W. B.**, Retinal and pineal hydroxyindole-*O*-methyltransferase activity in vertebrates, *Life Sci.*, 4, 983, 1965.
144. **Eichler, V. B. and Moore, R. Y.**, Studies on hydroxyindole-*O*-methyltransferase in frog brain and retina: enzymology, regional distribution and environmental control of enzyme levels, *Comp. Biochem. Physiol.*, 50C, 89, 1975.
145. **Van de Veerdonk, F. C. G.**, Demonstration of melatonin in amphibia, *Curr. Mod. Biol.*, 1, 1975, 1967.
146. **Gern, W. A. and Norris, D. O.**, Plasma melatonin in the neotenic tiger salamander *(Ambystoma tigrinum)*: effects of photoperiod and pinealectomy, *Gen. Comp. Endocrinol.*, 38, 393, 1979.
147. **Bagnara, J. T.**, Color change, in *Physiology of the Amphibia, Vol. 3*, Lofts, B., Ed., Academic Press, New York, 1976, 1.
148. **Charlton, H. M.**, The pineal gland and color change in *Xenopus laevis* Daudin, *Gen. Comp. Endocrinol.*, 7, 384, 1966.
149. **Bagnara, J. T. and Hadley, M. E.**, Endocrinology of the amphibian pineal, *Am. Zool.*, 10, 201, 1970.
150. **Wakahara, M.**, Daily variation in mitotic rate in tail-fin epidermis *Xenopus laevis* and its modification by pineal organ-subcommissural organ system and photoperiods, *Neuroendocrinology*, 9, 267, 1972.
151. **Adler, K.**, Pineal end organ: role in extraoptic entrainment of circadian locomotor rhythm in frogs, in *Biochronometry*, Menaker, M., Ed., Natl. Acad. Sci., Washington, D.C., 1971, 342.
152. **Adler, K.**, Extraocular photoreceptor in amphibians, *Photochem. Photobiol.*, 23, 275, 1976.
153. **Demian, J. J. and Taylor, D. H.**, Photoreception and locomotor rhythm entrainment by the pineal body of the newt, *Notophthalmus viridescens* (Amphibia, Urodela, Salamandridae), *J. Herpetol.*, 11, 131, 1977.

154. **Steyn, W.**, Three eyes: wider implications of a narrow specialty, *S. Afr. J. Sci.*, 62, 13, 1966.
155. **Ralph, C. L., Firth, B. T., Gern, W. A., and Owens, D. W.**, The pineal complex and thermoregulation, *Biol. Rev.*, 54, 41, 1979.
156. **Ralph, C. L., Firth, B. T., and Turner, J. S.**, The role of the pineal body in ectotherm thermoregulation, *Am. Zool.*, 19, 273, 1979.
157. **Hutchison, V. H., Black, J. J., and Erskine, D.**, Melatonin and chlorpromazine: thermal selection in the mudpuppy, *Necturus maculosus, Life Sci.*, 25, 527, 1979.
158. **Vullings, H. G. B.**, Influence of light and darkness on the hypothalamo-hypophysial system of *Rana esculenta* and the involvement of the pineal complex, *Z. Zellforsch. Mikrosk. Anat.*, 146, 491, 1973.
159. **Mazzi, V.**, The hypothalamus as a thermodependent neuroendocrine center in urodeles, in *The Hypothalamus*, Martini, L., Motta, M., and Fraschini, F., Eds., Academic Press, New York, 1970, 663.
160. **Werner, J. K.**, Temperature-photoperiod effects on spermatogenesis in the salamander *Plethodon cinereus, Copeia*, 1969, 592, 1969.
161. **Rastogi, R. K., Iela, L., Saxena, P. K., and Chieffi, G.**, The control of spermatogenesis in the green frog, *Rana esculenta, J. Exp. Zool.*, 196, 151, 1976.
162. **Disclos, P.**, Epiphysectomie chez le tetard d'Alytes, *C.R. Acad. Sci.*, 258, 3101, 1964.
163. **Biswas, N. M., Chakraborty, J., Chanda, S., and Sanyal, S.**, A basic experimental approach in perspective of pineal and melatonin involvement in photoperiod-induced alteration of spermatogenesis in toad *(Bufo melanostictus), Endokrinologie*, 71, 143, 1978.
164. **Juskiewicz, T. and Rakalska, Z.**, Anti-oestrogenic effects of bovine pineal glands, *Nature (London)*, 200, 1329, 1963.
165. **Juskiewicz, T. and Rakalska, Z.**, Lack of the effect of melatonin on the frog spermatogenic reaction, *J. Pharm. Pharmacol.*, 17, 189, 1965.
166. **John, T. M. and George, J. C.**, Diurnal variation in the effect of melatonin on plasma and muscle free fatty acid levels in the pigeon, *Endocrinol. Exp.*, 10, 131, 1976.
167. **Reiter, R. J., Blask, D. E., Johnson, L. Y., Rudeen, P. K., Vaughan, M. K., and Waring, P. J.**, Melatonin inhibition of reproduction in the male hamster; its dependency on time of day of administration and on an intact and sympathetically innervated pineal gland, *Neuroendocrinology*, 22, 107, 1976.
168. **O'Connor, J. M.**, Effect of melatonin on *in vitro* ovulation of frog oocytes, *Am. Zool.*, 19, 577, 1969.
169. **Ralph, C. L.**, The pineal gland and geographical distribution of animals, *Int. J. Biometeorol.*, 19, 289, 1975.
170. **Eakin, R. M.**, *The Third Eye*, Univ. California Press, Berkeley, 1973.
171. **Gundy, G. C., Ralph, C. L., and Wurst, G. Z.**, Parietal eyes in lizards: zoogeographical correlates, *Science*, 190, 671, 1975.
172. **Engbretson, G. A. and Lent, C. M.**, Parietal eye of the lizard: neuronal photoresponses and feedback from the pineal gland, *Proc. Natl. Acad. Sci. U.S.A.*, 73, 654, 1976.
173. **Oksche, A. and Kirschstein, H.**, Unterschiedlicher elektronmikroskopischer Feinbau der Sinneszellen in Parietalauge und in Pinealorgan (Epiphysis cerebri) der Lacertilia, *Z. Zellforsch. Mikrosk. Anat.*, 87, 159, 1968.
174. **Kappers, J. A.**, The sensory innervation of the pineal organ in the lizard, *Lacerta viridis*, with remarks on its position in the trend of pineal phylogenetic structural and functional evolution, *Z. Zellforsch. Mikrosk. Anat.*, 81, 581, 1967.
175. **Collin, J.-P. and Kappers, J. A.**, Synapses of the ribbon type in the pineal organ of *Lacerta vivipara* (Reptiles, Lacertilians), *Experientia*, 27, 1456, 1971.
176. **Petit, A. and Vivien-Roels, B.**, Présence de contacts neurosensoriels et de synapses d'un type nouveau dans l'épiphyse du Lézard des murailles (*Lacerta muralis*, Laurenti), *C.R. Acad. Sci.*, 284, 1911, 1977.
177. **Collin, J.-P., Juillard, M.-T., and Falcon, J.**, Localization of 5-hydroxytryptamine and protein(s) in the secretion granules of the rudimentary photoreceptor cells in the pineal of *Lacerta, J. Neurocytol.*, 6, 541, 1977.
178. **Kappers, J. A.**, The pineal organ: an introduction, in *The Pineal Gland*, Wolstenholme, G. E. W. and Knight, J., Eds., Churchill-Livingstone, Edinburgh, 1971, 3.
179. **Oksche, A., Ueck, M., and Rüdeberg, C.**, Comparative ultrastructural studies of sensory and secretory elements in pineal organs, *Mem. Soc. Endocrinol.*, 19, 7, 1971.
180. **Vivien-Roels, B.**, Ultrastructure, innervation et fonction de l'epiphyse chez les Chéloniens, *Z. Zellforsch. Mikrosk. Anat.*, 104, 429, 1970.
181. **Owens, D. W. and Ralph, C. L.**, The pineal-paraphyseal complex of sea turtles, *J. Morphol.*, 158, 169, 1978.
182. **Mehring, G.**, Licht-und electronen mikroskopische Untersuchung des Pineal Organs von *Testudo hermanni, Anat. Anz.*, 131, 184, 1972.

183. **Quay, W. B., Kappers, J. A., and Jongkind, J. F.**, Innervation and fluorescence histochemistry of monoamines in the pineal organ of a snake, *(Natrix natrix), J. Neuro-Visc. Relat.*, 31, 11, 1968.
184. **Vivien-Roels, B. and Petit, A.**, Relative roles of light and temperature in the control of pineal serotonin (5-HT) circadian and circannual variations in reptiles, *Gen. Comp. Endocrinol.*, 34, 77, 1978.
185. **Vivien-Roels, B. and Arendt, J.**, Variations circadiennes et cirannuelles de la mélatonine épiphysaire chez *Testudo hermanii* G. (Reptile-Chelonien) dans des conditions naturelles d'eclairement et de température, *Ann. Endocrinol. (Paris)*, 40, 93, 1979.
186. **Vivien-Roels, B., Arendt, J., and Bradtke, J.**, Circadian and circannual fluctuations of pineal indoleamines (serotonin and melatonin) in *Testudo hermanni:* Amelin (Reptilia, Chelonia), *Gen. Comp. Endocrinol.*, 37, 197, 1979.
187. **Quay, W. B., Jongkind, J. F., and Kappers, J. A.**, Localizations and experimental changes in monoamines of the reptilian pineal complex studied by fluorescence histochemistry, *Anat. Rec.*, 157, 304, 1967.
188. **Joss, J. M. P.**, Rhythmicity in the production of melatonin by *Lampropholas guichenoti* (Scincid lizard) and *Geotria australis* (Lamprey), in *Comparative Endocrinology*, Gaillard, P. J. and Boer, H. H., Eds., Elsevier/North-Holland, New York, 1978, 172.
189. **Quay, W. B., Stebbins, R. C., Kelley, T. D., and Cohen, N. W.**, Effects of environmental and physiological factors on pineal acetylserotonin and methyltransferase activity in the lizard *Sceloporus occidentalis, Physiol. Zool.*, 44, 241, 1971.
190. **Kennaway, D. J., Frith, R. G., Phillipou, G., Matthews, C. D., and Seamark, R. F.**, A specific radioimmunoassay for melatonin in biological tissue and fluids and its validation by gas chromatography-mass spectrometry, *Endocrinology*, 101, 119, 1977.
191. **Firth, B. T., Kennaway, D. J., and Rozenbilds, M. A. M.**, Plasma melatonin in the scincid lizard, *Trachydosaurus rugosus:* diel rhythm, seasonality, and the effect of constant light and constant darkness, *Gen. Comp. Endocrinol.*, 37, 493, 1979.
192. **Gern, W. A., Owens, D. W., Ralph, C. L., and Roth, J. J.**, Plasma melatonin from extra-pineal sites, *Am. Zool.*, 18, 670, 1978.
193. **Vivien-Roels, B. and Petit, A.**, Dosage spectrofluorimetrique de la sérotonine (5-hydroxytryptamine = 5-HT) épiphysaire chez les Reptiles. Etude des variations du taux de 5-HT en fonction du cycle saisonnier chez *Testudo hermanni, C.R. Acad. Sci.*, 280, 467, 1975.
194. **Owens, D. W., Gern, W. A., and Ralph, C. L.**, Melatonin in the blood and cerebrospinal fluid of the green sea turtle *(Chelonia mydas), Gen. Comp. Endocrinol.*, 40, 180, 1980.
195. **Underwood, H. and Menaker, M.**, Extraretinal photoreception in lizards, *Photochem. Photobiol.*, 23, 227, 1976.
196. **Graham, T. E. and Hutchison, V. H.**, Turtle diel activity: response to different regimes of temperature and photoperiod, *Comp. Biochem. Physiol.*, 63A, 299, 1979.
197. **Underwood, H.**, Circadian organization in lizards: the role of the pineal organ, *Science*, 195, 587, 1977.
198. **Underwood, H.**, Melatonin affects circadian rhythmicity in lizards, *J. Comp. Physiol.*, 130, 317, 1979.
199. **Licht, P.**, Regulation of the annual testis cycle by photoperiod and temperature in the lizard *Anolis carolinensis, Ecology*, 52, 240, 1971.
200. **Stebbins, R. C.**, The effect of parietalectomy on testicular activity and exposure to light in the desert night lizard *(Xantusia vigilis), Copeia*, 1970, 261, 1970.
201. **Stebbins, R. C. and Cohen, N. W.**, The effect of parietalectomy on the thyroid and gonads in free-living western fence lizards, *Sceloporus occidentalis, Copeia*, 1973, 662, 1973.
202. **Roth, J. J. and Ralph, C. L.**, Thermal and photic preferences in intact and parietalectomized *Anolis carolinensis, Behav. Biol.*, 19, 341, 1977.
203. **Clausen, H. J. and Poris, E. G.**, The effect of light upon sexual activity in the lizard, *Anolis carolinensis*, with special reference to the pineal body, *Anat. Rec.*, 69, 39, 1937.
204. **Licht, P. and Pearson, A. K.**, Failure of parietalectomy to affect the testes in the lizard *Anolis carolinensis, Copeia*, 1970, 172, 1970.
205. **Aron, E., Combescot, C., Demaret, J., and Guyon, L.**, Neurosécretion chez la tortue d'eau *Emys leprosa* apres destruction de la région épiphysaire, *C.R. Acad. Sci.*, 251, 1914, 1960.
206. **Combescot, C. and Demaret, J.**, Histophysiologie de l'épiphyse chez la Tortue d'eau *Emys leprosa* (Schw), *Ann. Endocrinol. (Paris)*, 24, 204, 1963.
207. **Levey, I. L.**, Effects of pinealectomy and melatonin injections at different seasons on ovarian activity in the lizard *Anolis carolinensis, J. Exp. Zool.*, 185, 169, 1973.
208. **Packard, M. J. and Packard, G. C.**, Antigonadotrophic effect of melatonin in male lizards *(Callisaurus draconoides), Experientia*, 33, 1665, 1977.
209. **Haldar, C. and Thaphiyal, J. P.**, Effect of pinealectomy on the annual testicular cycle of *Calotes versicolor, Gen. Comp. Endocrinol.*, 32, 395, 1977.

210. **Thapliyal, J. P. and Haldar, C. M.,** Effect of pinealectomy on the photoperiodic gonadal response of the Indian garden lizard, *Calotes versicolor, Gen. Comp. Endocrinol.,* 39, 79, 1979.
211. **Vivien, J. H.,** Signes de stimulation des activités sécrétoire des pinéalocytes chez la couleuvre *Tropidonotus natrix* L. traitée par des principes gonadotropes, *C.R. Acad. Sci.,* 260, 5371, 1965.
212. **Cardinali, D. P., Vacas, M. I., and Boyer, E. E.,** Specific binding of melatonin in bovine brain, *Endocrinology,* 105, 437, 1979.

Chapter 2

THE PINEAL AND REPRODUCTION IN BIRDS

Charles L. Ralph

TABLE OF CONTENTS

I. INTRODUCTION

A large portion of the metabolism, much of the behavior, and many morphological features in the Class Aves relate to reproduction.[1] Like reptiles and mammals, birds produce only a few progeny, in which considerable amounts of energy and material are invested. Courtship rituals often involve elaborate vocalization, display of colorful feathers and skin, and a variety of other behavioral maneuvers. The construction of nests, incubation, and the care of the young by one or both parents are energetically costly and require complex regulatory mechanisms. The reproductive organs of birds, and the regulating hormones and their neuroendocrine control mechanisms, have been well studied.[2] The pineal gland has been implicated as a component in the control system for avian reproduction.

The author published a comprehensive review of pineal structure and function of birds in 1970[3] which included works relating to the possible involvement of the pineal gland in reproductive processes. A brief summary of the subject also was incorporated in a review published in 1978.[4] To avoid redundancy, the present collation will deal mainly with works published in the last decade.

II. STRUCTURE OF THE AVIAN PINEAL GLAND

Comprehensive reviews dealing with the embryology, morphology, vascularization, innervation, and cytology of avian pineal glands are available.[3,5-9] Among species of birds, the morphology and cytology of the pineal gland are remarkably varied. In some Strigiformes (owls) and Procellariiformes (shearwaters, petrels) the pineal body is notably atrophic.[10] The Spheniscidae (penguins)[8,11] and Dromiceidae (emus),[8] in contrast, have very large pineal bodies, even when considered in relation to their relatively large body size. Cytologically, pineal bodies usually are compact, as in pigeons and chickens, or saccular, as in passerine birds.[3]

Recent microscopic studies of the pineal gland of the chicken (*Gallus domesticus*)[12] suggest that the organ is active and functional throughout the adult life, changing from a follicular configuration in the younger bird into cellular rosette formations in the older animal.[13] A histochemical examination of the pineal gland of the domestic fowl by Wight and MacKenzie[14] revealed abundant lipids, several enzymes (including alkaline and acid phosphatase), ATPase, lipase, and nonspecific esterase, as well as abundant RNA, but no glycogen, no Gomori-positive neurosecretory substances, and no calcium. There was PAS-positive material and globules in the lumina of some vesicles.

According to Collin,[15] the unique cell type in the avian pineal, the secretory pinealocyte, is evolutionarily derived from a cone-like receptor cell found among some reptiles and anamniotes. This regressed photoreceptor cell is, in Collin's terminology, a rudimentary photoreceptor cell.

III. PINEAL SECRETION AND PHOTIC RESPONSIVENESS

The regressed photoreceptor cell, or rudimentary photoreceptor cell,[15] may be a source of indoleamines, including melatonin. Collin et al.,[16] on the basis of an autoradiographic study, localized indoleamine precursors to dense-cored vesicles of the avian pinealocytes. They concluded that the secretory rudimentary photoreceptor cells synthesize serotonin and possibly other indoleamines as well. Using the Falck-Hillarp fluorescence technique for the demonstration of monoamines in the pineals of parakeets, Juillard et al.[17] provided further suggestive evidence that the rudimentary photoreceptor cells synthesize indoleamines.

A. Melatonin Rhythms

The first direct evidence that melatonin is present in the circulation was provided by Pelham et al.[18] who identified it by mass spectrometry in chicken blood. In all animals so far examined, the melatonin content of the pineal body, blood, and other tissues varies in a rhythmic manner with greater amounts present during nighttime than during daytime.[19,20] A precursor compound, serotonin (5-hydroxytryptamine), also varies rhythmically in its pineal content; in the Japanese quail the highest levels were found near the beginning of the light period and the lowest levels in the middle of the dark phase.[21]

The rhythmic variation in melatonin content of the pineal gland of chickens persists as a free-running, circadian rhythm for at least 2 weeks in continuous darkness,[22,23] but it disappears within 2 weeks in constant light.[23] The circadian melatonin oscillations appear to be phase-locked with the unique locomotor activity rhythm of the bird.[22] Following pinealectomy, the nocturnal rise in serum melatonin is abolished.[24]

When the pineal gland of chickens was autotransplanted to the anterior chamber of the eye, the melatonin content of both the brain and blood serum were lower than in normal, unoperated chickens; however, the ratio of brain to serum melatonin was the same as in normal birds (7:1).[25] This result was interpreted by Pang and Ralph[25] to mean that pineal melatonin is secreted into the blood and then taken up by the brain and concentrated there. The greatest amount of melatonin in the brain of chickens was found in the hypothalamic region.[26]

The autotransplanted pineal gland of chickens apparently released greater amounts of melatonin into the blood at night than during the day, since by bioassay melatonin could be detected there only when sampled at night.[25] It is paradoxical, then, that organ-cultured duck (*Anas platyrhinchos*) pineal glands respond to direct illumination by increasing the conversion of serotonin to melatonin and other products, according to Rosner et al.[27] Additionally, Hisano et al.[28] illuminated the "pineal region" of blinded ducks and obtained testicular stimulation. This latter result perhaps can best be explained as a response to stimulation of the photoreceptive region of the brain by light, which is known to mediate reproductive effects,[29] despite the finding that pinealectomy diminished the effects of the illumination.[28] The injury occasioned by the surgical removal of the pineal gland can affect the measured response in various ways, including increasing the opacity of the skull and underlying tissues.

B. Melatonin-Forming Enzymes

Two key enzymes for melatonin synthesis have been fairly well studied in bird pineals. These are hydroxyindole-*O*-methyltransferase (HIOMT) and *N*-acetyltransferase (NAT). HIOMT is slightly and variably influenced by ambient light. In Japanese quail (*Coturnix coturnix japonica*) pineal HIOMT has been reported to be inhibited by constant darkness.[30] HIOMT activity in developing chicks was found by Wainwright[31] to be higher in constant light than in constant darkness. Chickens[32] and *Coturnix*[33] that were maintained in 24-hr light-dark cycles had HIOMT activity slightly elevated during the dark phase. However, other studies generally failed to confirm a significant rhythmic change in pineal HIOMT of chickens.[34,35] Furthermore, HIOMT activity did not change consistently in chickens subjected to light-to-dark and dark-to-light transitions at atypical times.[36] In two species of penguins, sacrificed during the Antarctic summer (January) and sampled at 1200 and 2400 hr, neither pineal HIOMT nor melatonin levels varied significantly, although in every case the 2400 samples were slightly higher.[37]

The pineal glands of birds appear either to have two species of HIOMT or to have one HIOMT with a wider substrate specificity than that found in mammals. The chief metabolite of exogenous tryptophan by chick pineal glands in organ culture is melatonin, whereas exogenous serotonin is mainly converted to hydroxyindoleacetic acid according to Wainwright.[38] However, endogenously synthesized serotonin may enter a pool which is preferentially converted to melatonin.[38]

In 1970 interest shifted from HIOMT to NAT when the latter enzyme in rats was shown to be dramatically stimulated by norepinephrine and cyclic nucleotides and to be rapidly inactivated by light. Backstrom et al.[39] conjectured that NAT might be primarily responsible for the diurnal variation in pineal melatonin in the Japanese quail. In chicken pineals, the tenfold nocturnal rise in melatonin was found by Binkley et al.[34] to be phased identically with a 27-fold increase in NAT activity. A dark-to-light transition caused a rapid decrease (within 30 min) in pineal NAT activity and melatonin content.[36,40] Furthermore, the daily rhythms of both pineal NAT and melatonin persisted in phase in chickens that were kept in constant darkness but the rhythms disappeared under constant light.[22,23,40-42]

The shape of the pineal NAT rhythm is influenced by the pattern of the ambient light-dark cycle. In newly hatched chicks, NAT remained elevated longer when the dark phase of the photocycle was long (16 hr) and was elevated for a shorter period of time when the dark period was shorter (8 hr).[42] Explanted chick pineals in organ culture showed a marked increase in NAT when in the dark as compared to in the light.[43] The more closely the lighting conditions of in vitro chick pineal glands matched those under which the donor chicks had been raised, the closer was the similarity between cycles in levels of NAT activity in vitro and in vivo.[44]

Despite the rapid responsiveness of the avian pineal gland to lighting conditions, the normal diurnal cycle in NAT activity is regulated primarily by a circadian pacemaker. The nocturnal in vivo increase of activity begins before the start of the dark phase and the decline in activity is largely completed before the dark phase ends.[34,44] Most remarkably, entrainment of the cycle of pineal NAT in chickens to a light cycle does not require retinal photoreception, since it can be accomplished in blinded birds.[23,36]

Binkley et al.[45] have shown that the timing of the decline in NAT activity of the organ-cultured chick pineal gland is related to the light cycle to which the donor chick was previously exposed and not to the time the gland was placed into culture. Furthermore, repetitive cycles in levels of NAT activity persist in culture for at least 4 days under a diurnal cycle of illumination[44] and for at least 2 days in continuous darkness.[44-48]

Binkley et al.,[49] investigating the NAT of the retinae of chicks and sparrows (*Passer domesticus*), found that its activity varied daily, like pineal NAT, being greater during the dark phase than during light. Ocular NAT was depressed when chicks were exposed to continuous light and it was rapidly diminished in activity when they were given light during the normal dark time, in a manner similar to the responses of pineal NAT.

Barfuss and Ellis[50] collected free-living male *P. domesticus* throughout a year and assayed the pineal glands for HIOMT activity, noted beak color, and measured the size of the testes. They found HIOMT activity to be inversely related to testicular weights. Furthermore, in experimental situations, red light was found to be effective in reducing HIOMT activity and increasing testis size.

Alexander et al.[51] found that pineal HIOMT in female *Coturnix* increased from 25 to 30 days of age and then decreased during the period of rapid ovarian growth, remaining low through day 52. Preslock[52] likewise found a decline in pineal HIOMT activity during rapid sexual development in both male and female *Coturnix*. HIOMT activity rebounded in sexually mature *Coturnix* and was maintained at a high level until sexual activity declined. In contrast to HIOMT, pineal NAT was elevated in sexually maturing *Coturnix* and remained elevated in sexually mature males and females. In sexually inactive birds, however, NAT activity declined. Preslock speculates that gonadal steroids may stimulate NAT activity. Castration resulted in decreased HIOMT activity in both male and female *Coturnix*.[53,54] Administration of estrogens to females and androgens to males resulted in a return of HIOMT to levels similar to controls. NAT was unaffected by either castration or steroid treatment. Preslock[53] concluded that NAT is regulated primarily by photoperiod whereas HIOMT is regulated by photoperiod and gonadal steroids.

C. Other Biochemical Aspects

Only a few biochemical aspects of the avian pineal gland have been examined that do not relate to metabolism of indoles. The pineal gland of chickens appears to have the complex of enzymes necessary for sterol biosynthesis.[55] The diurnal uptake pattern of radioactive phosphorus (^{32}P) by the pineal gland of 3-day-old chicks is such that the greatest incorporation occurs during the light period.[56] Phosphorous is probably incorporated into phospholipid, nucleic acids, or proteins.

IV. EFFECTS OF PINEAL DENERVATION

The avian pineal gland is richly innervated by sympathetic afferent fibers.[3,57] These nerve fibers are adrenergic[58-60] and originate in the superior cervical ganglion.[59] Additionally, in the pineal gland of the chicken, histochemical methods suggest the presence of cholinergic fibers.[58] The origin of these fibers is unknown. Selective staining methods[3,61] plus ultrastructural features[60-62] have led some investigators to the conclusion that a few neurons are also present in bird pineal organs.

Most of the evidence, both cytological and electrophysiological, suggest that the avian pineal gland is not photoreceptive, at least not in the conventional sense.[3,4,57] However, according to Herbuté and Baylé,[63] illumination (light flashes) of the retina of Japanese quail inhibits spontaneous electrical activity (multiunit activity, MUA) in the pineal body. Blinding (by bilateral transection of the optic nerves) the quail reduces the inhibitory effect of light and results in spontaneous firing rates that are greater than in control birds. Quail maintained under short days (light 6 hr: dark 18 hr) had twice as much MUA as birds reared in long days (light 18 hr: dark 6 hr).[64] Superior cervical ganglionectomy significantly increased MUA, particularly in birds under short photoperiods. Castration of short-day birds increased MUA and testosterone administration suppressed firing rates. Testosterone was without effect in intact, short-day birds. Destruction of the habenular nuclei in quail by electrolytic lesions resulted in a marked increase in MUA in both long-day and short-day birds.[65] The suppressing effect of light flashes was attenuated by the habenular lesions. Thus, phasic photic information, detected by retinal receptors, may reach the pineal gland via the habenular nuclei and the pineal stalk in this species, but steady light may involve neither the habenular nor the cervical sympathetic routes.

Bilateral ablation of the superior cervical ganglia deprives the pineal body of gallinaceous birds of sympathetic innervation, as indicated by the disappearance of fluorescing amines from histological sections treated by the Falck-Owman technique.[59] Postganglionic sympathetic fibers have been demonstrated to be importantly involved in the control of melatonin synthesis in mammalian (rat) pineal glands.[66] Since melatonin had been shown to influence reproductive processes in mammals, it was surprising when Ralph et al.[67] failed to find any effect of superior cervical ganglionectomy on oviposition rates of Japanese quail. As further proof that the sympathetic innervation does influence indoleamine metabolism in the bird pineal, Hedlund et al.[21] demonstrated that superior cervical ganglionectomy of Japanese quail significantly reduced the serotonin content of the pineal gland at the time of expected peak levels. However, it was later shown that the daily changes in NAT activity and melatonin content in chicken pineal glands do not require the superior cervical ganglia.[23] (For these cycles to persist in constant darkness, however, the ganglia appear to be essential.) Neither blinding nor superior cervical ganglionectomy of chickens affects HIOMT activity.[23] As noted above, this is unlike the situation in the rat,[66] but, as also pointed out, bird pineal glands, again unlike most mammals, may be innerated by cholinergic fibers.[58] Perhaps the dual innervation is somehow related to the difference.

V. EFFECTS OF PINEALECTOMY

In the author's 1970 review,[3] some 30 reports on the effects of pinealectomy in birds were summarized. No general conclusion regarding the effect of pineal ablation could be drawn. In some studies pinealectomy resulted in gonadal stimulation and thus the pineal gland was alleged to be antigonadal. In others it has been considered to be progonadal, since pinealectomy retarded or inhibited gonadal function. In some reports the pineal gland was thought to be antigonadal during one stage of development but progonadal during another. In still other studies no effect at all was noted following pinealectomy. More recent studies continue to yield very mixed results.

Lesioning the pineals of house finches (*Carpodacus mexicanus*) failed to produce any effect on testis weight as compared to sham-operated birds.[68] In Harris's sparrows (*Zonotrichia querula*) pinealectomy did not prevent testicular regression nor eliminate photorefractoriness.[69] Pinealectomy had no detectable influence on the photoperiodic control of testicular growth in the white-crowned sparrow (*Zonotrichia leucophrys gambelii*).[62] When adult male ducks (*Anas platyrhinchos*) were pinealectomized in the spring, testicular growth and testosterone synthesis was inhibited during the breeding season, but a year later there was no difference among pinealectomized, sham-operated, and control ducks.[70] Cuello et al.[71] claimed that pinealectomy decreased the ovarian response to continuous light of enucleated *A. platyrhinchos*. However, no statistical analysis is given and the number of animals per group is small and variable, making their conclusion dubious. Menaker et al.[72] found that both the eyes and the pineal gland could be extirpated from house sparrows (*Passer domesticus*) and the testes still recrudesced normally in response to long photoperiods.

Pinealectomy of 1- to 2-day-old White Leghorn chicks had no effect on ovarian weights when examined 8 weeks later, either under continuous light or in 8-hr light: 16-hr dark cycles.[73] White Leghorn hens that were pinealectomized, optically blinded, or both pinealectomized and blinded all responded similarly, in terms of egg production to changes of photoperiod and spectrum.[74] Pinealectomy at 8 to 9 days of age, however, delayed normal sexual development and depressed plasma testosterone of White Leghorn cockerels.[75] Thus, in this last study, the pineal gland would appear to be normally progonadal, as would also appear to be the case for the duck (cited above[70,71]).

Pinealectomy of either blinded or intact male *Coturnix* did not produce a significant effect in the basic pattern of gonadal response to changes of photoperiod.[76] Neither the presence of the eyes nor the pineal gland are essential to gonadal maturation and function in male and female *Coturnix*.[77] Accelerated sexual maturity was observed after enucleation, in the males but pinealectomy had no effect on this response. Neither did pinealectomy influence the resistance of blinded birds of both sexes to photo-induced gonadal regression, nor did it influence gonadal changes induced by long or short photoperiods in intact birds.

Homma and Sakakibara[78] implanted small plates of solidified radioluminous paint along the fissura longitudinalis cerebri, adjacent to the hypothalamic area of pinealectomized *Coturnix* and observed significant testicular development. They concluded that the pineal does not play an important role as a photoreceptor or an endocrine organ in regulating gonadal development.

Indian weaver birds (*Ploceus philippinus*) became sexually precocious following pinealectomy.[79] When pinealectomy was combined with long photoperiods (18L:6D) the onset of the reproductive condition was accelerated more than by either treatment alone. Additionally, short photoperiods (8L:15D), which normally are nonstimulatory, became stimulatory in pinealectomized birds. The overall conclusion is that the pineal gland inhibits gonadal growth in juveniles and delays the onset of seasonal breeding in the adults of this species. A subsequent study,[80] involving pinealectomy of male Indian weaver birds during nonbreeding phase, resulted in an early recrudescence of the testis, increased hypothalamic luteinizing hormone-releasing hormone (LH-RH) and plasma luteinizing hormone (LH), and reduced pituitary LH. Pinealectomy in the breeding phase prevented gonadal regression, and there was reduced hypothalamic LH-RH and plasma LH. The pineal gland of Indian weaver birds, thus, appears to be antigonadal.

VI. EFFECTS OF ADMINISTERING PINEAL SUBSTANCES

A comprehensive literature review of a decade ago[3] revealed no truly convincing data to support the view that crude pineal preparations, melatonin or serotonin had consistently significant effects on the size or function of gonads or accessory reproductive organs in birds. More recent studies, however, tend to support the possibility that pineal-characteristic substances may alter reproductive function.

Balemans[81] reported that melatonin and 5-methoxytryptophol (5-MTP), a derivative of 5-hydroxytryptophan, have stimulatory effects on the testes and comb of juvenile White Leghorn chickens, but inhibitory actions on maturing and adult chickens. 5-MTP was more inhibitory than melatonin. Both compounds were administered in increasing concentrations (10 to 146 μg or 20 to 156 μg, three times weekly) for about 6 weeks. In a subsequent study, Balemans[82] reported that much lower doses of 5-MTP (0.1 to 35 μg) stimulated comb and testis growth in cockerels, but higher doses (2.5 to 350 μg) did not. When 5-MTP was administered three times weekly in increasing concentrations (1 to 39.25 μg) for about 5 weeks to adult females, Balemans[83] reported that ovarian and follicular growth were inhibited.

In an attempt to further verify the stimulatory effect of 5-MTP in juveniles and the inhibitory effect in older birds, White Leghorn hens that were 6 weeks old, 4 months old or 1 year old were given 5-MTP.[84] Again, 5-MTP was given thrice weekly and the amount was increased over time. The juveniles received 0.1 to 3.9 μg per injection, the intermediate group 0.1 to 9.55 μg and the oldest group 1 to 30.25 μg. Comb development was retarded only in the juveniles but, paradoxically, they had an increase in ovarian weight as compared to controls. In the maturing group, ovarian growth was retarded but the adult group was unaffected. However, the size of the follicles weighing more than 2 g each was decreased in adult birds but in the maturing birds the indole did not affect follicle mass.

In still another study by Balemans et al.,[85] White Leghorn cockerels were treated in a manner similar to the females just described. Birds were sorted on the basis of comb size, since this is commonly regarded as an index of testicular function. When melatonin was administered to cockerels having a comb index (length × height/2) of 2 or less there was no effect on testis size. However, when given to cockerels with indices between 2.0 and 2.45, melatonin administration resulted in a decrease in testis size. 5-MTP, regardless of the comb index when injections were begun, accelerated the development of the gonads. Paradoxically, when melatonin was given in combination with 5-MTP the stimulatory effect was exaggerated. In summary, melatonin would appear to be inhibitory to testicular development whereas 5-MTP is stimulatory.

All the above experiments by Balemans were conducted under such variable conditions that it is difficult to interpret the results. They were done at different times of the year, even within one set of experiments, and under varying, natural photoperiods. The amount of the indoles also injected varied with time (presumably, in part, to keep pace with the changing size of the animal). In addition, there were no tests for significant differences. Thus, the potential value of these data is unfortunately diminished by the complexity of the experimental designs and lack of statistical treatment of the data.

Sackman[86] observed that a 20 μg injection of LH depressed ^{32}P uptake (an indicator of metabolic rate) by the pineal gland of 13-day-old White Leghorn cockerels. However, in birds 3- or 10-days-old, LH was without effect on uptake of the isotope. This result was interpreted to mean that the metabolism of the pineal gland is not responsive to LH administration until the birds are more than 11-days-old. A lack of an effect of LH on pineal gland in vitro led Sackman to conclude that the response observed in the oldest birds is mediated by the testes.

Sackman[87] injected melatonin (100 μg) daily (at 1700 hr; 11.5 hr after lights on) into White Leghorn cockerels from the second to thirteenth day of life. Throughout the experiment the birds were exposed to 14 hr of light per day. On day 14, the experimental birds were given a single injection of 2.5 μg LH-RH at 30, 90, or 120 min before sacrifice. At 2 hr before sacrifice the birds were also given 2.0 μCi ^{32}P each. Testis weight and ^{32}P uptake, as well as pituitary weight and gonadotrophin content, were determined. Neither testis nor pituitary gland weight was affected by any of the treatments. However, LH-RH significantly increased ^{32}P uptake by the testes in birds not given melatonin, the effect increasing with the interval of time between LH-RH injection and sacrifice. Melatonin-injected birds failed to accumulate ^{32}P. LH-RH in birds that did not receive melatonin caused an initial decrease (90 min) in pituitary gonadotrophin content followed by an increase (120 min). Surprisingly, melatonin appeared to enhance the effect of LH-RH on pituitary gonadotrophin, the 90 min levels being lower and the 120 min levels being higher than in those cockerels given only LH-RH.

Mature female *Coturnix* that were implanted weekly for 4 weeks with melatonin (10 μg or 1 μg) in beeswax pellets exhibited significantly reduced pineal HIOMT and NAT activities, according to Wright and Preslock.[88] Oviductal weights tended to be higher with increasing quantity of melatonin. The investigators suggest that melatonin exerts a feedback inhibition on its synthesizing enzymes.

Melatonin injections can affect plasma levels of fatty acids in pigeons, depending upon the time of day it is administered.[89] Melatonin (1.25 mg/kg body weight) injected intravenously into pigeons during midphotophase caused an increase in plasma growth hormone but not during the scotophase, according to McKeown et al.[90] A higher dose (5 mg/kg body weight) did not cause any significant change in plasma growth hormone. However, the higher dose allegedly caused plasma glucose to be elevated when given during both midphotophase and midscotophase. These results are paradoxical and also questionable because of the very large amounts of melatonin administered. They may suggest, nevertheless, that melatonin has actions on the brain that can result in influences on any of several hypothalamic-hypophysial axes.

Melatonin has been detected in the brain of the chicken[26,91] and, as noted above, it is highly probable that the brain accumulates melatonin from the blood,[25] especially the region that includes the hypothalamus. Melatonin (1 to 2 mg), when injected into chickens, altered the electroencephalogram (EEG) within 2 min.[92] Also, chickens that had been pinealectomized 10 to 12 weeks earlier had atypical EEGs.[92] Melatonin, appears to act directly on the central nervous system to augment ketamine-induced catatonia in 13- to 22-day-old chicks.[93] Thus, it is quite logical to presume that melatonin acts through the central nervous system, a speculation previously entertained.[4,20]

VII. SUMMARY AND CONCLUSIONS

The pineal gland is clearly implicated in reproductive processes in some mammals (Reiter, Hoffmann this volume). Since there are parallels between birds and mammals in regard to the chemistry and photoresponsiveness of their pineal glands,[66] one might expect them to serve similar functions. Both birds and mammals* synthesize and secrete melatonin in a rhythmic pattern, with greater amounts being released at night. In both, the phasing of the rhythms is set by the ambient photoperiod and the rhythms persist in continuous darkness for several days. However, there also are differences between avian and mammalian pineal glands.[66] The avian gland can be driven by light that is perceived by nonretinal (cephalic) photoreceptors;[29] the mammalian pineal gland appears to be only responsive to retinally perceived light. Both the avian and mammalian glands are innerated by adrenergic fibers from the superior cervical ganglia,[59-62] but the pineal gland of the bird also may have a second set of fibers that are cholinergic.[58] Superior cervical ganglionectomy in gallinaceous birds, unlike the situation in mammals, has no apparent effect on pineal HIOMT activity or on the daily rhythmic changes in NAT activity or on melatonin synthesis.[66] Again, unlike the mammalian pineal gland, the avian gland in vitro does not respond to norepinephrine by increasing its synthesis of melatonin.[66]

Despite these differences, it is tempting to believe that, as in mammals, the pineal may have a controlling role in the reproductive physiology of birds. Birds are remarkably responsive, in terms of reproductive activity to day length. They are also seasonal breeders. The pineal gland, so very responsive to light in terms of melatonin production, would seem to be a logical candidate as a transducer of photic information for the reproductive system. However, the supporting evidence for such a role for the pineal gland among birds is disappointingly weak.

Changes in the activity of melatonin-forming enzymes do correlate in a somewhat logical way with sexual activity. HIOMT activity has been reported to decline during times of sexual activity or sexual maturation.[50-54] However, HIOMT can be at a high level in sexually-mature birds as can NAT activity. What such changes signify remains questionable.

Denervation of the pineal gland by superior cervical ganglionectomy failed to affect egg production in Japanese quail.[67] Pinealectomy caused either no changes (passerines,[62,68,69,72] chicken,[74] and *Coturnix*[76-78]) in reproductive function or minor (chicken[75]) and transient (duck[70]) effects. More dramatic effects are reported for Indian weaver birds[79,80] where pinealectomy resulted in sexual precocity and made the bird responsive to short photoperiods, which normally are not stimulatory.

Injections into gallinaceous birds of substances characteristic of pineal glands — melatonin and 5-methoxytryptophol (5MTP) — have produced intriguing effects. 5-MTP was generally stimulatory to sexual development in juvenile chickens but inhibitory in older birds. Melatonin tended to be inhibitory for sexual development. However, the experiments (Balemans[81-85]) on which these conclusions are based, are complicated in design and execution, not statistically analyzed, and difficult to interpret due to incoherent results.

Thus, the assembled evidence does not provide convincing evidence that the pineal gland is of primary importance in the control of reproduction or sexual activity among birds. There are tantalizing bits of data which suggest that the pineal gland does have certain actions in the control of reproductive processes, but collectively they fail to make a solid case.

* It would be more accurate to say chickens and rats in place of birds and mammals, respectively, since most of what is known about melatonin in the two classes is derived from these two species.

The nonreproductive functions of the avian pineal gland also remain obscure. In four species of the Family Fringillidae (sparrows) of the Order Passeriformes, the pineal gland is essential for maintenance of a circadian locomotor rhythm; in the Family Sturnidae (starling) of the same order, it is equivocal whether or not the pineal body is essential for the rhythm; in the Order Galliformes (chicken and *Coturnix*) it is not.[94] Under certain conditions, the pineal gland of sparrows, pigeons, and chickens appears to participate in body temperature regulation.[95] The administration of melatonin to sparrows and pigeons depresses the body temperature.[95] The specific effect of such manipulations commonly depends upon the time of day (daytime vs. nighttime) when they are performed.[95,96] Thus, perhaps there is a relationship between the pineal's role in rhythmicity and in thermoregulation.

The function of the avian pineal gland, thus, remains frustratingly enigmatic. The available information base is small, the number of species so far examined is few, and the experimental manipulations are limited and even questionable in some cases. We know a fair amount about the pineals of birds, especially in regard to melatonin synthesis — the enzymes, the rhythmic pattern, and the effects of light and darkness. What one would most like to know is how pineal chemistry relates to the regulations and responses of birds as they function daily and annually. This largely remains to be discovered.

REFERENCES

1. **Welty, J. C.,** *The Life of Birds,* 2nd ed., W. B. Saunders, Philadelphia, 1975.
2. **Murton, R. K. and Westwood, N. J.,** *Avian Breeding Cycles,* Clarendon Press, Oxford, 1977.
3. **Ralph, C. L.,** Structure and alleged functions of avian pineals, *Am. Zool.,* 10, 217, 1970.
4. **Ralph, C. L.,** Pineal control of reproduction: nonmammalian vertebrates, *Prog. Reprod. Biol.,* 4, 30, 1978.
5. **Quay, W. B. and Renzoni, A.,** Comparative and experimental studies of pineal structure and cytology in passeriform birds, *Riv. Biol.,* 56, 363, 1963.
6. **Quay, W. B. and Renzoni, A.,** The diencephalic relations and variably bipartite structure of the avian pineal complex, *Riv. Biol.,* 60, 9, 1967.
7. **Menaker, M. and Oksche, A.,** The avian pineal organ, in *Avian Biology,* Vol. 4, Farner, D. S., King, J. R., and Parkes, K. C., Eds., Academic Press, New York, 1974, 79.
8. **Ralph, C. L.,** The pineal gland and geographical distribution of animals, *Int. J. Biometeor.,* 19, 289, 1975.
9. **Ralph, C. L.,** Cytology of the pineal gland: changes produced by various treatments, *J. Neural Transm.,* Suppl. 13, 25, 1978.
10. **Quay, W. B.,** Influence of pineal atrophy among birds and its relation to nocturnality, *Condor,* 74, 33, 1972.
11. **Piezzi, R. S. and Gutierrez, L. S.,** Electron microscopic study on the pineal gland of the antarctic penguin *(Pygoscelis papua), Cell Tissue Res.,* 164, 559, 1975.
12. **Omura, Y.,** Ultrastructural study of embryonic and post-hatching development in the pineal organ of the chicken (brown leghorn), *Gallus domesticus), Cell Tissue Res.,* 183, 255, 1977.
13. **Boya, J. and Calvo, J.,** Evolution of the pineal gland in the adult chicken, *Acta Anat.,* 104, 104, 1979.
14. **Wight, P. A. L. and MacKenzie, G. M.,** The histochemistry of the pineal gland of the domestic fowl, *J. Anat.,* 198, 261, 1971.
15. **Collin, J.-P.,** Differentiation and regression of the cells of the sensory line in the epiphysis cerebri, in *The Pineal Gland,* Wolstenholme, G. E. W. and Knight, J., Eds., Churchill Livingstone, Edinburgh, 1971, 79.
16. **Collin, J.-P., Calas, A., and Julliard, M. T.,** The avian pineal organ. Distribution of exogenous indoleamines. A quantitative study of the rudimentary photoreceptor cells by electron microscopic radioautography, *Exp. Brain Res.,* 25, 15, 1976.

17. **Julliard, M. T., Hartwig, H. G., and Collin, J.-P.,** The avian pineal organ. Distribution of endogenous monoamines; a fluorescence microscopic, microspectrofluorimetric and pharmacological study of the parakeet, *J. Neural Transm.*, 40, 269, 1977.
18. **Pelham, R. W., Ralph, C. L., and Campbell, I. M.,** Mass spectral identification of melatonin in blood, *Biochem. Biophys. Res. Commun.*, 46, 1236, 1972.
19. **Lynch, H. J.,** Diurnal oscillations in pineal melatonin content, *Life Sci.*, 10, 791, 1971.
20. **Ralph, C. L.,** Correlation of melatonin content in pineal gland, blood, and brain of some birds and mammals, *Am. Zool.*, 16, 35, 1976.
21. **Hedlund, L., Ralph, C. L., Chepko, J., and Lynch, H. J.,** A diurnal serotonin cycle in the pineal body of Japanese quail: photoperiod phasing and the effect of superior cervical ganglionectomy, *Gen. Comp. Endocrinol.*, 16, 52, 1971.
22. **Ralph, C. L., Pelham, R. W., MacBride, S. E., and Reilly, D. R.,** Persistent rhythms of pineal and serum melatonin in cockerels in continuous darkness, *J. Endocrinol.*, 63, 319, 1974.
23. **Ralph, C. L., Binkley, S., MacBride, S. E., and Klein, D.,** Regulation of pineal rhythms in chickens: effects of blinding, constant light, constant dark and superior cervical ganglionectomy, *Endocrinology*, 97, 1373, 1975.
24. **Pelham, R. W.,** A serum melatonin rhythm in chickens and its abolition by pinealectomy, *Endocrinology*, 96, 543, 1975.
25. **Pang, S. F. and Ralph, C. L.,** Mode of secretion of pineal melatonin in the chicken, *Gen. Comp. Endocrinol.*, 27, 125, 1975.
26. **Pang, S. F., Ralph, C. L., and Reilly, D. P.,** Melatonin in the chicken brain: its origin, diurnal variation, and regional distribution, *Gen. Comp. Endocrinol.*, 22, 499, 1974.
27. **Rosner, J. M., Denari, J. H., Nagle, C. A., Cardinali, D. P., de Pérez Bedés, G. D., and Orsi, L.,** Direct action of light on serotonin metabolism and RNA biosynthesis in duck pineal explants, *Life Sci.*, 11, 829, 1972.
28. **Hisano, N., Cardinali, D. P., Rosner, J. M., Nagle, C. A., and Tramezzani, J. H.,** Pineal role in the duck extraretinal photoreception, *Endocrinology*, 91, 1318, 1972.
29. **Menaker, M. and Underwood, H.,** Extraretinal photoreception in birds, *Photochem. Photobiol.*, 23, 299, 1976.
30. **Alexander, B., Dowd, A. J., and Wolfson, A.,** Effect of continuous light and darkness on hydroxyindole-*O*-methyltransferase and 5-hydroxytryptophan decarboxylase activities in the Japanese quail, *Endocrinology*, 86, 1441, 1970.
31. **Wainwright, S. D.,** Effects of changes in environmental lighting upon levels of hydroxyindole-*O*-methyltransferase activity in the developing-chick pineal gland, *Can. J. Biochem.*, 53, 438, 1975.
32. **Pelham, R. W. and Ralph, C. L.,** Pineal hydroxyindole-*O*-methyltransferase (HIOMT) in the chicken: effect of diurnal lighting and substrate concentration, *Life Sci.*, 11, 51, 1972.
33. **Preslock, J. P.,** Photoperiodic and gonadal steroid regulation of pineal hydroxyindole-*O*-methyltransferase and *N*-acetyltransferase in *Coturnix* quail, *Life Sci.*, 17, 1227, 1975.
34. **Binkley, S., MacBride, S. E., Klein, D. C., and Ralph, C. L.,** Pineal enzymes: regulation of avian melatonin synthesis, *Science*, 181, 273, 1973.
35. **Binkley, S. and Geller, E. B.,** Pineal enzymes in chickens: development of daily rhythmicity, *Gen. Comp. Endocrinol.*, 27, 424, 1975.
36. **Binkley, S., MacBride, S. E., Klein, D. C., and Ralph, C. L.,** Regulation of pineal rhythms in chickens: refractory period and nonvisual light perception, *Endocrinology*, 96, 848, 1975.
37. **Benelaz, G. A., Piezzi, R. A., and Lynch, H. J.,** Hydroxyindole-*O*-methyltransferase (HIOMT) and melatonin in the pineal gland of the Antarctic penguins (*Pygoscelis adeliae* and *P. papua*), *Gen. Comp. Endocrinol.*, 30, 43, 1976.
38. **Wainwright, S. D.,** Metabolism of tryptophan and serotonin by the chick pineal gland in organ culture, *Can. J. Biochem.*, 55, 415, 1977.
39. **Backstrom, M., Hetta, J., Wahlstrom, G., and Wetterberg, L.,** Enzyme regulation of melatonin synthesis in the pineal gland of Japanese quail, *Life Sci.*, 12, 493, 1972.
40. **Fraser, I. H. and Wainwright, S. D.,** The influence of lighting conditions upon the level and course of increase in specific activity of serotonin *N*-acetyltransferase in the developing chick pineal gland, *Can. J. Biochem.*, 54, 103, 1976.
41. **Binkley, S. and Geller, E. B.,** Pineal *N*-acetyltransferase in chickens: rhythm persists in constant darkness, *J. Comp. Physiol.*, 99, 67, 1975.
42. **Binkley, S., Stephens, J. L., Riebman, J. B., and Reilly, K. B.,** Regulation of pineal rhythms in chickens: photoperiod and dark-time sensitivity, *Gen. Comp. Endocrinol.*, 32, 411, 1977.
43. **Wainwright, S. D. and Wainwright, L. K.,** Regulation of the diurnal cycle in activity of serotonin acetyltransferase in the chick pineal gland, *Can. J. Biochem.*, 56, 685, 1978.
44. **Wainwright, S. D. and Wainwright, L. K.,** Chick pineal serotonin acetyltransferase: a diurnal cycle maintained *in vitro* and its regulation by light, *Can. J. Biochem.*, 57, 700, 1979.

45. **Binkley, S., Riebman, J. B., and Reilly, K. B.**, Timekeeping by the pineal gland, *Science*, 197, 1181, 1977.
46. **Binkley, S. A., Riebman, J. B., and Reilly, K. B.**, The pineal gland: a biological clock *in vitro*, *Science*, 202, 1198, 1978.
47. **Kasal, C. A., Menaker, M., and Perez-Polo, J. R.**, Circadian clock in culture: *N*-acetyltransferase activity of chick pineal glands oscillates *in vitro*, *Science*, 203, 656, 1979.
48. **Deguchi, T.**, Circadian rhythm of serotonin *N*-acetyltransferase activity in organ culture of a chicken pineal gland, *Science*, 203, 1245, 1979.
49. **Binkley, S., Hryshchyshyn, M., and Reilly, K.**, *N*-acetyltransferase activity responds to environmental lighting in the eye as well as in the pineal gland, *Nature (London)*, 281, 479, 1979.
50. **Barfuss, D. W. and Ellis, L. C.**, Seasonal cycles in melatonin synthesis by the pineal gland as related to testicular function in the house sparrow *(Passer domesticus)*, *Gen. Comp. Endocrinol.*, 17, 183, 1971.
51. **Alexander, B., Dowd, A. J., and Wolfson, A.**, Pineal hydroxyindole-*O*-methyltransferase (HIOMT) activity in female Japanese quail *(Coturnix coturnix japonica)*, *Neuroendocrinology*, 6, 236, 1970.
52. **Preslock, J. P.**, Pineal hydroxyindole-*O*-methyltransferase and *N*-acetyltransferase during sexual maturation of *Coturnix* quail, *Life Sci.*, 15, 1805, 1974.
53. **Preslock, J. P.**, Photoperiodic and gonadal steroid regulation of pineal hydroxyindole-*O*-methyltransferase activity and *N*-acetyltransferase in *Coturnix* quail, *Life Sci.*, 17, 1227, 1975.
54. **Preslock, J. P.**, Regulation of pineal enzymes by photoperiod, gonadal hormones and melatonin in *Coturnix* quail, *Hormone Res.*, 7, 108, 1976.
55. **Wells, J. R.**, Steroids in the pineal gland of the domestic fowl (*Gallus domesticus*): cholesterol and its biosynthesis, *Comp. Biochem. Physiol.*, 40, 723, 1971.
56. **Sackman, J. W.**, Circadian rhythm of pineal uptake of ^{32}P in domestic fowl, *J. Endocrinol.*, 73, 99, 1977.
57. **Ralph, C. L.**, The pineal complex: a retrospective view, *Am. Zool.*, 15 (Suppl. 1), 105, 1975.
58. **Wight, P. A. L. and MacKenzie, G. M.**, Dual innervation of the pineal of the fowl, *Gallus domesticus*, *Nature (London)*, 228, 474, 1970.
59. **Hedlund, L.**, Sympathetic innervation of the avian pineal body, *Anat. Rec.*, 166, 406, 1970.
60. **Ueck, M.**, Weitere Untersuchungen zur Feinstruktur und Innervation des Pineal organs von *Passer domesticus* L., *Z. Zellforsch. Mikrosk. Anat.*, 105, 276, 1970.
61. **Calvo, J. and Boya, J.**, Development of the innervation in the chicken pineal gland, *Acta Anat.*, 103, 212, 1979.
62. **Oksche, A., Kirschstein, H., Kobayashi, H., and Farner, D. S.**, Electron microscopic and experimental studies of pineal organ in the white-crowned sparrow *Zonotrichia leucophyrys gambelii*, *Z. Zellforsch. Mikrosk. Anat.*, 124, 247, 1972.
63. **Herbuté, S. and Baylé, J. D.**, Multiple-unit activity in the pineal gland of the Japanese quail: spontaneous firing and responses to photic stimulations, *Neuroendocrinology*, 16, 52, 1974.
64. **Herbuté, S. and Baylé, J. D.**, Influence of castration and testosterone on pineal multiunit activity in quail, *J. Physiol. (Paris)*, 72, 949, 1976.
65. **Herbuté, S. and Baylé, J. D.**, Suppression of pineal multiunit response to flash after habenular lesion in quail, *Am. J. Physiol.*, 233, E293, 1977.
66. **Binkley, S.**, Comparative biochemistry of the pineal glands of birds and mammals, *Am. Zool.*, 16, 57, 1976.
67. **Ralph, C. L., Lynch, H. J., Gundy, G. C., and Hedlund, L.**, Pineal function and oviposition in Japanese quail: superior cervical ganglionectomy and photoperiod, *Science*, 170, 995, 1970.
68. **Hamner, W. M. and Barfield, R. J.**, Ineffectiveness of pineal lesions on the testis cycle of a finch, *Condor*, 72, 99, 1970.
69. **Donham, R. S. and Wilson, F. E.**, Photorefractoriness in pinealectomized Harris' sparrow, *Condor*, 72, 101, 1970.
70. **Cardinali, D. P., Cuello, A. E., Tramezzani, J. H., and Rosner, J. M.**, Effects of pinealectomy on the testicular function of the adult male duck, *Endocrinology*, 89, 1082, 1971.
71. **Cuello, A. C., Hisano, N., and Tramezzani, J. H.**, The pineal gland and the photosexual reflex in female ducks, *Gen. Comp. Endocrinol.*, 18, 162, 1972.
72. **Menaker, M., Roberts, R., Elliott, J., and Underwood, H.**, Extraretinal light perception in the sparrow, III: the eyes do not participate in photoperiodic photoreception, *Proc. Nat. Acad. Sci. U.S.A.*, 67, 320, 1970.
73. **Godkhindi, S. S., Mithuji, G. F., Nephade, M. S., and Memon, G. N.**, Effect of light and pinealectomy on body growth and endocrine glands of white leghorn chicks, *Indian J. Exp. Biol.*, 9, 508, 1971.
74. **Harrison, P. C.**, Extraretinal photocontrol of reproductive responses of leghorn hens to photoperiods of different length and spectrum, *Poultry Sci.*, 51, 2060, 1972.

75. **Cogburn, L. A. and Harrison, P. C.**, Retardation of sexual development in pinealectomized single comb White Leghorn cockerels, *Poultry Sci.*, 56, 876, 1977.
76. **Homma, K., Wilson, W. O., and Siopes, T. D.**, Eyes have a role in photoperiodic control of sexual activity of *Coturnix, Science*, 178, 421, 1972.
77. **Siopes, T. D. and Wilson, W. O.**, Extraocular modification of photoreception in intact and pinealectomized *Coturnix, Poultry Sci.*, 53, 2035, 1974.
78. **Homma, K. and Sakakibara, Y.**, Encephalic photoreceptors and their significance in photoperiodic control of sexual activity in Japanese quail, in *Biochronometry*, Menaker, M., Ed., National Academy of Sciences, Washington, D.C., 1971, 333.
79. **Balasubramanian, K. S. and Saxena, R. N.**, Effect of pinealectomy and photoperiodism in the reproduction of Indian weaver birds, *Ploceus philippinus, J. Exp. Zool.*, 185, 333, 1973.
80. **Saxena, R. N., Malhotra, L., Kant, R., and Baweja, P. K.**, Effect of pinealectomy and seasonal changes on pineal antigonadotropic activity of male Indian weaver bird *Ploceus phillipinus, Indian J. Exp. Biol.*, 17, 732, 1979.
81. **Balemans, M. G. M.**, Age-dependent effects of 5-methoxytryptophol and melatonin on testes and comb growth of the White Leghorn (*Gallus domesticus* L.,), *J. Neural Transm.*, 33, 179, 1972.
82. **Balemans, M. G. M.**, The stimulatory effects of several concentrations of 5-methoxytryptophol on testicular growth in the White Leghorn (*Gallus domesticus* L.), *J. Neural Transm.*, 34, 49, 1973.
83. **Balemans, M. G. M.**, The inhibitory effect of 5-methoxytryptophol on ovarian weight, follicular growth and egg production of adult White Leghorn hens (*Gallus domesticus* L.), *J. Neural Transm.*, 34, 159, 1973.
84. **Balemans, M. G. M., van de Veerdonk, F. C. G., and van de Kamer, J. C.**, The influence of 5-methoxytryptophol, a pineal compound, on comb growth, follicular growth and egg production of juvenile, maturing and adult White Leghorn hens (*Gallus domesticus* L.), *J. Neural Transm.*, 41, 37, 1977.
85. **Balemans, M. G. M., van de Veerdonk, F. C. G., and van de Kamer, J. C.**, Effect of injecting 5-methoxy indoles, pineal compounds, on testicular weight of White Leghorn cockerels (*Gallus domesticus* L.), *J. Neural Transm.*, 41, 47, 1977.
86. **Sackman, J. W.**, The effect of luteinizing hormone on chicken testis and pineal gland uptake of ^{32}P, *Experientia*, 31, 1367, 1975.
87. **Sackman, J. W.**, The effect of melatonin on LH-RH induced changes in pituitary and gonadal function of immature domestic fowl, *Biol. Reprod.*, 17, 718, 1977.
88. **Wright, N. and Preslock, J. P.**, Melatonin inhibition of pineal enzymes in *Coturnix* quail, *Neuroendocrinology*, 19, 177, 1975.
89. **John, T. M. and George, J. C.**, Diurnal variation in the effect of melatonin on plasma and muscle free fatty acid levels in the pigeon, *Endocrinol. Exp.*, 10, 131, 1976.
90. **McKeown, B. A., John, T. M., and George, J. C.**, Diurnal variation in effects of melatonin on plasma growth hormone and glucose in the pigeon, *Endocrinol. Exp.*, 9, 263, 1975.
91. **Pang, S. F., Brown, G. M., Grota, L. J., Chambers, J. W., and Rodman, R. L.**, Determinations of *N*-acetylserotonin and melatonin activities in the pineal gland, retina, Hardarian gland, brain and serum of rats and chickens, *Neuroendocrinology*, 23, 1, 1977.
92. **Pang, S. F., Ralph, C. L., and Petrozza, J. A.**, Effects of melatonin administration and pinealectomy on the electroencephalogram of the chicken (*Gallus domesticus*) brain, *Life Sci.*, 18, 961, 1976.
93. **Lakin, M. L., Giedt, W. R., and Winters, W. D.**, The effect of pineal indolamines on the response to ketamine in the pinealectomized chick, *Proc. West. Pharmacol. Soc.*, 21, 41, 1978.
94. **Takahashi, J. S. and Menaker, M.**, Physiology of avian circadian pacemakers, *Fed. Proc. Fed. Am. Soc. Exp. Biol.*, 38, 2583, 1979.
95. **Ralph, C. L., Firth, B. T., Gern, W. A., and Owens, D. W.**, The pineal complex and thermoregulation, *Biol. Rev.*, 54, 41, 1979.
96. **Cogburn, L. A., Harrison, P. C., and Balsbaugh, R. K.**, Temperature preferendum of pinealectomized cockerels during their light-dark cycle, *Proc. Soc. Exp. Biol. Med.*, 161, 425, 1979.

Chapter 3

REPRODUCTIVE EFFECTS OF THE PINEAL GLAND AND PINEAL INDOLES IN THE SYRIAN HAMSTER AND THE ALBINO RAT

Russel J. Reiter

TABLE OF CONTENTS

I. INTRODUCTION

In a comprehensive monograph on the pineal published in 1919, Tilney and Warren* closed with the following comments:

The pineal gland cannot be considered a vestige in the light of the histological evidence, since the tendency toward specialization is definitely in the interest of glandular function in ophidians, chelonians, birds, and mammals. Ontogenetically, in two forms at least, in *Felis domestica* and man, the development of the pineal body follows the general lines of glandular differentiation. The pineal body is, therefore, a glandular structure and as such, is necessary in some way to metabolism.
The histology of the organ gives clear evidence that the epiphyseal complex of vertebrates possesses a pluripotentiality whose fundamental inherent tendency is in the interest of glandular differentiation, but in a few instances, as in cyclostomes, amphibia, and in primitive reptiles, the pineal organ may become further differentiated in the interest of a highly specialized sensory mechanism which has, or has had, visual function. As a gland, it may in some cases, contribute its secretion to the cerebrospinal fluid, but in higher vertebrates, as in ophidians, chelonians, birds, and mammals, it is an endocrinic (sic) organ, contributing the products of its secretion to the blood stream.

The firm conclusion that the pineal had differentiated into a glandular structure in higher vertebrates was based heavily on the observation that the organ was extremely vascular, a well-known characteristic of organs of internal secretion. However, besides the vascularity feature the authors also described in great detail the neuronal and glial elements of the organ as well as the prominent acervuli present in the pineal of some species. In addition to Tilney and Warren, other morphologists[2] throughout the first half of the twentieth century were advocating the idea that the pineal had to have an internal secretory function. About the only facet of the pineal on which the pioneering authors did not comment was the chemical nature of the active pineal constituents.

Interestingly, 2 years before Tilney and Warren[1] made their claims, McCord and Allen[3] had published the results of an important set of experiments on the physiological actions of an extract of the bovine pineal gland. The extract, when fed to amphibians, lightened their skin by causing the aggregation of the pigment granules in the melanocytes. This discovery proved to be a milestone in pineal research since the lighting substance was later identified as *N*-acetyl-5-methoxytryptamine or melatonin,[4,5] a compound which has been shown to have other potent actions as well. Tilney and Warren[1] apparently did not recognize the significance of the earlier discovery of McCord and Allen since they never referenced the work in their 432 citations.

Despite the repeated proclamations of these early morphologists, the scientific community generally ignored the pineal as a functional component of the endocrine system. Although most scientists didn't deny the morphological data, the negativism was prompted by equivocal physiological data. For example, it seems for every three experiments performed involving the administration of pineal extracts, one showed a positive effect of the pineal on the endocrine system,[6] a second showed a negative relationship,[7] while the third would indicate that the pineal constituents completely lacked any endocrine capability.[8] Similar variable changes often were observed after surgical removal of the pineal gland.[9,10]

In the mid-1950s two monographs appeared which summarized the literature of the world on pineal-endocrine interactions. One of these was written by Kitay and Altschule[9] in the United States while the second appeared in Europe under the authorship of Thieblot and LeBars.[10] The message of both reviews was the same, namely, that the pineal is very likely an organ of internal secretion and that it exerts a controlling influence over the reproductive organs. Indeed, both groups of authors were so con-

* From Tilney, F. and Warren, L. F., *The Morphology and Evolutional Significance of the Pineal Body*, Wistar Press, Philadelphia, 1919. With permission.

vinced of its endocrine capabilities that in the titles of their respective books they referred to the structure as a gland rather than an organ. Although this may seem like a minor point, the nomenclature of this structure is even disputed today.[11] After all, the phrases anterior pituitary organ or adrenal organ have long been obsolete because of the proven endocrine functions of these glands. The same should apply to the pineal, especially when referring to this structure in mammals.

The present résumé will only deal in a very cursory manner with the older equivocal data. The bulk of what will be discussed herein will be related to advances within the last 15 years. Although during this interval there has been the usual amount of healthy controversy and disagreement, it is also the era when some of the functions of the pineal gland have been uncovered. This survey will be concerned primarily with the influence of the pineal on reproduction, however, it also has many other endocrine and extra-endocrine effects.

II. ITS CAPABILITIES

A. End Organ Responses

While examining the influence of various environmental parameters on the hibernatory potential of the Syrian hamster in the mid-1960s, Hoffman and colleagues[12] noted that when this species was maintained under conditions of restricted photopeiods their testes underwent pronounced atrophy. Although even before this time it was generally considered that short days were suppressive to the reproductive physiology of rodents, the changes observed in the reproductive organs of the hamster were especially marked. Thus, the testes of male hamsters exposed to short days within several weeks exhibited a five- to sixfold drop in weight (Figure 1) and they became aspermatogenic; the animals were obviously sexually incompetent. In search of an explanation for the observed changes, the pineal gland was considered as a possible intermediary between the environmental photoperiod and the neuroendocrine-reproductive system for the following two reasons: it was theoretically capable of producing gonad-inhibiting compounds[9,10] and secondly, there were indications that the activity of the organ was under the influence of the light:dark cycle.[13-18] Thus, we speculated that when Syrian hamsters were placed in short daily photoperiods, the antigonadal activity of the pineal gland was proportionately increased and gonadal atrophy ensued. This led to the obvious question, i.e., would the reproductive systems of hamsters exposed to short days atrophy if their pineal gland was surgically removed before their exposure to restricted day lengths? Before conducting this experiment it was necessary to develop a reliable technique of surgical pinealectomy which would ensure minimal damage to the adjacent neural structures and maximal survival of the operated animals. This problem was solved with the design of a special trephine that fit into a standard handheld dental drill which was used on animals mounted in a head holder.[19] The trephine was fitted with an adjustable collar which allowed for the proper depth of cut into the skull without damage to the subjacent vascular and neural structures. The technique allowed hamsters to be successfully pinealectomized in less than 5 min. When, in fact, adult male pinealectomized hamsters were exposed to light:dark (LD) cycles of 1:23 (in hours) neither their testes nor their accessory sex glands (seminal vesicles and coagulating glands) underwent regression. Indeed, surgical removal of the pineal gland, which in the hamster weighs on the order of 300 to 500 μg, prevented short days from inducing reproductive organ collapse.[20,21] Similar observations were made in female Syrian hamsters. Dark exposure was followed by a cessation of estrous cyclicity and ovarian and uterine involution.[22,23] Following pinealectomy, however, the estrous cycles and the reproductive organs of the dark exposed animals were indistinguishable from those of hamsters kept under long daily photoperiods. These observations con-

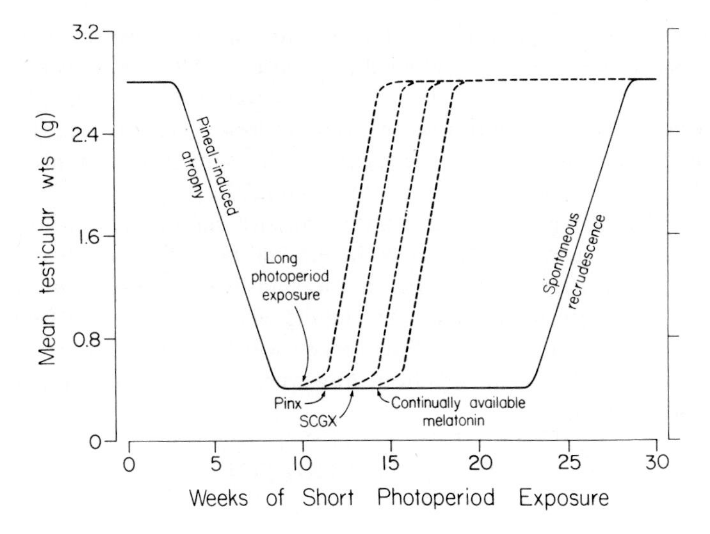

FIGURE 1. Gonadal involution in male hamsters exposed to short daily photoperiods. Testes of pineal-intact sexually mature hamsters usually weigh about 3000 mg; however, when they are exposed to short days, the gonads regress to an infantile state (roughly 500 mg) within 8 to 10 weeks. Pinealectomized hamsters never experience darkness-mediated gonadal involution. Not only is the intact pineal required for darkness to cause regression of the reproductive organs, in addition, once they are atrophic the pineal gland must remain intact and sympathetically innervated for them to be maintained in an infantile condition. Thus, if dark-exposed hamsters with involuted gonads are either pinealectomized (PINX) or superior cervical ganglionectomized (SCGX), testicular recrudescence is initiated and complete restoration is achieved within approximately 8 weeks. Likewise, moving the hamsters either to long daily photoperiods or placing melatonin pellets subcutaneously (so the indole is continually available) also results in the regrowth of the reproductive organs. If hamsters are kept for prolonged periods under conditions of short days their testes and accessory sex glands undergo what has been referred to as spontaneous or endogenous recrudescence.

vincingly established that the pineal gland, presumably because of its endocrine capabilities, was a critical intermediary between the prevailing photoperiod and the neuroendocrine-reproductive axis in this species.

Concurrent with the establishment of an interaction of the pineal gland with reproductive physiology in hamsters were studies on the synthesis of pineal compounds, especially melatonin, within the pineal gland of the rat.[24,25] The constituent of greatest interest at the time was the enzyme hydroxyindole-*O*-methyltransferase (HIOMT), which at that juncture was believed to be rate limiting in melatonin formation (Figure 2). It had also been shown in rats that melatonin was antigonadotrophic.[26-28] Thus, it was assumed that any experimental parameter which would exaggerate or increase pineal melatonin synthesis would also lead to a suppression of the reproductive organs. Certainly, surgical removal of the orbital globes bilaterally, i.e., blinding, was reported to cause a high sustained activity of HIOMT and, therefore, presumably melatonin production.[29] Blinding also was suppressive to reproduction. Likewise, destroying the postganglionic sympathetic innervation of the pineal gland by removal of the superior cervical ganglia allegedly prevented periods of light from inhibiting HIOMT activity[30,31] and, therefore, this procedure was believed to produce the same effects on the neuroendocrine-gonadal axis as did surgical removal of the eyes.[32] Because of this

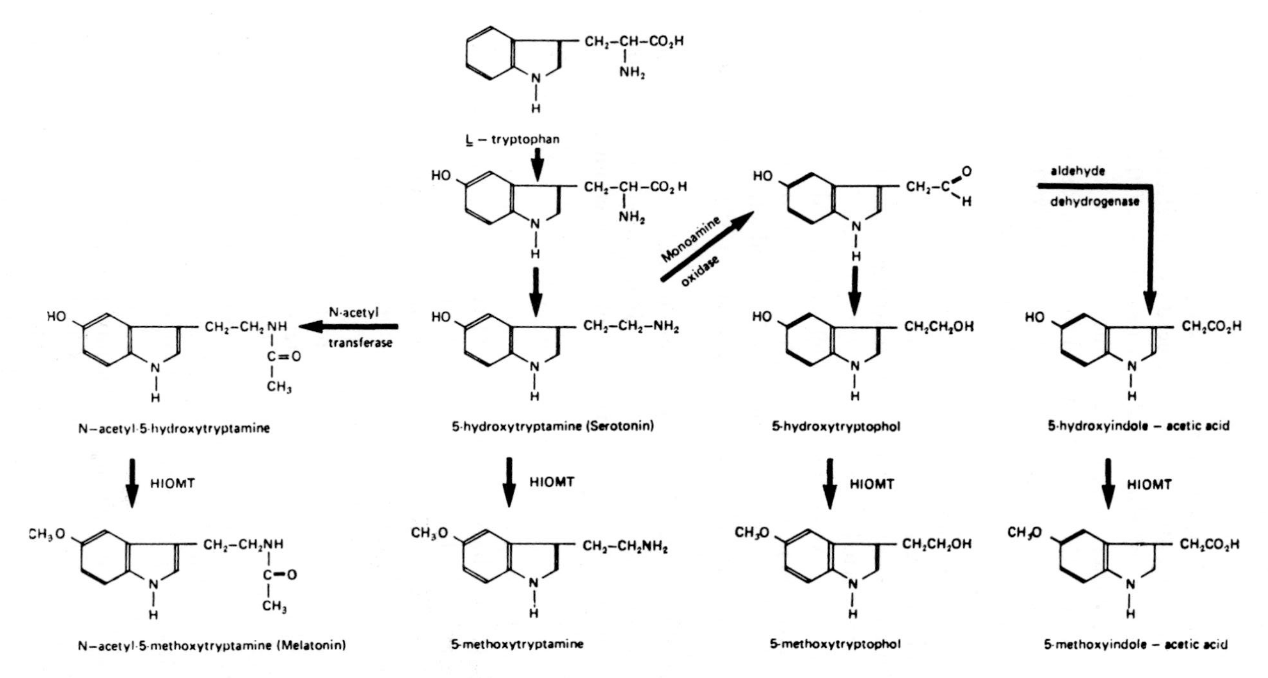

FIGURE 2. Indole metabolism within the pineal gland. All the constituents shown are derived from the amino acid tryptophan. There is ample evidence that melatonin is an important secretory product of the gland. 5-Methoxytryptophol apparently also is produced in a rhythmic manner within the mammalian pineal. Hydroxyindole-*O*-methyltranferase (HIOMT) is required for the synthesis of at least four compounds and was initially thought to be rate limiting in melatonin production. Presently, the activity of *N*-acetyltransferase is felt by many to determine the quantity of melatonin produced. The serotonin concentration in the pineal exceeds that of any other organ. How many of these compounds actually have physiological activity remains to be determined.

we decided to compare the effects of blinding and superior cervical ganglionectomy (SCGX) on gonadal size and function in the hamster.

Rather than having effects similar to blinding, however, SCGX altered reproductive physiology in the same manner as did pinealectomy.[33] Thus, either pinealectomy or SCGX negated the inhibitory effects of blinding on the testes and accessory sex organs of males and the ovaries and uterus of females. The data were unequivocal and established that destruction of the sympathetic innervation to the pineal gland severely impairs its endocrine capabilities. This has been routinely documented in subsequent studies.[34,35] Hence in terms of the end organs response, rather than being equivalent to blinding, SCGX produces changes identical to those induced by pinealectomy.

One of the more perplexing aspects of pineal-mediated involution of the neuroendocrine-reproductive system of hamsters maintained in short days or otherwise deprived of photoperiodic information is the interval required for the gonads to involute. In papers published in 1965, we reported total reproductive collapse within 4 weeks after male hamsters were deprived of light.[20,21] Three years subsequent to these initial observations, a detailed study was conducted to determine the actual rate of the ponderal changes in the testes. In the study in question, hamsters were surgically blinded after adulthood was attained and thereafter groups were sacrificed at weekly intervals for 8 weeks.[36] Although within 4 weeks significant gonadal atrophy was recorded, the degree of reproductive regression reported 3 years earlier was not attained until 6 weeks after light deprivation. In more recent years, the time interval required for darkness to induce gonadal atrophy is becoming progressively longer so that, as of this writing, it may require 10 to 12 weeks of darkness to cause total reproductive collapse.[37] Whereas both the rate of gonadal atrophy (once it is initiated) and the degree of change have remained the same, the interval after the onset of light deprivation to the onset of gonadal involution has been prolonged. It has been speculated that this may be related to the perpetual inbreeding of the Syrian hamster.[37] Hence, commercial breeders, in their quest for highly prolific specimens, may be selectively eliminating the highly photoperiodic animals from their breeding stock.

Other potential explanations include the improved nutritional status of the animals and yet unidentified factors which may override the inhibitory influence of the pineal gland. Only occasional investigations have been directed at elucidating such interactions. It was once thought, for example, that replacing the normal dry laboratory chow with fresh vegetables during the period of light restriction might at least retard gonadal involution. In support of this supposition, Negus and Berger[38] claimed that certain constituents in green plants suppress HIOMT activity within the pineal gland. When tested, however, fresh vegetables had no beneficial effect in curtailing gonadal involution in blinded male hamsters.[39] Optically enucleated hamsters were given fresh lettuce, carrots, and carrot greens daily; despite this their testes underwent atrophy at the normal rate (Figure 3). Since these data are obviously limited in terms of the food stuffs examined and the species tested, the findings do not definitely prove that dietary factors are incapable of modifying either the biochemical or endocrine activity of the pineal gland.

Another exogenous factor which might partially determine the rapidity with which the testes of male hamsters involute is the presence of the female, particularly females, exhibiting normal estrous cycles. Certainly, in at least some species pheromonal[40,41] and olfactory[42,43] information have major effects on reproductive capability. To test for such a possibility, blinded males were caged with either ovariectomized or hysterectomized females; the former were obviously acyclic while the latter exhibited normal estrous cyclicity but could not become pregnant. The males were sacrificed at 2-week intervals and the ponderal changes in their gonads were recorded; the testes were also histologically evaluated.[39] The presence of either normally cycling or ovariectomized

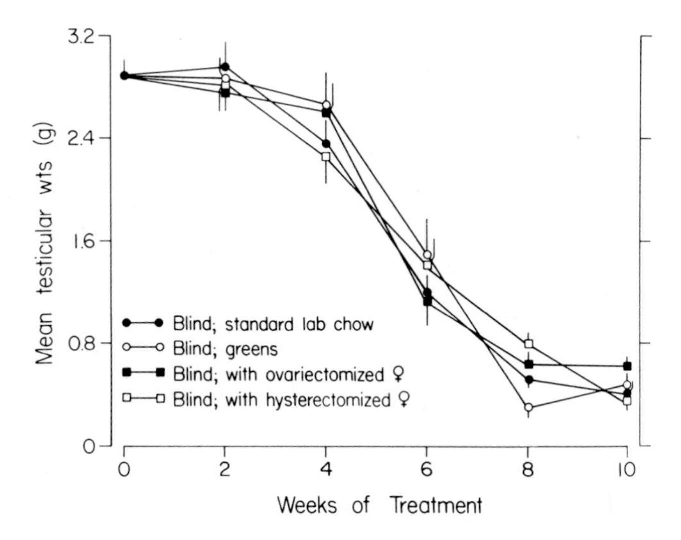

FIGURE 3. Mean testicular weight changes in blinded male hamsters. Animals were either fed standard dry commercial hamster chow or their diet was supplemented with lettuce, carrots, and carrot greens; despite the presence of the standard chow the animals consumed abundant amounts of the vegetables. Gonadal involution was not impeded in animals whose diet was supplemented with vegetables. Other groups of blinded males were caged (three males with two females) with either ovariectomized or hysterectomized females. Again, the presence of the females did not retard testicular involution in the light restricted males.

females did not impede testicular degeneration (either grossly or histologically) in the light-deprived males (Figure 3). From this it seems apparent that olfactory stimuli are without a major impact on pineal-induced testicular involution following light deprivation due to surgical removal of the orbital globes. Whether ovarian involution in light-restricted females would proceed unabated in the presence of reproductively functional males has not been determined.

Undoubtedly, the gonadal weight changes experienced by male hamsters after alterations in the photoperiod or manipulations of the pineal gland are very striking. Because of the obvious impact of the photoperiodic environment on testicular size, gross photographs of the end organs have sometimes been provided to convince the reader of the magnitude of the observed changes.[36,37,44] In long term studies where repeated measurements are desirable on the same group of animals, the testicular index (TI) has sometimes been used in lieu of testicular weights.[45-47] The TI is determined by multiplying the maximal testicular length (in millimeters) by the maximal testicular width (in millimeters) (determined with the use of calipers) and dividing the product by the body weight (in grams) of the animals (Figure 4). Initially, the testes were exposed by means of a laparotomy and the measurements were taken. Some workers have subsequently resorted to measurements taken through the scrotal sac with only the testicular width being recorded.[48] It is this reviewer's experience that repeated surgical exposure of the gonads is associated with adhesions and occasionally with testicular damage, both of which may lead to at least partial gonadal regression. There are times, however, when repeated measurements are essential and the use of the TI is useful. However, when this parameter is employed, it is much more reliable when the gonads are exteriorized and measured directly. The use of testicular width only, when measured through the scrotal sac, would seem to be the least reliable method for as-

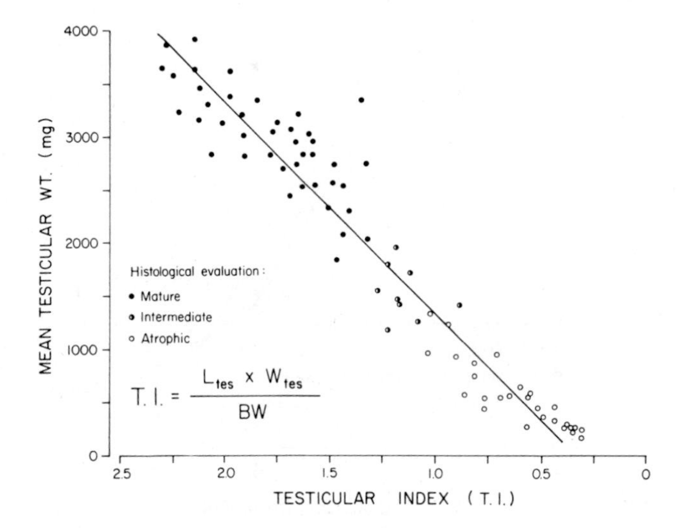

FIGURE 4. Testicular index (TI) in adult male golden hamsters. To determine the TI, the testes of anesthetized hamsters were surgically exposed and measured to the nearest millimeter. Both the maximal length (L_{tes}) and maximal width (W_{tes}) are recorded; the product of these two measurements is divided by the body weight (BW) in grams. Generally, gonads with a TI of less than 1.0 were grossly and histologically involuted while those above 1.25 were microscopically normal in appearance. When the TI fell between 1.0 and 1.25, the testes were often classified as being in an intermediate state. Small TIs were produced by placing the hamsters in short daily photoperiods.

sessing the status of the reproductive system. The testicular index in hamsters has been shown to be highly correlated with the spermatogenic and steroidogenic activities of the testes.[49,50] This is supported by the observations summarized in Figure 4.

Convenient endpoints by which to overtly monitor the reproductive state in female hamsters include vaginal smears[51] and the postovulatory vaginal discharge;[52] both have been effectively utilized to continually assess the sexual status of female hamsters after manipulations of the pineal and/or perturbations of the photoperiodic environment. Intact female hamsters maintained under long daily photoperiods (12.5 or more hours of light per day[53,54]), exhibit highly regular estrous cycles with an estrous smear occurring every 4 days. This smear is characterized by the presence of elongated and oval nucleated cells. When females are placed under conditions where their exposure to light is reduced, the percentage of estrous smears decreases greatly and eventually the animals become acyclic.[55] The vaginal smears at this time contain cells typical of both the metestrous and diestrous states and have been classified as intermediate in type.[55] Such smears are invariably associated with infantile uteri and anovulatory ovaries. Vaginal cycles of pinealectomized animals remain unaltered even under conditions where the photoperiods are greatly reduced in length, again indicating that the effects of short days on the reproductive state of the highly photoperiodic Syrian hamster involves an alteration of pineal biosynthetic and secretory activity.[55,56]

The postovulatory vaginal discharge also provides a simple and accurate means to determine ovarian and vaginal cyclicity in the hamster.[52] The absence of the discharge is indicative of animals whose hormonal state is not compatible with ovulation at 4-day intervals. After experimental manipulation of the pineal or following the administration of pineal hormones, the vaginal discharge pattern may reflect reduced ovulatory processes and acyclic features.[57,58] Again, the changes induced by the photoperiod are negated in hamsters that are pinealectomized.

Not all common laboratory species are as photosensitive as the hamster in terms of the responses of their reproductive organs. For example, the neuroendocrine-gonadal axis of the rat is relatively less responsive to alterations in the daily LD cycle. Rats may be inherently less reactive or their photoinsensitivity may be related to the fact that they have been inbred for numerous generations for the expressed purpose of maximal reproductive performance.[37] Regardless of the explanation, they are much less reliable in terms of the magnitude of the changes for studies involved with pineal-reproductive interactions.

Often when the influence of the pineal gland on reproduction in rats is studied, maturing animals are used as the experimental subjects. When deprived of light during prepubertal development, the growth of the gonads and adnexa in both males and females characteristically is retarded.[59-61] Hence, the growth of the testes, accessory sex organs, and ventral prostates is slowed in males while in females vaginal introitus is delayed, as is the normal enlargement of the ovaries and uterus. Even with total or near total light restriction, however, pubertal onset is not severely retarded, rather it is only slightly delayed. The effects of darkness on puberty in the rat is in marked contrast to the situation in the hamster. Despite the fact that the hamster is so exquisitely sensitive to photoperiodic manipulation after adulthood, during prepubertal development darkness is incapable of delaying pubertal onset; however, once adulthood is attained, their gonads totally regress (provided the short-day exposure is continued).[62-64] In both rats and hamsters, pinealectomy negates any effects of darkness on sexual regression.

There are several experimental perturbations which increase the sensitivity of the reproductive systems of rats to the pineal secretory products. These include early neonatal treatment of the animals with androgens or estrogens, surgically induced anosmia, and a quantitative reduction in food intake.[65,66] Because they indirectly exaggerate the antigonadotrophic potential of the pineal gland they are generally referred to as potentiating agents. The first of these to be uncovered was early neonatal steroid treatment. In the late 1960s Kordon and Hoffmann[67] and Hoffman and colleagues[68] reported that the early treatment of male and female rats with testosterone proprionate (TP) (1 mg at 5 days of age) combined with blinding at 25 days of age, severely retarded the growth of the gonads such that in early adulthood the reproductive organs were only a fraction of their normal size. These workers did not, however, realize the importance of the pineal gland in these responses. It was later found that when TP-treated, blinded rats had their pineal gland surgically removed the gonads of the animals grew much more rapidly. From these studies it was obvious that pinealectomy prevented the effects of light deprivation although it was without influence in reversing the consequences of early neonatal steroid treatment in either male[69] or female[70] rats. Besides testosterone, other steroids including estradiol benzoate[71] and diethylstilbesterol (a synthetic estrogen) given early after birth exaggerates the ability of the pineal gland in light deprived rats to curtail the development of the reproductive system. Again, pineal removal nearly completely negates the ability of the combined treatments to retard sexual development.

The second potentiating factor to be considered is surgically induced anosmia. Like blinding, anosmia alone has only a slight inhibitory influence on the growth of the reproductive organs of rats. However, when the procedures are combined, the depressant action on gonadal maturation is much greater than the sum of the two individual treatments. Testes and accessory sex organs of males[73] and ovaries and uteri[74] of blinded, olfactory bulbectomized females mature at a very slow pace unless the animals are additionally pinealectomized. Likewise, if intact rats receive the surgical procedures after adulthood is attained the reproductive system undergoes regression.[75,76] As with early androgen treatment, it is believed that anosmia has no direct stimulatory influence on the pineal gland but rather it potentiates the inhibitory action of the pineal antigonadotrophic agent at its site of action, wherever that may be. Certainly, if serotonin *N*-acetyltransferase (NAT) activity[77] and bioassayable levels of pineal melatonin[78] are any index, pineal biosynthetic activity is not augmented in anosmic rats. As in hamsters, superior cervical ganglionectomy seriously jeopardizes the ability of the pineal to inhibit reproduction in blinded, anosmic rats.[79]

The final potentiating factor to be discussed is underfeeding. Whereas a 50% reduction in food intake slowed the growth of the testes and adnexa, when combined with blinding these organs remained infantile until the rats were at least 59 days of age. Pinealectomy at an early age overcame the effects of food restriction and blinding on the growth of the reproductive organs.[80,81]

It should be reemphasized that the potentiating factors are believed not to augment the production or secretion of the antigonadotrophin by the pineal gland, rather they supposedly increase the sensitivity of the inhibited site to the pineal product.[77,78] This judgement is, of course, made with the knowledge that the only indices of pineal metabolism that have been measured in anosmic rats are those concerned with indole metabolism, namely, HIOMT activity,[66] NAT activity[77] (the alleged rate-limiting enzyme in melatonin production[82]), and melatonin levels.[78] There may be other pineal gonad-inhibiting factors, the production and discharge of which are greatly augmented following olfactory bulbectomy.

B. Hormonal Responses

1. Males

Considering the marked changes in the testes and accessory sex organs which follow the exposure of hamsters to short days, an associated drop in circulating gonadotrophins almost seemed inevitable. In male hamsters kept under naturally short days, radioimmunoassayable luteinizing hormone (LH) values were found to be significantly lower than those in pinealectomized control animals maintained under the same conditions.[83] A reduction in serum LH levels after the maintenance of hamsters in reduced photoperiods of the laboratory was confirmed soon thereafter[49,84,85] although the reversal of the blood levels of LH with pineal removal was not tested. The first of these reports[49] claimed that circulating LH titers fell to barely detectable limits within 60 days after the maintenance of males in constant darkness. Less clear-cut reductions in LH were observed in hamsters kept under LD cycles of 10:14[84] or 8:16.[85] It has become apparent that blood levels of LH may not always be statistically significantly depressed in male hamsters suffering from darkness-induced testicular atrophy.[37] Similar observations have been made relative to follicle stimulating hormone (FSH) levels. Although one would expect a great reduction in this gonadotrophin after exposing hamsters to short days, there are times when the changes are somewhat less dramatic than anticipated.[84,86] However, when the results of all the published reports are summarized it can be stated with confidence that a drop in serum gonadotrophins after the exposure of male hamsters to short days is the rule rather than the exception.

The relative importance of diminished levels of LH and FSH in preventing gonadal involution in male hamsters was investigated by injecting the animals daily with synthetic luteinizing hormone-releasing hormone (LRH) which, in hamsters, is known to release both LH and FSH from the anterior pituitary gland.[87-89] The daily subcutaneous administration of LRH into males maintained under light:dark cycles of 1:23 was incapable of even slightly retarding the involution of the reproductive system normally observed in hamsters kept under short photoperiods.[90] A rise in circulating immunoreactive LH was documented in these animals; no attempt was made to measure FSH titers. These somewhat unexpected findings at the time could be interpreted in one of several ways. They could mean, for example, that the pineal normally interferes with the action of the gonadotrophins at the testicular level and, thus, even though substantial amounts of LH and FSH were available they were simply incapable of stimulating testicular growth and function. Another possibility was that the daily rises in circulating LH and FSH levels following LRH administration were shortlived and thereby incapable of significantly stimulating the testes. Finally, the failure of LRH to promote testicular growth in this case could have meant that a hormone (or hormones) in addition to LH and FSH are required for the maintenance of a high level of testicular function.

The other hormone of interest in this response is prolactin. The present reviewer has adamantly maintained for a number of years that the regulatory influence of the pineal on prolactin in hamsters is apparently greater (in terms of the magnitude of the changes) than that for LH and FSH.[37,91-93] Indeed, prolactin levels in the blood are dramatically decreased after exposure of male hamsters to short daily photoperiods (Figure 5). This change did not go unnoticed; very soon following the initial reports, Bartke and colleagues[94] began examining the potential of prolactin in testicular growth and circulating testosterone levels in hamsters exposed to short days. Exogenously injected ovine prolactin stimulated both the testicular weights and testosterone production in animals kept under light:dark cycles of 5:19. However, prolactin did not totally restore the gonadal size to that seen in hamsters housed under long day conditions.

Further experimentation showed that the ameliorative effects of prolactin on the testes of dark-maintained hamsters may be related to the ability of the hormone to stimulate or activate LH receptors in the gonads.[95-97] In lieu of daily or twice daily prolactin injections, subcapsular renal homografts have been utilized to further document the essential role of prolactin in promoting normal testicular function in the hamster. Such homografts, presumably because of their copious secretion of prolactin, have been shown to at least partially negate the inhibitory influence of darkness and the pineal secretory products on reproduction.[98-100]

In the most recent paper in this series it was convincingly shown that all three gonadally active hormones must be available in adequate amounts to cause full testicular development in male hamsters.[101] The expressed purpose of the investigation was to determine whether the combination of LH, FSH, and prolactin could totally overcome pineal-induced gonadal involution in male hamsters. Adequate circulating prolactin levels were ensured by placing two pituitary homografts under the capsule of one kidney while LH and FSH secretion were augmented by giving twice daily injections of 1 μg LRH. In surgically blinded hamsters, the pituitary homografts by themselves had a very modest stimulatory action on the testes and accessory sex organs. Likewise, twice daily LRH administration by itself essentially was without stimulatory action on the growth of the gonads and adnexa in the light-deprived animals. Conversely, combined pituitary transplants and LRH administration fully restored gonadal size and function suggesting that all three hormones are equally important in promoting the spermatogenic and steroidogenic activity of the testes in this species. Increases in the blood levels of all three hormones were verified by specific radioimmunoassay. These findings are explicable in terms of the theory presented earlier,[95,97] namely, that prolactin acts by promoting either the production or increase of the sensitivity of the LH, and possibly the FSH,[101] receptors in the testes. By itself, prolactin failed to appreciably stimulate the end organs; likewise, LH and FSH by themselves were similarly ineffective in augmenting testicular maturation. However, when LH, FSH, and prolactin were available in ample quantities, normal testicular function followed. Thus, it seems safe to conclude that the pineal gland of light-deprived male hamsters depresses sexual physiology by influencing the metabolism of all three hormones.[37]

Changes in plasma levels of hormones induced by the activated pineal gland are mirrored by alterations in pituitary concentrations of these constituents (Figure 6). As

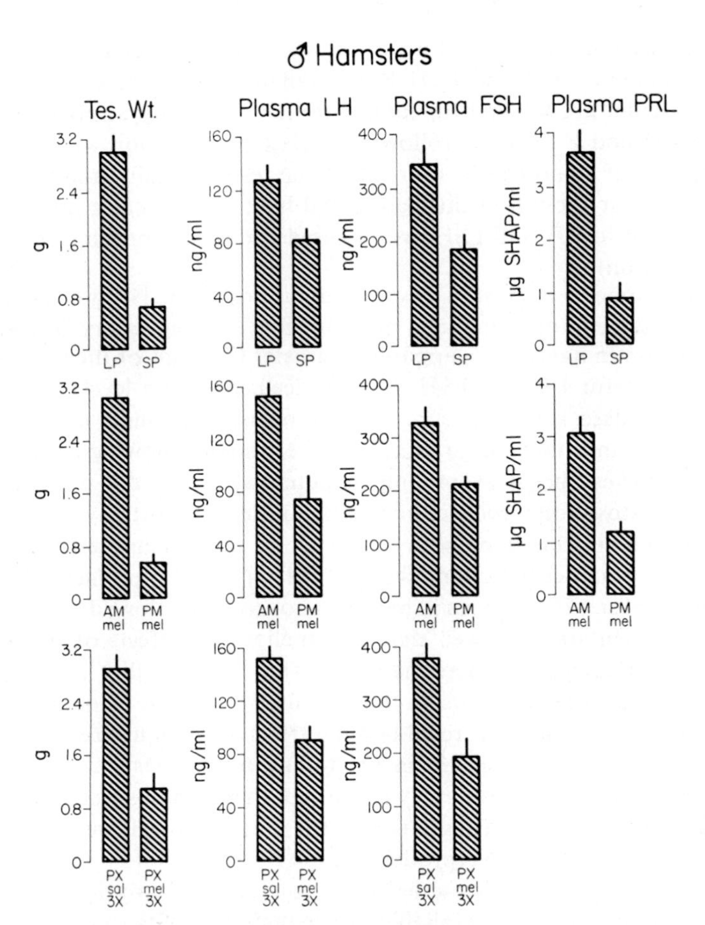

FIGURE 5. Testicular weights and radioimmunoassayable plasma LH, FSH, and prolactin titers in male hamsters after various experimental treatments. Some animals were kept under either long (LP) or short daily photoperiods (SP) while others, kept under long-day conditions, were treated with melatonin daily either early (AM mel) or late (PM mel) in the light period. Finally, pinealectomized (PX) hamsters were treated thrice daily with either saline (sal 3X) or melatonin (mel 3X). In terms of the reproductive parameters that were measured, the responses were the same for the animals exposed to short photoperiods, given afternoon melatonin administration, and given thrice daily melatonin injections into animals lacking their pineal gland. Vertical lines from the top of the bars signify standard errors. All values differ with a p value of at least 0.02. The most profound changes are in terms of the testicular weights and the circulating prolactin levels.

the titers of LH, FSH, and prolactin in the blood drop so do the quantities of these hormones in the anterior lobe to the pituitary gland.[91,102] Indeed, the lower levels of circulating gonadotrophins undoubtedly reflect the lower releasable reserves within the pituitary. Again, prolactin levels are usually more severely affected by the pineal gland than are those of LH and FSH.

In tests of the biological activity, it was found that the hypothalami of male hamsters suffering from pineal-induced gonadal atrophy possessed reduced quantities of LH and FSH releasing activity.[103] At least the FSH releasing activity of the extracted hypothalamic fragments was reversed in the pinealectomized control animals whereas the

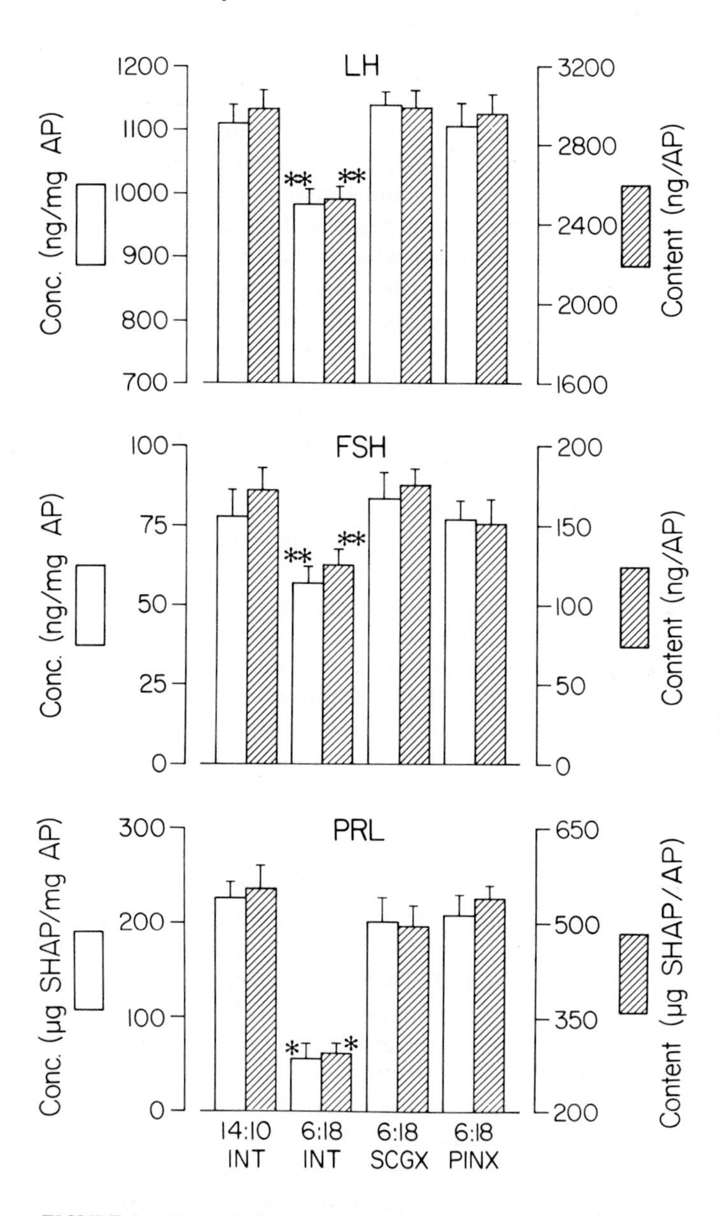

FIGURE 6. Mean pituitary concentration and content of LH, FSH, and prolactin in male hamsters kept in either long (light:dark cycles of 14:10) or short (8:16) days. Intact (INT) hamsters kept in short day lengths experienced a drop in the levels of all three hormones. These changes were prevented if the animals were either bilaterally superior cervical ganglionectomized (SCGX) or pinealectomized (PINX). Prolactin levels were more severely affected than were LH and FSH. Vertical lines from top of bars signify standard errors(** $p<0.05$ and * $p<0.001$ VS animals in long days).

LH releasing activity seemed to remain depressed under such conditions. These data are in seeming conflict with more recent findings in which the hypothalamic content of LRH was estimated using radioimmunoassay. Packard and Silverman[104] noted no difference in the content of LRH between hypothalami collected from hamsters maintained under either LD 14:10 or LD 8:16; the latter group had involuted reproductive systems whereas the former were sexually mature animals. In this study the influence of pineal removal on hypothalamic LRH content was not examined. Thus, the hypothalamic changes associated with pineal-mediated regression of the pituitary-testicular axis in male hamsters kept under short-day conditions remains unresolved. This is an important question since the information could provide a hint as to the site of action of the pineal antigonadotrophic factor(s). A measurement of the turnover of LH and FSH releasing activity of the hypothalamus under such conditions may be more informative than simply an estimate of the content of these constituents.

In male rats, the consequences of pineal activation produce similar hormonal imbalances as those seen in light-deprived male hamsters. As noted above, the most reliable way in which to cause substantial changes in the reproductive system of rats is to combine pineal activation (by light deprivation) with an experimental perturbation which indirectly exaggerates the antigonadotrophic potential of the pineal gland. The three potentiating agents that have been defined to date include neonatal androgen (or estrogen) treatment, anosmia, and underfeeding.[56,66,105,106] The hormonal patterns accompanying involution of the testes in such animals have been described in greatest detail in the blind, anosmic rat; however, even here the information available on gonadotrophin levels is meager compared to that published for hamsters with a pineal altered reproductive state. Combined blinding and anosmia in male rats leads to either no change[107] or a reduction[108] in circulating levels of immunoreactive LH. Prolactin titers in the blood under such conditions are consistently depressed.[107,108] After early neonatal treatment with testosterone and subsequent light deprivation, circulating levels of FSH are reportedly subnormal.[66] Pituitary hormone levels undergo similar changes. The hypophyseal content of LH and prolactin decline in male rats deprived of both their olfactory sense and photic information.[107,108]

2. Females

Although in both female and male hamsters the reproductive organs regress to an infantile state under the influence of the activated pineal gland, the associated changes in gonadotrophin regulation are quite different for the two sexes. As noted above, when female Syrian hamsters are exposed to short daily photoperiods they are rendered acyclic within a several-week period. The vaginal smears under these conditions contain cells characteristic of both the metestrous and diestrous stages of the cycle.[55,56] Also, rather than becoming smaller, the ovaries may actually enlarge due to the marked proliferation of the interstitial tissue.[117,118] This change in ovarian morphology is representative of female hamsters with ample LH.[119,120] Thus, it was surmised in 1969 that, when eventually measured, at least the LH levels in light restricted hamsters would be either normal or possibly even elevated.[118] Indeed, when circulating LH and FSH titers were actually estimated by means of radioimmunoassay both were observed to exhibit daily afternoon surges.[121,122] In normally cycling female hamsters, of course, the surge occurs once every 4 days; in this case the rise in gonadotrophins are associated with the shedding of ova by the ovary.[123] However, in the acyclic female the daily surge in the gonadotrophins is not releated to ovulation even though the midafternoon rise in LH and FSH occurs at about the same time of day as the normal proestrous preovulatory release of these hormones in cycling females.

In the case of LH, the magnitude of the daily afternoon rise in light-restricted females is on the order of 15- to 20-fold over baseline levels observed in the morning. For FSH, the afternoon surge is roughly twofold.[124] Even when short day exposed females are ovariectomized they continue to exhibit a daily afternoon surge of LH and FSH. What, in effect, these results mean is that even though the gonads of female hamsters kept under restricted photoperiods are nonfunctional in terms of ovulation and the production of ovarian steroids, their mean levels of gonadotrophins are actually elevated by virtue of the daily rises in these constituents in the blood. This contrasts with the male where it is generally considered that circulating LH and FSH levels are depressed. The significance of the daily afternoon spurts of LH and FSH in light-restricted female hamsters remains unknown.

These results could be interpreted to mean that the pineal antigondotrophic substance normally inhibits the action of LH and FSH directly at the level of the ovaries. However, this is not the only explanation which is compatible with these changes. Besides the gonadotrophins, circulating prolactin levels may also be altered in hamsters placed under short day conditions. Usually prolactin titers in the blood are considered to be diminished although a careful analysis of the levels of this hormone throughout a 24-hour period has never been accomplished. If, in fact, prolactin levels drop and if, as in the male hamster, prolactin in the female is involved with the maintenance of LH and FSH receptors at the gonadal level,[94-97] then the elevated LH and FSH titers may be incapable of stimulating the ovaries because of the reduced availability of receptors. The relative importance of the alterations in these three hormones in causing darkness-induced gonadal regression in females obviously requires additional experimentation.

In line with the elevated LH titers in the blood of females with pineal mediated ovarian atrophy, the pituitary stores of this gonadotrophin also increase. In some cases the levels of LH within the pituitary gland of the female may actually double in response to light deprivation.[125,126] Pituitary FSH values after female hamsters are exposed to darkness have virtually gone uninvestigated. The most remarkable change in the anterior pituitary gland of female hamsters suffering from pineal mediated gonadal involution is the drop in prolactin reserves. Prolactin levels may decrease by as much as 60 to 80% after intact females are exposed to daily short photoperiods.[125-127] The reserves of this hormone drop to extremely low levels and, as noted above, this portends an important role for this constituent in the gonadal atrophy which ensues as a consequence of the maximally activated pineal gland. Nothing is known concerning possible changes in the hypothalamic content of releasing or inhibiting factors in the female hamster as a consequence of alterations in the status of the pineal gland.

The hormones governing the function of the ovaries and the uterus in female rats also change after surgical or environmental manipulations which alter the physiology of the pineal relative to the pituitary-ovarian axis. Dual sensory deprivation, i.e., blinding and olfactory bulbectomy, when done either before[74] or after[79] puberty greatly suppresses the function of the ovaries with respect to ovulatory processes[76,128] and the secretion of steroid hormones.[129] Most commonly plasma LH[130,131] and FSH[132] levels are subnormal under such conditions. Circulating prolactin levels vary according to the experiment in which they are examined; in some studies they were found to be elevated[131,133] while on another occasion they were within the normal range.[130] The content of LH within the anterior pituitary gland of blind, anosmic female rats is invariable increased[130,131] while prolactin levels exhibit a uniform and marked drop.[130,134]

Studies on the releasing and inhibitory factor activity of the hypothalami of blind, anosmic female rats suggests that the release of LRH may normally be inhibited by the pineal hormone resulting in its accumulation in the median eminence.[131] The alleged augmented prolactin secretion could be due to a suppression of prolactin inhibiting factor release and/or to a concomitant stimulation of prolactin releasing factor.[131] Pineal removal from blind, anosmic females reverses the effects of these procedures at the hypothalamic, the pituitary, and the gonadal level.[135-137] In other words, the changes observed in the hypothalamo-pituitary-ovarian axis of dual sensory deprived female rats is directly attributable to a change in the activity of the pineal gland.

Under long photoperiodic conditions, pineal removal from otherwise intact female rats causes rather minimal or no changes in the hormones associated with reproductive physiology.[138-141] There may be some exceptions to this generalization inasmuch as Trentini and colleagues[142] reported rather substantial alterations in ovulation in adult female rats that were kept under long days. The usual shortcomings of pinealectomizing rats kept in long days and expecting dramatic changes in the function of the sexual organs have been summarized elsewhere.[37,143]

III. ITS HORMONES

There are many potential hormones produced in and secreted from the pineal gland. Indeed, the number of constituents being proposed as active pineal fractions is increasing rapidly.[144-152] Often these judgments are made on the basis of scanty experimental evidence. A substance does not qualify as an internal secretory product merely because it can be isolated from pineal tissue. It is essential that individuals working in the field exercise caution in ascribing important endocrine capabilities or hormonal properties to partially purified substances which have been shown to possess endocrine activity only under a few limited conditions. This precaution does not mean to imply that the compounds in question are not hormonally active under normal circumstances, it's simply that this reviewer desires a more cautious approach to the assignation of the title "hormone" to possible fragments of larger molecules which are often not in pure form and which have not been shown to fulfill the physiological criteria of a hormonal product.

There are two major categories of pineal substances that are considered in all discussions of pineal secretory products, i.e., methoxyindoles and proteins or polypeptides. More information is available on the chemical structure of the former than of the latter. Likewise, the indoles have probably been tested under a greater variety of physiological conditions than have the peptides. This is perhaps primarily related to their commercial availability and to the fact that they are relatively inexpensive. Melatonin is the best known of the indole compounds whereas arginine vasotocin (AVT), a compound a number of scientists now seriously question as being a pineal product,[153,154] is probably the most thoroughly studied polypeptide.

A. Melatonin

Melatonin is present in the pineal gland of both the hamster[155-158] and the rat.[18,159-161] In both species its synthesis exhibits a night time rise which is directly related to the attendant period of darkness. There is presumed to be a close relationship between plasma levels of melatonin and the pineal content of the indole,[162] although this has not been proven for the Syrian hamster.

The actions of melatonin are very complex. It can either inhibit or promote gonadal growth and function.[163] Since short days had been shown to produce such potent inhibitory effects on reproduction, it was assumed that melatonin, an alleged pineal hormone, would also cause suppression of sexual physiology. However, the early studies were never very successful in this regard.[106]

In these early studies, male and female hamsters were characteristically kept under long day conditions (>12.5 hr light/day) and injected with melatonin once per day. Invariably, the indole was administered early in the light phase of the light:dark cycle. This treatment was surprisingly devoid of any inhibitory effect in terms of the reproductive system.[106] However, there seemed to be a logical explanation for the lack of any inhibitory effect of daily morning injections of melatonin. It seemed likely that an injection of melatonin merely caused a rapid rise in the blood levels of this hormone with an equally rapid disappearance from the blood. Thus, animals injected with melatonin in this manner were presumably only exposed to the indole for a brief period every morning; presumably so brief that reproductive involution did not ensue. In an attempt to resolve this problem, melatonin was mixed with either beeswax or cholesterol and the resulting pellets were placed subcutaneously. It was reasoned that the subcutaneously placed pellets would continually release melatonin and, as a consequence, the gonads would involute. However, even utilizing this means of administration of the indole, the reproductive organs were not adversely influenced.[106] As a consequence of these observations, some doubt was raised as to the antigonadotrophic capability of melatonin.

It was not until the mid1970s that this problem was resolved. Remarkably, it was found that merely shifting the time of the melatonin injection from the morning to the afternoon was sufficient to cause the pituitary-gonadal axis of Syrian hamsters to involute.[57,164] Thus, daily injections of melatonin (25 μg) in the morning (early in the light phase of the light:dark cycle) had, as previously shown,[106] no suppressive influence on reproduction. Conversely, the same dose given daily in the afternoon (late in the light phase) caused an atrophic response of the pituitary-gonadal axis equivalent to that which is seen in hamsters exposed to short days.

Confirmation of the findings of Tamarkin and colleagues[57,164] was soon forthcoming.[165] Invariably, afternoon as opposed to morning injections of melatonin are potently antigonadotrophic (Figure 7, cf., groups 3 and 4) with the resulting hormonal imbalance being reminiscent of that caused by the exposure of hamsters to restricted photoperiods (Figure 5). In fact, the similarity of the response to short days and afternoon melatonin injections suggests that the indole may be the compound which mediates pineal-induced gonadal involution.[37] Careful scrutiny of the data indicates that the melatonin must be injected at least 6.5 hr after lights on in order for it to be effective in its antigonadotrophic capacity. The gonadal responses to afternoon melatonin injections are not unique to the males but occur in females as well. Some of these findings are summarized in Figure 8. Adult female Syrian hamsters were kept under long daily photoperiods (light:dark cycles of 14:10) and they were given daily injections of melatonin at either 0800, 1000, 1200, 1400, 1600, or 1800 hr. Only injections of melatonin given at 1600 or 1800 hr produced vaginal acyclicity in 100% of the animals. Melatonin administered daily at 1400 hr prevented the postovulatory discharge in about 50% of the hamsters whereas earlier injections did not perceptibly alter the vaginal cycles. The minimal effect dose of melatonin required to cause gonadal atrophy is on the order of 1.6 μg daily.[158]

It is now clear that of the known pineal indoles that have been tested, only melatonin injected daily by this precise scheme is capable of causing a pronounced regression of the sexual organs.[166] Whereas 5-methoxytryptophol reportedly has a very modest suppressive influence on the testicular and accessory sex organ weights of hamsters, *N*-acetylserotonin (the immediate precursor of melatonin) and 5-hydroxytryptophol (the immediate precursor of 5-methoxytryptophol) completely lack gonad-inhibiting activity when the same experimental paradigm is employed.

As a result of these findings, it now seems convincingly established that melatonin is an antigonadotrophic compound and that it could be the pineal factor which mediates gonadal involution normally attendant upon the exposure of hamsters to darkness. There are, however, claims that neutralizing circulating melatonin levels by the administration of antibodies against melatonin does not prevent darkness-induced atrophy of the hypothalamo-pituitary-gonadal axis.[169-171] However, these reports failed to establish whether in fact all the endogenously produced melatonin is neutralized by the antibodies. Since this is critical to the argument that melatonin is not the essential antigonadotrophin, the validity of the procedures employed remains questionable.

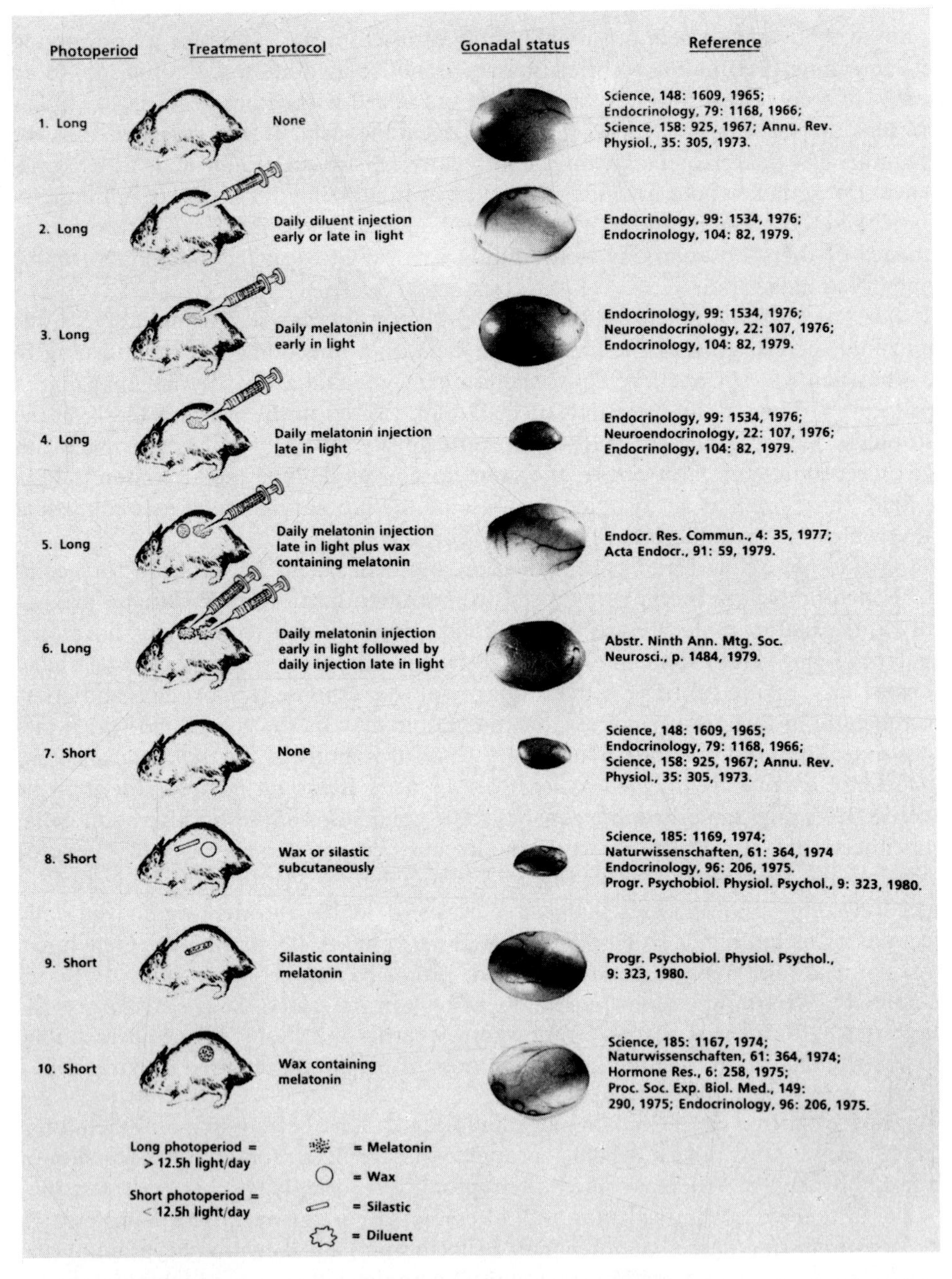

FIGURE 7. Responses of the reproductive systems of hamsters to either long (>12.5 hr light daily) or short (<12.5 hr light daily) days and to melatonin administered by any of several means. When injected subcutaneously, melatonin is either dissolved in alcoholic saline or in oil. When continuous-release subcutaneous depots are implanted, the melatonin is either mixed with beeswax or placed in silastic capsules. Similar responses occur in both males and females. The size of the testes indicates the general reproductive state of the animals. Injected melatonin, when properly administered, induces total reproductive collapse. On the contrary, melatonin continually available from subcutaneous pellets prevents darkness from inducing gonadal involution, i.e., it acts in a counter antigonadotrophic manner. Both the antigonadotrophic and the counter antigonadotrophic actions of melatonin are explicable in terms of the sensitivity of the melatonin receptor to the indole. For more details the original references can be consulted.

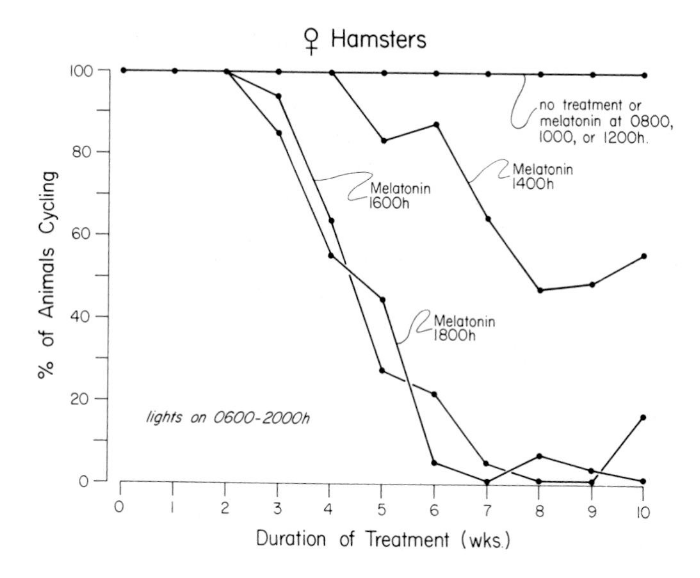

FIGURE 8. Responses of the vaginal cycles of hamsters to melatonin (25 μg daily) given at either 0800, 1000, 1200, 1400, 1600, or 1800 hr. The animals were kept in 14 hr of light daily. Morning injections were ineffective in altering vaginal cyclicity in the animals while late afternoon injections (i.e., 1600 and 1800 hr) interrupted vaginal cyclicity in 100% of the hamsters within about 2 months. The administration of melatonin in the early afternoon (1400 hr) interfered with the vaginal cycles in about half of the animals. Vaginal cycles were monitored by means of the postovulatory discharge.

There are other seemingly paradoxical effects of melatonin in hamsters which could theoretically militate against it being an antigonadotrophic agent. These are the so-called counter antigonadotrophic (or progonadal) influences of the indole. In 1974 it was unexpectedly found that continually available melatonin from a subcutaneous reservoir prevented gonadal involution in dark-exposed male hamsters.[172] This observation was contrary to virtually everything that was known about the ability of melatonin to influence reproduction up until this time. In the experiment, melatonin was mixed with beeswax and the resulting pellets (which contained 1 mg melatonin in 24 mg wax) were surgically implanted beneath the skin of male hamsters that were exposed to short days. Surprisingly, the subcutaneous melatonin reserves were as effective as pinealectomy itself in preventing reproductive involution which normally follows light restriction. In other words, the pellets caused a functional pinealectomy (Figure 7, groups 9 and 10). Females, as well as males, were found to respond to subcutaneously placed melatonin depots in the same manner.[173,174] Also, not only did continually available melatonin prevent gonadal involution due to dark exposure, if hamsters with already atrophic reproductive organs received an implant of melatonin in beeswax, their reproductive systems were restored to a functionally mature state within 6 to 8 weeks (Figure 1).[175,176] Although the counter antigonadotrophic action of melatonin may be shared by 5-methoxytryptophol,[90] neither *N*-acetylserotonin nor 5-hydroxytryptophol[177] are capable of negating the inhibitory influence of the pineal gland on the physiology of the sexual organs. The counter antigonadotrophic action of melatonin has been confirmed following the subcutaneous implantation of melatonin-filled silastic capsules.[178]

When the initial observation was made that melatonin from a subcutaneous reservoir leads to a functional pinealectomy, it was tentatively surmised that this was a bogus reaction of the system to a pharmacologic dose of the compound. In many of the studies reviewed above, 1 mg melatonin was placed in a subcutaneous location every 2 weeks. If the indole was totally dissolved during this interval, this translates into a daily dose of 143 μg melatonin. Thus, studies were designed to lower the daily dose to which the animals were exposed. This was done by decreasing both the quantity of melatonin per beeswax pellet and the frequency of pellet implantation. Using these methods, it was established that continual exposure of adult hamsters to somewhat less than 3.6 μg per day is still sufficient to prevent darkness and the pineal from causing involution of the reproductive organs in light-restricted male hamsters.[179] Hence, it was assumed that the counter antigonadotrophic activity of melatonin was probably not a result of the administration of supraphysiological amounts of melatonin.

About the same time another rather startling observation was made. Although it seemed like an unusual experiment the following rationale seemed sound. Both short day exposure and daily afternoon melatonin injections induce gonadal involution. Since the reproductive collapse resulting from short days is prevented by the subcutaneous placement of melatonin reservoirs, could it be that depots of melatonin under the skin would also prevent daily-melatonin injections from influencing the pituitary-gonadal axis, i.e., could chronically available melatonin negate the action of acute daily melatonin injections? When put to the test it was indeed found that if hamsters possessing subcutaneous melatonin-beeswax pellets are injected every afternoon with melatonin, their gonads are resistant to inhibition; thus, melatonin inhibited its own action.[180]

These diametrically opposed actions of melatonin seemed to defy explanation, however, one theory has been presented which is compatible with the majority of the findings. The theory relies on the assumption that the sensitivity of the melatonin receptors is regulated by melatonin itself.[37,181] Thus, when melatonin is normally synthesized and secreted from the pineal gland at night, it acts on its receptors and simultaneously renders them transiently insensitive to the indole for a period of time, i.e., it down regulates its own receptors. Thus, daily melatonin injections in the morning shortly after lights on are ineffective in suppressing reproduction because the receptors are in a down regulated state. By late in the light period the receptors restore their capacity to respond to melatonin and, as a consequence, melatonin given daily at this time induces gonadal involution. When melatonin deposits are placed under the skin of hamsters, the continual availability of the indole keeps the receptors in a down regulated state thereby preventing even daily afternoon injections of melatonin from influencing reproductive physiology.

If this theory has any validity we reasoned that we could alter the sensitivity of the receptors with appropriately timed exogenously administered melatonin. For this study adult male Syrian hamsters were kept under light:dark cycles of 14:10 (lights on at 0600 hr). Animals that received a 1 mg subcutaneous injection of melatonin at 1100 hr did not experience gonadal involution presumably because the receptors for the indole were in a down regulated state.[182] However, the 1 mg injection at 1100 hr completely overcame the ability of the 25 μg injections at 1700 hr from inducing reproductive atrophy (Figure 7, cf., groups 4 and 5). Control hamsters that received saline at 1100 hr and melatonin later in the day (at 1700 hr) exhibited the usual involution of the hypothalamo-pituitary-testicular system. Thus, injecting a large dose of melatonin in the morning somehow alters the responsiveness of the animal to later melatonin injections; this is theoretically related to the fact that the morning injection of the indole prolongs the insensitivity of the receptors and, as a consequence, melatonin later in the day is ineffective as an antigonadotrophic agent.[37,182,183]

Melatonin receptors tentatively have been identified but generally very little is known concerning their physiology. According to Cohen et al.,[184] cytoplasmic melatonin receptors are present in the ovaries collected from rats, hamsters, and humans. Conversely, another group claims that the receptors are related to membrane fractions of the medial basal hypothalamus of rats and hamsters.[185,186] Finally, Niles and colleagues[187] provide evidence that melatonin receptors are associated with cytosol fractions of the hypothalamus, hippocampus, striatum, and midbrain. None of the studies have been confirmed or denied by other laboratories even though this is undoubtedly a subject of intense research interest. In terms of the down regulation theory of the melatonin receptor presented above, it is noteworthy that in a recent paper by Vacas and Cardinali[188] the claim was made that there are larger numbers of melatonin receptors detectable in the hamster brain when the animals are killed at the end of the light period compared to those killed earlier in the day; this, of course, coincides with the time the intact hamster is maximally sensitive to the indole (Figure 7).[57,164,165] Down regulation of the melatonin receptors (as occurs early in the light phase of the light:dark cycle) may be manifested as a reduced number of available receptor cites. The observations of Vacas and Cardinali[188] are important and warrant confirmation.

Finally, the point should be made that the down regulation theory of melatonin action as described above was formulated in an attempt to explain the effects of melatonin in the Syrian hamster and should not generally be applied to all mammals or even to other rodents. Whereas it may have widespread applicability, with the information at hand the more conservative approach seems judicious.

Not all published data on the antigonadotrophic effects of melatonin in the hamster can be explained by the down regulation theory of melatonin action. Work from a single laboratory indicates that continually available melatonin from a subcutaneous melatonin-filled silastic capsule is capable of inducing regression of the sexual organs in hamsters kept in long daily photoperiods.[178,189,190] The findings, however, are somewhat variable and inconsistent. For example, in the initial paper Turek and colleagues[178] reported that the implantation of melatonin-filled silastic capsules of either 50, 100, 150, or 200 mm length eventually resulted in decreased weights of the testes and accessory sex organs. In a later publication they compared melatonin from two different sources, i.e., Regis and Sigma. In this study only the 300-mm melatonin-filled capsules (from both sources) consistently depressed paired testes weights; additionally, the melatonin from Regis was effective as an antigonadotrophin in about half the animals that received 50-mm-length capsules. We have attempted, without success, on a number of occasions to confirm that continually available melatonin from subcutaneous depots will lead to the cessation of normal reproductive organ growth. For one experiment, the melatonin-filled silastic capsules were supplied by one of the authors (C. Desjardins) of the papers cited above. Also, he advised us as to the details of the experimental design. As was our previous experience, melatonin released from the silastic capsules was ineffective as an antigonadotrophic agent in hamsters kept under long photoperiods (Table 1). There is also a supposed advantage to using melatonin-filled silastic capsules as opposed to melatonin-beeswax pellets; namely, the capsules can be weighed before and after use and, hence, the daily dose of melatonin to which the animals are exposed can be calculated. Reportedly, 0.5 μg melatonin is released per day for each mm-length capsule.[188] When we have weighed capsules before and after use, the variability has been so great that the results have been of little value. Finally, the before and after capsule weights as an indication of the amount of melatonin that has escaped was recently criticized (at an International Symposium on Pineal Function, Thredbo, Australia, February, 1980) as being undependable by G. B. Ellis who, at the time, was a graduate student with Dr. Turek.

The reason Turek and colleagues have been able to force gonadal involution in Syrian hamsters with continually available melatonin while others have not is unclear. Since they have had their own breeding colony, possibly the animals they used were

Table 1
ABSOLUTE AND RELATIVE MEAN (± STANDARD ERROR) TESTICULAR AND ACCESSORY SEX ORGAN (SEMINAL VESICLES AND COAGULATING GLANDS) WEIGHTS IN HAMSTERS THAT RECEIVED SUBCUTANEOUS IMPLANTS OF EITHER EMPTY OR MELATONIN-FILLED SILASTIC CAPSULES

Silastic capsule length	No. of Animals	Testes		Accessory organs	
		mg	mg/100g BW	mg	mg/100g BW
50 mm (empty)	6	3016±136	1946±91	290±12	187±9
50 mm (melatonin filled)	5	2750±49	1936±35	306±20	209±14
100 mm (melatonin filled)	5	2681±151	1915±124	284±21	202±15
200 mm (melatonin filled)	5	2981±127	1743±87	286±24	167±16
300 mm (melatonin filled)	4	2704±189	1511±112	298±19	166±14
400 mm (melatonin filled)	4	2993±47	1672±40	306±22	171±17

Note: The treatment period was 77 days. Throughout the experiment the animals were maintained under a light:dark cycle of 14:10. There are no statistically significant differences between any of the groups. BW = body weight.

genetically slightly different, i.e., a different substrain, than those obtained from commercial breeders. Alternatively, there may be something different (temperature, food, etc.) in the execution of the experiments which changes the responsiveness of their animals to melatonin. Hopefully, the reason for some of the discrepant findings will be forthcoming in the near future.

Like the reproductive system of rats being relatively less responsive to the activated pineal than that of hamsters, so it is less responsive to melatonin treatment. Very few, if any, studies have been conducted since melatonin first became commercially available which show that the indole can seriously jeopardize the ability of rats to successfully reproduce. Nevertheless, the rat continues to be a common model in which to check for the antigonadotrophic capability of the pineal methoxyindoles.

Shortly after its discovery, melatonin was found to delay ovarian growth and to lower the incidence of estrous smears in rats.[27,28] A number of years later it was reported that the implantation of melatonin directly into either the median eminence or the midbrain reticular formation depresses pituitary and plasma levels of bioassayable LH in castrated male rats.[115,116,191] This group of workers also reemphasized the possibility that pineal melatonin may normally be secreted directly into the cerebrospinal fluid rather than into the vascular system, a concept which presently has little support among pinealologists. Almost concurrent with these discoveries of Fraschini and colleagues[115,116,191] were the elegant studies of Kamberi et al.,[192-194] wherein it was shown that the intraventricular infusion of small quantities of melatonin depressed circulating levels of radioimmunoassayable LH and FSH while elevating prolactin titers in anesthetized adult rats. Conversely, when the indole was infused directly into the anterior lobe of the pituitary, LH and FSH secretion continued unabated. In view of this latter finding, it was somewhat surprising that a number of years later it was observed that when anterior pituitary glands of rats were incubated in the presence of synthetic LRH, melatonin inhibited the action of the hypothalamic hormone on the release of LH.[195,196] It was later found that this was a phenomenon unique to the pituitaries of neonatal rats and does not apply when glands of adult animals are placed in culture.[197] Thus, the findings are not in conflict with the earlier observations of Kamberi and coworkers.[192-194] Also in rats, peripherally administered melatonin has been shown to depress ovulation[198] and the associated release of LH[199] in immature rats treated with pregnant mare's serum gonadotrophin; these results are also consistent with a central level of action of melatonin.

Of special interest concerning the actions of melatonin is a recent publication by Kao and Weisz[200] in which the perifusion of medial basal hypothalami of rats with melatonin (1.0 m*M*) resulted in a highly significant release of radioimmunoassayable gonadotrophin releasing hormone. There have been no follow-up studies on this interesting observation and how the results can be reconciled with the findings discussed above remains to be answered. It is possible that the endocrine milieu of the blood normally perfusing the hypothalamus and pituitary may be important in determining the responses of these organs to melatonin and other pineal fractions. Hence, the results of in vitro studies should be interpreted with more than the usual degree of caution.

With the exception of a few isolated reports,[201] there is virtually a consensus that melatonin normally acts in an inhibitory manner on the neuroendocrine-reproductive axis of the rat.[105,144,202] However, the point of continued debate seems to be the site at which the indole intervenes to influence sexual physiology. Whereas the bulk of the findings are consistent with the action of melatonin at the hypothalamo-pituitary unit, other workers argue that it mediates the effects of LH or FSH directly at the gonadal level[203,204] (see also Blask, this volume) or possibly it may even influence reproduction via alterations in the metabolic activity of the liver.[205,206]

The point has already been made that the reproductive system of rats can be sensitized to the antigonadotrophic influence of the pineal gland by any of several potentiating factors including anosmia, neonatal androgen (or estrogen) treatment, and a quantitative reduction in food intake.[66] An important question is whether the potentiating factors would also increase the sensitivity of the neuroendocrine axis to daily melatonin injections. This problem was resolved by the results of recent investigations. Blask and colleagues[207] found that reducing the food intake by 50% of maturing male rats combined with daily afternoon injections of melatonin led to a marked curtailment of the growth of the testes, seminal vesicles, and ventral prostate, and also reduced prolactin levels in the pituitary. The afternoon melatonin injection schedule was selected because of the observation that hamsters are increasingly sensitive to the indole as the light phase continues.[57] Similar observations were made by the same group[208] in reference to anosmic rats. Melatonin injections 11 hr after lights on severely depressed the maturation of the reproductive system and pituitary prolactin levels in male rats that were incapable of receiving olfactory stimuli due to bilateral removal of the olfactory bulbs.

Although melatonin obviously exhibited antigonadotrophic activity in these studies,[207,208] since it was only injected in the afternoon it could not be determined whether the animals were more or less responsive to the indole at other times of the day. It does appear, however, from a study in anosmic female rats that daily afternoon (12 hr after lights on) injections of melatonin are somewhat more potent in reducing the maturation of the pituitary-ovarian axis than are injections given earlier in the day (3 hr after lights on).[209] Hence, rats seem to be much like hamsters inasmuch as they probably exhibit a changing sensitivity to the pineal indole melatonin.

The final point to be covered in reference to melatonin in rats relates to its counter antigonadotrophic action in this species. Again, the studies that have been designed to test this possibility have utilized rats in which the growth of the reproductive system was held in check by a combination of an activated pineal gland and a potentiating factor. The growth retardation of the sexual organs experienced by neonatally steroid-treated rats that were also blinded was for the most part overcome in animals that were given a biweekly subcutaneous implant of melatonin in beeswax.[210] Likewise, blind, anosmic male rats which usually mature sexually at a very low pace were restored essentially to normal by continually available melatonin from a subcutaneous depot.[107,108] Unquestionably, melatonin, when placed as a deposit under the skin from which it can be released, continually acts in rats, as in hamsters, as a counter antigonadotrophic agent. Because of the many similarities in the responses of rats and hamsters to melatonin administered by various means, the scheme, described above, concerning the changing sensitivity of the melatonin receptor may apply to both species.

B. Other Indoles

In a few studies, the physiological consequences of other indoles have been checked as to their ability to modify the functioning of the neuroendocrine-reproductive axis. However, in only an occasional isolated case have they been shown to be as active as melatonin.[113,148] One compound that has received some investigative effort is 5-methoxytryptophol (5-Mtol); it, like melatonin, is an end product of serotonin metabolism in the pineal gland.[160,211] The interest in 5-Mtol was reinforced by the recent demonstration that it exhibits a rhythm in the blood of rats which is highly correlated with the level of this substance in the pineal gland.[212] The experimenters used a highly specific gas-chromatography mass-spectrometry method for the quantification of 5-Mtol.

In physiological studies, 5-Mtol is reportedly more effective than melatonin in altering the estrous cycles of rats[213] while in hamsters[166] 5-Mtol is considerably less potent than melatonin, when both are injected in the afternoon, in suppressing the growth of the testes. Other studies revealed that 5-Mtol curtails the rise in circulating FSH levels which follows castration in male rats.[214] Whereas Fraschini and Martini[116] consider this indole to be a specific inhibitor of FSH this concept is questioned by others.[215] None of the changes induced by 5-Mtol have been particularly dramatic. However, if the sensitivity of the receptors for 5-Mtol vary with time of day and if the cycle is not in phase with the alleged rhythm in sensitivity of the melatonin receptors, then 5-Mtol could be very potent as an antigonadotrophin but it simply may never have been administered at the most propitious time.

The precursor of 5-Mtol, 5-hydroxytryptophol (5-Htol) has been used infrequently as a modifier of reproduction. It has been reported to inhibit the percentage of immature rats that ovulate in response to pregnant mare's serum gonadotrophin.[216] According to Fraschini, Collu, and Martini,[191] 5-Htol acts like melatonin and specifically influences LH synthesis in the anterior pituitary of the rat. The small amount of data on the hormonal properties of 5-Htol make it difficult to draw any firm conclusions concerning its endocrine role.

There are at least two other indoles, i.e., *N*-acetylserotonin and 5-methoxytryptamine, which could be secretory products of the pineal gland. The results obtained with the use of these compounds, however, have either been inconsequential or contradictory.[148]

It is obvious that, other than melatonin, relatively little information is available on the reproductive sequelae of indole administration. This is a fertile field for investigation of which few researchers are presently taking advantage. This frontier may well be thoroughly exploited in the decade of the 1980s.

C. Polypeptides

Inasmuch as the role of pineal polypeptides in controlling reproductive physiology have been discussed by other reviewers in this volume (see chapters by Vaughan and Benson), this subject is considered beyond the scope of the present survey. Besides the reviews noted above, the reader may wish to consult one or more of the following articles.[148,150,152,217,218]

IV. ITS NICHE

Organs and organ systems evolve because they afford some physiological advantage to the organism. Thus the question arises, of what advantage is it to mammals to possess an organ that secretes hormones in response to changes in the photoperiodic environment? Obviously, it is essential that certain functions of any organism which lives under natural environmental conditions must be in proper synchrony with the seasons of the year. This is particularly true of animals which inhabit the temperate and polar regions of the earth. One of the most important of these functions is reproduction. It would be physiological suicide for animals to deliver their young during seasons that would not maximally promote the survival of the newborns. Thus, under natural environmental conditions, virtually all species have their young in the spring of the year. Young delivered during other seasons would have a greater probability of dying because of the harshness of the environment. To ensure spring birth, mammals have evolved a variety of different mechanisms whereby breeding and subsequent delivery of the young are prohibited from occurring indiscriminantly. Because of its annual reproducibility, many animals have come to rely on the changing photoperiod to signal the appropriate time for sexual behavior. The pineal gland is the organ which responds to the ratio of light:darkness and determines when some animals can and cannot breed. The interrelationships of the photoperiod, the pineal gland, and the annual cycle of reproduction were put into a theoretical scheme in the early 1970s.[220,221]

In this theory, the photoperiod acting by way of the pineal gland induces reproductive involution in photosensitive species during the fall and winter months. The scheme presented arises from experimental data obtained with the use of the Syrian hamster, however, it is likely that it applies to other seasonally breeding animals which rely on the light:dark cycle to determine their annual patterns of sexual activity.

The annual cycle of reproduction for photosensitive long day breeding rodents, has been divided into the following four phases: inhibition phase, sexually quiescent phase, restoration phase, and the sexually active phase.[37,222,223] Each of these phases coincides with a particular season or seasons of the year. For example, in the fall, hibernatory species such as Syrian hamsters, experience progressively decreasing day lengths and additionally, the animals spend increasingly more time in subterranean burrows in darkness. This increased duration of time spent in the dark maximally stimulates the antigonadotrophic activity of the pineal gland and reproductive regression ensues. The interval during which the gonads are regressing is known as the inhibition phase of the annual cycle. Reproductive involution normally follows rapidly after the exposure of hamsters to naturally decreasing day lengths in the fall of the year.[83,224] The morphological and physiological changes which occur in the sexual organs of hamsters under naturally short days are identical with those observed in animals subjected to artificially reduced day lengths under laboratory conditions.[37] The involution of the reproductive organs in the fall of the year ensures long term, deep hibernation[225-229] and it prevents the animals from breeding.[230] Any young born in the late fall or winter would likely die because of the adverse weather conditions associated with these seasons. This is particularly true for animals that inhabit the temperate and polar regions of the earth. Indeed, it is in these areas where seasonal reproduction is probably most essential. At the end of the inhibition phase, the animals are sexually incompetent; this signifies the onset of the sexually quiescent phase.[223]

During the period of sexual quiescence, bouts of hibernation are interrupted by occasional periods of arousal from the toporous state.[225] In animals maintained in short day conditions with infantile reproductive organs, gonadal recrudescence can be initiated by either pinealectomy,[71] or by moving the animals to long day conditions.[231] Likewise, as subcutaneous pellets of melatonin prevent gonadal involution in laboratory-maintained hamsters kept in short days so also under naturally short photoperiods, subcutaneous reserves of melatonin will prevent gonadal atrophy.[232] Also, if hamsters with totally involuted reproductive organs receive a subcutaneous deposit of melatonin in beeswax, gonadal recrudescence follows shortly (Figure 1). Since none of these experimental paradigms occur in the natural state, during the sexually quiescent phase under field conditions, male hamsters characteristically exhibit very low sperm production[227,233] and females do not ovulate.[14] Furthermore, reduced gonadotrophins and sexual steroids probably render the animals behaviorally incapable of mating, etc. This combination of circumstances, coupled with the fact that hibernating hamsters probably do not cohabitate, prevent sexual reproduction and guarantee successful hibernation.[228]

Proof that pinealectomized hamsters, even under winter photoperiodic and temperature conditions can breed, comes from the observation that if pinealectomized males are caged with pinealectomized females at midwinter, the females do become pregnant.[230] Pregnancy and delivery of the young seem to proceed normally. However, within a few days after delivery all of the young die, presumably because of the prevailing low temperatures at this time of the year. Thus, to prevent the wastage of enormous amounts of energy it is obviously essential that these species are incapable of sexual activity during the winter months.

As spring approaches, but while the hamsters are still hibernating in dark underground burrows, the gonads of both males and females begin to recrudesce. This regrowth is light independent and has been described as being spontaneous or endogenous.[37,234] The initiation of recrudescence identifies the onset of the restoration phase of the annual reproductive cycle. One investigator has suggested the regrowth of the gonads is a consequence of the increased sensitivity of the reproductive system to the stimulatory effect of light.[235] This is probably not correct since in fact complete restoration of the gonads can occur in total darkness.[118] Rather, the regrowth theoretically occurs because the hypothalamo-pituitary-gonadal axis becomes refractory to the antigonadotrophic influence of the pineal gland. The refractoriness explanation is also supported by the observation that the production of at least one pineal antigonadotrophin, melatonin, is not diminished during the restoration phase of the cycle.[236]

The onset of refractoriness is an important phenomenon in the seasonally breeding hibernatory hamster inasmuch as it allows the gonads to achieve adult size and function during the later stages of hibernation. Thus, when these animals finally terminate the hibernatory state, they are immediately capable of successful reproduction. If refractoriness did not set in, the regrowth of the gonads would have to await the exposure of the animals to long days after the time of emergence from hibernation. This would delay reproduction by about 8 weeks and would effectively reduce the length of the breeding season. The restoration phase of the annual cycle is complete when the gonads are fully recrudesced and the animals terminate hibernation.

As animals appear above ground in the spring of the year they begin mating, etc. This represents the onset of the sexually active phase.[37] Since the gestation period for the animals is quite short (16 days) the young are born early in the spring. Female hamsters probably have several litters throughout the summer months. The refractory period which initiated the restoration phase extends into the sexually active phase in the summer. Thus, if animals are deprived of light in the spring, immediately after they have experienced a period of short days during which the gonads degenerated and subsequently spontaneously regenerated, their reproductive systems do not undergo atrophy.[231] This raises an important question concerning the control of the annual reproductive cycle by the light:dark cycle. Does light have any importance in cueing the annual cycle? In the case of feral animals it is the virtual absence (or reduced) light in the fall that initiates the inhibition phase of the cycle. Likewise, it is the absence of light which allows the sexually quiescent phase to proceed uninterrupted. Also, the restoration phase is light independent and, finally, during the sexually active phase the reproductive systems of hamsters remain refractory to darkness, i.e., light is seemingly not required at this time.

Some light is required during the summer sexually active phase for the purpose described below. Recall that surgically blinded hamsters, i.e., animals deprived of all photic input, experience gonadal atrophy and subsequent spontaneous regeneration of their reproductive organs. Thereafter, the reproductive systems never again regress.[118] On the contrary, the reproductive systems of hamsters kept under seasonally changing photoperiods involute each winter. Apparently, light during the sexually active phase, although not required to support high gonadal activity, is necessary to interrupt the refractory period so darkness can again induce sexual involution during the subsequent fall and winter. This was experimentally borne out in experiments from at least two separate laboratories.[223,237] Thus, both light and darkness are critical for the proper synchronization of the phases of the annual reproductive cycle with the appropriate seasons of the year.[37,222,223]

Inasmuch as the mechanisms whereby the pineal gland induces gonadal atrophy are reviewed in another section of this volume (see Blask), it is considered beyond the scope of this survey to review in detail the data relating to this problem. However, certain papers do deserve mention at this point. It has been proposed that pineal secretory products increase the sensitivity of the feedback centers of the brain to circulating gonadal steroids such that even small amounts of these compounds strongly inhibit the secretion of LH and FSH and, as a consequence, gonadal involution results.[238,239] This is a worthwhile hypothesis, however, it does not totally explain pineal induced gonadal involution since it does not account for the very important role played by prolactin in the atrophic response. It is apparent that even under conditions of a relatively low output of LH and FSH, prolactin can promote substantial testicular growth in hamsters kept under short days.[97-99]

In this review considerable emphasis has been placed on melatonin as an important pineal antigonadotrophin. Despite this, the present reviewer remains open minded as to other inhibitory factors of pineal origin. It is apparent, however, that melatonin can mediate many of the effects of the pineal on reproduction in the hamster. The point has already been made that both darkness and melatonin produce similar degrees of atrophy of the reproductive organs and similar hormonal imbalances (Figure 5). Recently accumulated data indicate that properly timed melatonin injections can also duplicate the annual reproductive cycle of the hamster. Certainly, afternoon melatonin injections, like short days, cause gonadal involution.[57] If the injections are continued, the gonads eventually spontaneously regenerate just as they do in hamsters kept in darkness (Figure 9). After regeneration occurs, the gonads are refractory to the inhibitory effects of both darkness and melatonin.[231,240,241] Finally, both the refractory periods to darkness and melatonin are interrupted by the exposure of animals to long days for prolonged periods of time.[234,237,241] Thus, as noted above, melatonin by itself could mediate seasonal reproductive changes which normally rely on darkness acting by way of the pineal gland.[37]

V. CONCLUDING REMARKS

The mysteries of the pineal gland have been very difficult to unravel because in many respects the pineal is an unorthodox organ of internal secretion. For example, its biosynthetic and secretory activity depend very heavily on its noradrenergic sympathetic innervation. This is not the case with most endocrine glands which function at virtually the normal pace after denervation. Another unusual feature is that the pineal products

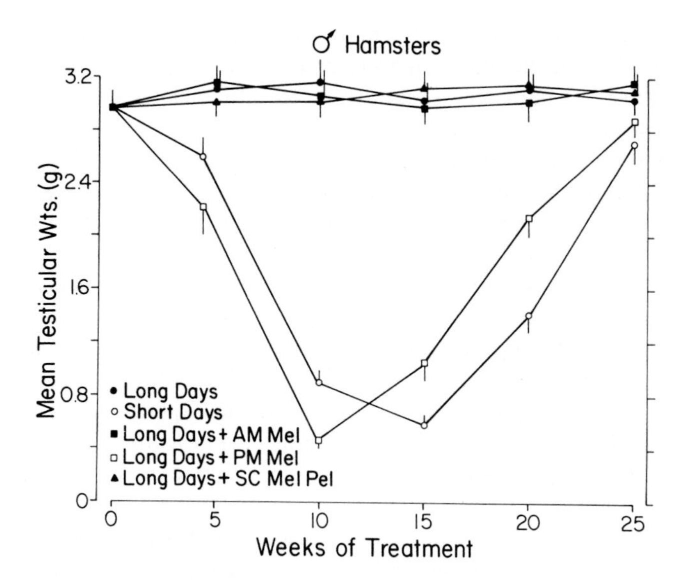

FIGURE 9. Mean testicular weights in hamsters given various treatments. Both short-day exposure and PM-melatonin injections (PM Mel) induce a transient inhibitory effect on the testes. Eventually however, the neuroendocrine-gonadal axis becomes refractory to both types of treatment. Continually available melatonin from a subcutaneous reserve (SC Mel Pel) prevents short day-induced gonadal atrophy presumably by down regulating its own receptors. AM melatonin injections (AM Mel) are without measurable influence on reproduction in the Syrian hamster.

are primarily inhibitory. Most hormones promote some aspect of metabolism. Indeed, by conventional definition the term hormone implies stimulation. Finally, despite the fact that it has been amply demonstrated, many scientists outside the immediate field seem reluctant about accepting the pineal as a legitimate endocrine structure because it has suffered from a long term vestigiality complex.

The chemical nature of the pineal hormone(s) remains an area of considerable controversy. There could be one or many hormonal products from the pineal gland. Certainly, the methoxyindole melatonin has captured a great deal of interest. In terms of the reproductive system, there are many similarities between pineal and melatonin induced gonadal atrophy in hamsters. Because of this it is tempting to speculate that melatonin is the hormone which mediates the antigonadotrophic action of the pineal in the Syrian hamster and possibly in other species. For example, (1) the degree of atrophy and the hormonal alterations produced by the two treatments are similar; (2) the time period required for these procedures to induce total reproductive collapse are roughly equivalent; (3) eventually the gonads of hamsters become refractory to the influence of the pineal and melatonin; (4) the refractory period to the pineal and melatonin are interrupted by exposure of the animals to long photoperiods for a matter of weeks; (5) subcutaneous melatonin depots prevent both pineal and afternoon melatonin-induced gonadal atrophy and; (6) any procedure which interferes with the nocturnal rise in melatonin production also prevents pineal induced gonadal involution. On the basis of these findings it is difficult to envisage that melatonin is not somehow involved with pineal mediated inhibition of reproduction in the Syrian hamster (and probably the albino rat). Even if melatonin is proven to be the hormone which accounts for pineal induced gonadal atrophy, research on other pineal constituents should continue. It is becoming increasingly apparent that the pineal is a complex secretory organ which probably interacts, in one way or another, with most organs in the body. This being the case, it would be naive to assume it secretes but a single product.

It should again be emphasized that the present review deals with pineal-reproductive interrelationships as we believe them to exist in the Syrian hamster and albino rat. What is stated herein may or may not apply to other species. The pineal and its hormones often seem to work differently in various species. Thus, generalizations should be made with extreme caution. The actions of the pineal envoys are diverse and complex and anyone entering the field should bear this in mind when attempting to investigate this multifaceted organ of internal secretion.

ACKNOWLEDGMENTS

Work by the author was supported by research grants from the National Science Foundation and from the National Institutes of Health.

REFERENCES

1. **Tilney, F. and Warren, L. F.**, *The Morphology and Evolutional Significance of the Pineal Body*, Wistar Press, Philadelphia, 1919.
2. **Gladstone, R. J. and Wakely, C. P. G.**, *The Pineal Organ*, Bailliere, Tindall and Cox, London, 1940.
3. **McCord, C. P. and Allen, F. P.**, Evidences associating pineal gland function with alterations in pigmentation, *J. Exp. Zool.*, 23, 207, 1917.
4. **Lerner, A. B., Case, J. D., Takahashi, Y., Lee, T. H., and Mori, W.**, Isolation of melatonin, the pineal factor that lightens melanocytes, *J. Am. Chem. Soc.*, 80, 2587, 1958.
5. **Lerner, A. B., Case, J. D., and Heinzelman, R. V.**, Structure of melatonin, *J. Am. Chem. Soc.*, 81, 6084, 1959.
6. **Milco, S. M.**, L'action de l'extrait épiphysaire sur le cycle oestral de la rate, *Bull. Soc. Roum. Endocrinol.*, 7, 86, 1941.
7. **Engel, P.**, Zirbeldrüse und gonadotropes Hormon, *Zschr. Ges. Exp. Med.*, 94, 333, 1934.
8. **den Hartog Jager, W. A. and Heil, J. F.**, Uber die Epiphysefrage, *Acta Brevia Neerl. Physiol. Pharmacol. Microbiol.*, 5, 32, 1935.
9. **Kitay, J. I. and Altschule, M. D.**, *The Pineal Gland*, Harvard University Press, Cambridge, 1954.
10. **Thieblot, L. and Le Bars, H.**, *La Glande Pinâle ou Épiphyse*, Librairie Maloine S. A., Paris, 1955.
11. **Reiter, R. J.**, *The Pineal-1977*, Eden Press, Montreal, 1977, chap. 1.
12. **Hoffman, R. A., Hester, R. J., and Townes, C.**, Effect of light and temperature on the endocrine system of the golden hamster (*Mesocricetus auratus* Waterhouse), *Comp. Biochem. Physiol.*, 15, 525, 1965.
13. **Quay, W. B.**, Volumetric and cytologic variation in the pineal body of *Peromycus leucopus* (Rodentia) with respect to sex, captivity and daylength, *J. Morphol.*, 98, 471, 1956.
14. **Mogler, R. K.- H.**, Das Endokrine System des Syrischen Goldhamster unter Berücksichtigung des Natürlichen Winterschlafs, *Zeit. Morphol. Oekol. Tiere*, 47, 267, 1958.
15. **Quay, W. B.**, Reduction in mammalian pineal weight and lipid content during continuous light, *Gen. Comp. Endocrinol.*, 1, 211, 1960.
16. **Quay, W. B.**, Circadian rhythm in rat pineal serotonin and its modification by estrous cycle and photoperiod, *Gen. Comp. Endocrinol.*, 3, 473, 1963.
17. **Wurtman, R. J., Axelrod, J., and Phillips, L. S.**, Melatonin synthesis in the pineal gland: control by light, *Science*, 142, 1071, 1963.
18. **Quay, W. B.**, Circadian and estrous rhythms in pineal melatonin and 5-hydroxyindole-3-acetic acid, *Proc. Soc. Exp. Biol. Med.*, 115, 710, 1964.
19. **Hoffman, R. A. and Reiter, R. J.**, Rapid pinealectomy in hamsters and other small rodents, *Anat. Rec.*, 153, 19, 1965.
20. **Hoffman, R. A. and Reiter, R. J.**, Pineal gland: influence on gonads of male hamsters, *Science*, 148, 1609, 1965.

21. **Hoffman, R. A. and Reiter, R. J.,** Influence of compensatory mechanisms and the pineal gland on dark-induced gonadal atrophy in male hamsters, *Nature (London),* 203, 658, 1965.
22. **Hoffman, R. A. and Reiter, R. J.,** Responses of some endocrine organs of female hamsters to pinealectomy and light, *Life Sci.,* 5, 1147, 1966.
23. **Reiter, R. J., Hoffman, R. A., and Hester, R. J.,** The effects of thiourea, photoperiod and the pineal gland on the thyroid, adrenals and reproductive organs of female hamsters, *J. Exp. Zool.,* 162, 263, 1966.
24. **Axelrod, J. and Wurtman, R. J.,** The formation, metabolism and some actions of melatonin, a pineal substance, in *Endocrines and the Central Nervous System,* Levine, R., Ed., Williams & Wilkins, Baltimore, 1966, 200.
25. **Wurtman, R. J., Axelrod, J., and Kelly, D. E.,** *The Pineal,* Academic Press, New York, 1968, chap. 2.
26. **Kappers, J. A.,** Melatonin, a pineal compound. Preliminary investigations of its function in the rat, *Gen. Comp. Endocrinol.,* 2, 16, 1962.
27. **Wurtman, R. J., Axelrod, J., and Chu, E. W.,** Melatonin, a pineal substance: effect on the rat ovary, *Science,* 141, 277, 1963.
28. **Chu, E. W., Wurtman, R. J., and Axelrod, J.,** An inhibitory effect of melatonin of the estrous phase of the estrous cycle of the rodent, *Endocrinology,* 75, 238, 1964.
29. **Wurtman, R. J., Axelrod, J., and Fischer, J. E.,** Melatonin synthesis in the pineal gland: effect of light mediated by the sympathetic nervous system, *Science,* 143, 1328, 1964.
30. **Wurtman, R. J. and Axelrod, J.,** The pineal gland, *Sci. Am.,* 213, 5, 1965.
31. **Axelrod, J. and Wurtman, R. J.,** Photic and neural control of indoleamine metabolism in the rat pineal gland, *Adv. Pharmacol.,* 6(A), 157, 1968.
32. **Wurtman, R. J., Axelrod, J., Chu, E. W., and Fischer, J. E.,** Mediation of some effects of illumination on the rat estrous cycle by the sympathetic nervous system, *Endocrinology,* 75, 266, 1964.
33. **Reiter, R. J. and Hester, R. J.,** Interrelationships of the pineal gland, the superior cervical ganglia and the photoperiod in the regulation of the endocrine systems of hamsters, *Endocrinology,* 79, 1168, 1966.
34. **Reiter, R. J. and Fraschini, F.,** Endocrine aspects of the mammalian pineal gland: a review, *Neuroendocrinology,* 5, 219, 1969.
35. **Reiter, R. J.,** Physiologic role of the pineal gland, in *The Action of Hormones,* Foa, P. P., Ed., Charles C Thomas, Springfield, Ill., 1971, 283.
36. **Reiter, R. J.,** Morphological studies on the reproductive organs of blinded male hamsters and the effects of pinealectomy or superior cervical ganglionectomy, *Anat. Rec.,* 160, 13, 1968.
37. **Reiter, R. J.,** The pineal and its hormones in the control of reproduction in mammals, *Endocr. Rev.,* 1, 109, 1980.
38. **Negus, N. C. and Berger, P. J.,** Pineal weight response to a dietary variable in *Microtus montanus, Experientia,* 27, 215, 1971.
39. **Reiter, R. J.,** *The Pineal Gland: A Report of Some Recent Physiological Studies,* Edgewood Arsenal Technical Report, Edgewood Arsenal, Maryland, 1967.
40. **Beauchamp, G. K., Doty, R. L., Moutton, D. G., and Mugford, R. A.,** The phermone concept in mammalian chemical communication: a critique, in *Mammalian Olfaction, Reproductive Processes, and Behavior,* Doty, R. L., Ed., Academic Press, New York, 1976, 144.
41. **Richmond, M. and Stehn, R.,** Olfaction and reproduction in microtine rodents, in *Mammalian Olfaction, Reproductive Processes, and Behavior,* Doty, R. L., Ed., Academic Press, New York, 1976, 198.
42. **Macrides, F.,** Olfactory influences on neuroendocrine function in mammals, in *Mammalian Olfaction, Reproductive Processes, and Behavior,* Doty, R. L., Ed., Academic Press, New York, 1976, 29.
43. **Murphy, M. R.,** Olfactory impairment, olfactory bulb removal, and mammalian reproduction, in *Mammalian Olfaction, Reproductive Processes, and Behavior,* Doty, R. L., Ed., Academic Press, New York, 1976, 96.
44. **Reiter, R. J.,** The effect of pineal grafts, pinealectomy, and denervation of the pineal gland on the reproductive organs of male hamsters, *Neuroendocrinology,* 2, 138, 1967.
45. **Rusak, B. and Morin, L. P.,** Testicular responses to photoperiod are blocked by lesions of the suprachiasmatic nuclei in golden hamsters, *Biol. Reprod.,* 15, 366, 1976.
46. **Zucker, I. and Morin, L. P.,** Photoperiodic influences on testicular regression, recrudescence and the induction of scotorefractoriness in male golden hamsters, *Biol. Reprod.,* 17, 493, 1977.
47. **Turek, F. W. and Losee, S. H.,** Melatonin-induced testicular growth in golden hamsters maintained in short days, *Biol. Reprod.,* 18, 299, 1978.
48. **Turek, F. W. and Campbell, C. S.,** Photoperiodic regulation of neuroendocrine-gonadal activity, *Biol. Reprod.,* 20, 32, 1979.

49. **Berndtson, W. E. and Desjardins, C.**, Circulating LH and FSH levels and testicular function in hamsters during light deprivation and subsequent photoperiodic stimulation, *Endocrinology,* 95, 195, 1974.
50. **Desjardins, C., Ewing, L. L., and Johnson, B. H.**, Effects of light deprivation upon the spermatogenic and steroidogenic elements of the hamster testes, *Endocrinology,* 89, 791, 1971.
51. **Kent, C. G., Jr.**, Physiology of reproduction, in *The Golden Hamster,* Hoffman, R. A., Robinson, P. F., and Magalhaes, H., Eds., Iowa State Univ. Press, Ames, 1968, 119.
52. **Orsini, M. W.**, The external vaginal phenomena characterizing the stages of the estrous cycle, pregnancy, pseudopregnancy, lactation, and the anestrous hamster, *Mesocricetus auratus,* Waterhouse, *Proc. Animal Care Panel,* 11, 193, 1961.
53. **Gaston, S. and Menaker, M.**, Photoperiod control of hamster testis, *Science,* 158, 925, 1967.
54. **Elliott, J.**, Circadian rhythms and photoperiodic time measurement, *Fed. Proc. Fed. Am. Soc. Exp. Biol.,* 35, 2339, 1976.
55. **Sorrentino, S., Jr. and Reiter, R. J.**, Pineal-induced alteration of estrous cycles in blinded hamsters, *Gen. Comp. Endocrinol.,* 15, 39, 1970.
56. **Reiter, R. J.**, The role of the pineal in reproduction, in *Reproductive Biology,* Balin, H. and Glasser, S., Eds., Excerpta Medica, Amsterdam, 1972, 71.
57. **Tamarkin, L., Westrom, W. K., Hamill, A. I., and Goldman, B. D.**, Effect of melatonin on the reproductive systems of male and female Syrian hamsters: a diurnal rhythm in sensitivity to melatonin, *Endocrinology,* 99, 1534, 1976.
58. **Andre, J. S. and Parrish, J.**, Inhibition of estrous cyclicity in golden hamsters by melatonin administration on the day of proestrus, *J. Exp. Zool.,* 207, 161, 1979.
59. **Browman, L. G.**, Light and its relation to activity and estrous rhythms in the albino rat, *J. Exp. Zool.,* 75, 375, 1937.
60. **Browman, L. G.**, The effect of optic enucleation on the albino rat, *Anat. Rec.,* 78, 59, 1940.
61. **Eayrs, J. T. and Ireland, K. L.**, The effect of total darkness on the growth of the newborn albino rat, *J. Endocrinol.,* 6, 386, 1950.
62. **Reiter, R. J.**, The pineal gland and gonadal development in male rats and hamsters, *Fertil. Steril.,* 19, 1009, 1968.
63. **Reiter, R. J.**, Growth of the endocrine organs of female hamsters blinded at 25 days of age, *Experientia,* 25, 751, 1969.
64. **Reiter, R. J., Sorrentino, S., Jr., and Hoffman, R. A.**, Early photoperiodic conditions and pineal antigonadal function in male hamsters, *Int. J. Fertil.,* 15, 163, 1970.
65. **Reiter, R. J. and Sorrentino, S., Jr.**, Reproductive effects of the mammalian pineal, *Am. Zool.,* 10, 247, 1970.
66. **Reiter, R. J. and Sorrentino, S., Jr.**, Factors influential in determining the gonad-inhibiting activity of the pineal gland, in *The Pineal Gland,* Wolstenholme, G. E. W. and Knight, J., Eds., Churchill Livingstone, London, 1971, 329.
67. **Kordon, C. and Hoffman, J.**, Mise en évidence d'un èffet fortement gonadostimulant de la lumiere chez le rat mâle prétraité par une injection postnatale de testostérone, *C. R. Seances Soc. Biol. Paris,* 161, 1262, 1967.
68. **Hoffmann, J. C., Kordon, C., and Benoit, J.**, Effect of different photoperiods and blinding on ovarian and testicular functions in normal and testosterone-treated rats, *Gen. Comp. Endocrinol.,* 10, 109, 1968.
69. **Reiter, R. J., Hoffmann, J. C., and Rubin, P. H.**, Pineal gland: influence on gonads of male rats treated with androgen three days after birth, *Science,* 160, 420, 1968.
70. **Reiter, R. J., Rubin, P. H., and Richert, J. R.**, Pineal-induced ovarian atrophy in rats treated neonatally with testosterone, *Life Sci.,* 7, 299, 1968.
71. **Reiter, R. J.**, Pineal-gonadal relationships in male rodents, in *Progress in Endocrinology,* Gual, C., Ed., Excerpta Medica, Amsterdam, 1969, 631.
72. **Orbach, J. and Kling, J.**, Effect of sensory deprivation on onset of puberty, mating, fertility and gonadal weights in rats, *Brain Res.,* 3, 141, 1966.
73. **Reiter, R. J., Klein, D. C., and Donofrio, R. J.**, Preliminary observations on the reproductive effects of the pineal gland in blinded, anosmic male rats, *J. Reprod. Fertil.,* 19, 563, 1969.
74. **Reiter, R. J. and Ellison, N. M.**, Delayed puberty in blinded anosmic female rats: role of the pineal gland, *Biol. Reprod.,* 2, 216, 1970.
75. **Sorrentino, S., Jr., Reiter, R. J., Schalch, D. S., and Donof rio, R. J.**, Role of the pineal gland in growth restraint of adult male rats by light and smell deprivation, *Neuroendocrinology,* 8, 116, 1971.
76. **Reiter, R. J., Sorrentino, S., Jr., and Donofrio, R. J.**, Interaction of photic and olfactory stimuli in mediating pineal-induced gonadal regression in adult female rats, *Gen. Comp. Endocr.,* 15, 326, 1970.

77. **Klein, D. C., Reiter, R. J., and Weller, J. L.,** Pineal *N*-acetyltransferase activity in blinded and anosmic male rats, *Endocrinology,* 89, 1020, 1971.
78. **Reiter, R. J., Sorrentino, S., Jr., Ralph, C. L., Lynch, H. J., Mull, D., and Jarrow, E.,** Some endocrine effects of blinding and anosmia in adult male rats with observations on pineal melatonin, *Endocrinology,* 88, 895, 1971.
79. **Reiter, R. J., Sorrentino, S., Jr., and Jarrow, E. L.,** Central and peripheral neural pathways necessary for pineal function in the adult female rat, *Neuroendocrinology,* 8, 321, 1971.
80. **Sorrentino, S., Jr., Reiter, R. J., and Schalch, D. S.,** Interactions of the pineal gland, blinding, and underfeeding on reproductive organ size and radioimmunoassayable growth hormone, *Neuroendocrinology,* 7, 105, 1971.
81. **Sorrentino, S., Jr., Schalch, D. S., and Reiter, R. J.,** Environmental control of growth hormone and growth, in *Growth and Growth Hormone,* Müller, E., Ed., Excerpta Medica, Amsterdam, Amsterdam, 1972, 330.
82. **Klein, D. C., Weller, J. L., and Moore, R. Y.,** Melatonin metabolism: neural regulation of pineal serotonin *N*-acetyltransferase activity, *Proc. Nat. Acad. Sci. U.S.A.,* 68, 3107, 1971.
83. **Reiter, R. J.,** Pineal control of a seasonal reproductive rhythm in male golden hamsters exposed to natural daylight and temperature, *Endocrinology,* 92, 423, 1973.
84. **Tamarkin, L., Hutchinson, J. S., and Goldman, B. D.,** Regulation of serum gonadotropins by photoperiod and testicular hormone in the Syrian hamster, *Endocrinology,* 99, 1528, 1976.
85. **Turek, F. W., Alvis, J. D., Elliott, J. A., and Menaker, M.,** Temporal distribution of serum levels of LH and FSH in adult male golden hamsters, *Biol. Reprod.,* 14, 630, 1976.
86. **Bridges, R., Tamarkin, L., and Goldman, B.,** Effects of photoperiod and melatonin on reproduction in the Syrian hamster, *Ann. Biol. Anim. Biochem. Biophys.,* 16, 399, 1976.
87. **Arimura, A., Matsua, H., Baba, Y., and Schally, A. V.,** Ovulation induced by synthetic luteinizing hormone-releasing hormone in the hamster, *Science,* 174, 511, 1971.
88. **Arimura, A., Debeljuk, L., and Schally, A. V.,** LH release by LH-releasing hormone in golden hamsters at various stages of the estrous cycle, *Proc. Soc. Exp. Biol. Med.,* 140, 609, 1972.
89. **Turek, F. W., Alvis, J. D., and Menaker, M.,** Pituitary responsiveness to LRF in castrated male hamsters exposed to different photoperiodic conditions, *Neuroendocrinology,* 24, 140, 1977.
90. **Reiter, R. J., Vaughan, M. K., Blask, D. E., and Johnson, L. Y.,** Pineal methoxyindoles: new evidence concerning their function in the control of pineal-mediated changes in the reproductive physiology of male golden hamsters, *Endocrinology,* 96, 206, 1975.
91. **Reiter, R. J. and Johnson, L. Y.,** Depressant action of the pineal gland on pituitary luteinizing hormone and prolactin in male hamsters, *Hormone Res.,* 5, 311, 1974.
92. **Reiter, R. J.,** Reproductive involution in male hamsters exposed to naturally increasing daylengths after the winter solstice, *Proc. Soc. Exp. Biol. Med.,* 163, 264, 1980.
93. **Reiter, R. J.,** Regulation of pituitary gonadotrophins by the mammalian pineal gland, in *Neuroendocrine Regulation of Fertility,* Anand Kumar, T. C., Eds., S. Karger, Basel, 1976, 215.
94. **Bartke, A., Croft, B. T., and Dalterio, S.,** Prolactin restores plasma testosterone levels and stimulates testicular growth in hamsters exposed to short day-length, *Endocrinology,* 97, 1601, 1975.
95. **Bex, F. J. and Bartke, A.,** Testicular LH binding in the hamster: modification by photoperiod and prolactin, *Endocrinology,* 100, 1223, 1977.
96. **Bartke, A., Goldman, B. D., Bex, F. J., and Dalterio, S.,** Mechanism of reversible loss of reproductive capacity in a seasonally-breeding mammal, *Int. J. Androl.,* Suppl. 2, 345, 1978.
97. **Bex, F., Bartke, A., Goldman, B. D., and Dalterio, S.,** Prolactin, growth hormone, luteinizing hormone receptors, and seasonal changes in testicular activity in the golden hamster, *Endocrinology,* 103, 2069, 1978.
98. **Matthews, M. J., Benson, B., and Richardson, D. L.,** Partial maintenance of testes and accessory organs in blinded hamsters by homoplastic anterior pituitary grafts or exogenous prolactin, *Life Sci.,* 23, 1131, 1978.
99. **Bartke, A., Smith, M. S., and Dalterio, S.,** Reversal of short photoperiod-induced sterility in male hamsters by ectopic pituitary homografts, *Int. J. Androl.,* 2, 257, 1979.
100. **Reiter, R. J. and Ferguson, B. N.,** Delayed reproductive regression in male hamsters bearing intrarenal pituitary homografts and kept under natural winter photoperiods, *J. Exp. Zool.,* 209, 175, 1979.
101. **Chen, H. J. and Reiter, R. J.,** The combination of twice daily luteinizing hormone releasing factor administration and renal pituitary homografts restores normal reproductive organ size in male hamsters with pineal-mediated gonadal atrophy, *Endocrinology,* 106, 1382, 1980.
102. **Reiter, R. J.,** Pineal-mediated reproductive events, in *Novel Aspects of Reproductive Physiology,* Spilman, C. H. and Wilks, J. W., Eds., Spectrum, New York, 1978, 369.
103. **Blask, D. E., Reiter, R. J., Vaughan, M. K., and Johnson, L. Y.,** Differential effects of the pineal gland on LH-RH and FSH-RH activity in the medial basal hypothalamus of the male golden hamster, *Neuroendocrinology,* 28, 36, 1979.

104. **Pickard, G. E. and Silverman, A. J.**, Effects of photoperiod on hypothalamic luteinizing hormone releasing hormone in the male hamster, *J. Endocrinol.*, 83, 421, 1980.
105. **Reiter, R. J.**, Pineal-anterior pituitary gland relationships, in *MTP International Review of Science — Physiology Series One, Vol. 5, Endocrine Physiology*, McCann, S. M., Ed., Butterworths, London, 1974, 277.
106. **Reiter, R. J.**, Pineal regulation of the hypothalamo-pituitary axis: gonadotrophins, in *Handbook of Physiology, Endocrinology IV, Part 2*, Knobil, E. and Sawyer, W. H., Eds., American Physiological Society, Washington, D.C., 1974, 519.
107. **Reiter, R. J., Blask, D. E., and Vaughan, M. K.**, A counter antigonadotrophic effect of melatonin in male rats, *Neuroendocrinology*, 19, 72, 1975.
108. **Chen, H. J. and Reiter, R. J.**, Influence of subcutaneous deposits of melatonin on the antigonadotrophic effects of blinding and anosmia in male rats. A dose response study, *Neuroendocrinology*, 30, 169, 1980.
109. **Clementi, F., de Virgilis, G., Fraschini, F., and Mess, B.**, Modifications of pituitary morphology following pinealectomy and the implantation of the pineal body in different areas of the brain, in *Proc. Sixth Int. Cong. Electron Microscopy*, Marzuen Co., Tokyo, 1966, 539.
110. **Clementi, F., de Virgilis, G., and Mess, B.**, Influence of pineal gland principles on gonadotrophin-producing cells of the rat anterior pituitary gland. An electron microscopic study, *J. Endocrinol.*, 44, 241, 1969.
111. **Girod, C., Cure, M., Czyba, J. C., et Durand, N.**, Influence de l'epiphysectomie sur les cellules gonadotropes antéhypophysaires du Hamster doré (*Mesocricetus auratus* Waterh.), *C. R. Soc. Biol. Paris*, 158, 1636, 1964.
112. **Mess, B.**, Endocrine and neurochemical aspects of pineal function, *Int. Rev. Neurobiol.*, 11, 171, 1968.
113. **Reiter, R. J., Vaughan, M. K., Vaughan, G. M., Sorrentino, S., Jr., and Donofrio, R. J.**, The pineal gland as an organ of internal secretion, in *Frontiers of Pineal Physiology*, Altschule, M. D., Ed., MIT Press, Cambridge, 1975, 54.
114. **Motta, M., Fraschini, F., and Martini, L.**, Endocrine effects of pineal gland and melatonin, *Proc. Soc. Exp. Biol. Med.*, 126, 431, 1967.
115. **Fraschini, F.**, The pineal gland and the control of LH and FSH secretion, in *Progress in Endocrinology*, Gual, C., Ed., Excerpta Medica, Amsterdam, 1969, 637.
116. **Fraschini, F. and Martini, L.**, Rhythmic phenomena and pineal principles, in *The Hypothalamus*, Martini, L., Motta, M., and Fraschini, F., Eds., Academic Press, New York, 1970, 529.
117. **Reiter, R. J.**, Changes in the reproductive organs of cold-exposed and light-deprived female hamsters *(Mesocricetus auratus), J. Reprod. Fertil.*, 16, 217, 1968.
118. **Reiter, R. J.**, Pineal function in long term blinded male and female hamsters, *Gen. Comp. Endocrinol.*, 12, 460, 1969.
119. **Greenwald, G. S.**, Histologic transformation of the ovary of the lactating hamster, *Endocrinology*, 77, 641, 1965.
120. **Greenwald, G. S.**, Luteotrophic complex in the hamster, *Endocrinology*, 80, 118, 1967.
121. **Seegal, R. F. and Goldman, B. D.**, Effects of photoperiod on cyclicity and serum gonadotropins in the Syrian hamster, *Biol. Reprod.*, 12, 223, 1975.
122. **Bridges, R. S. and Goldman, B. D.**, Diurnal rhythms in gonadotropins and progesterone in lactating and photoperiod induced acyclic hamsters, *Biol. Reprod.*, 13, 617, 1975.
123. **Bex, F. J. and Goldman, B. D.**, Serum gonadotropins and follicular development in Syrian hamster, *Endocrinology*, 96, 928, 1975.
124. **Goldman, B. D. and Brown, S.**, Sex differences in LH and FSH patterns in hamsters exposed to short photoperiod, *J. Steroid Biochem.*, 11, 531, 1979.
125. **Reiter, R. J. and Johnson, L. Y.**, Pineal regulation of immunoreactive luteinizing hormone and prolactin in light-deprived female hamsters, *Fertil. Steril.*, 25, 958, 1974.
126. **Reiter, R. J. and Johnson, L. Y.**, Elevated pituitary LH and depressed pituitary prolactin levels in female hamsters with pineal-induced gonadal atrophy and the effects of chronic treatment with synthetic LRF, *Neuroendocrinology*, 14, 310, 1974.
127. **Reiter, R. J.**, Changes in pituitary prolactin levels of female hamsters as a function of age, photoperiod, and pinealectomy, *Acta Endocrinol.*, 79, 431, 1975.
128. **Reiter, R. J.**, Effects of the pineal gland on reproductive organ growth and fertility in dual sensory deprived female rats, *Endocrinol. Exp.*, 6, 3, 1972.
129. **Vaughan, G. M., Reiter, R. J., Siler-Khoder, T. M., Sackman, J. W., Allen, J., Vaughan, M. K., Johnson, L. Y., and Starr, P.**, Influence of pinealectomy on serum estrogen and progesterone levels in blind-anosmic female rats, *Experientia*, 34, 1378, 1978.
130. **Blask, D. E. and Reiter, R. J.**, Pituitary and plasma LH and prolactin levels in female rats rendered blind and anosmic: influence of the pineal gland, *Biol. Reprod.*, 12, 329, 1975.

131. **Blask, D. E. and Reiter, R. J.**, The pineal gland of the blind-anosmic female rat: its influence on medial basal hypothalamic LRH, PIF and/or PRF activity *in vivo*, *Neuroendocrinology*, 17, 362, 1975.
132. **Rønnekliev, O. K. and McCann, S. M.**, Effects of pinealectomy, anosmia and blinding alone or in combination on gonadotrophin secretion and pituitary and target gland weights in intact and castrated male rats, *Neuroendocrinology*, 19, 97, 1975.
133. **Shiino, M., Arimura, A., and Rennels, E. G.**, Effects of blinding, olfactory bulbectomy, and pinealectomy on prolactin and growth hormone cells of the rat with special reference to ultrastructure, *Am. J. Anat.*, 139, 191, 1974.
134. **Donofrio, R. J. and Reiter, R. J.**, Depressed pituitary prolactin levels in blinded anosmic female rats: role of the pineal gland, *J. Reprod. Fertil.*, 31, 159, 1972.
135. **Blask, D. E., Reiter, R. J., and Johnson, L. Y.**, The influence of pineal activation, removal or denervation on hypothalamic luteinizing hormone-releasing hormone activity: an *in vitro* study, *Neurosci. Lett.*, 1, 327, 1975.
136. **Blask, D. E., Reiter, R. J., and Johnson, L. Y.**, Pineal-induced alterations in reproductive function and pituitary prolactin in the female rat: the effects of bilateral superior cervical ganglionectomy and nervi conarii transection, *J. Neurosci. Res.*, 3, 127, 1977.
137. **Blask, D. E. and Reiter, R. J.**, Pineal removal or denervation: effects on hypothalamic PRF activity in the rat, *Mol. Cell. Endocrinol.*, 11, 243, 1978.
138. **Blake, C. A.**, Effects of pinealectomy on the rat oestrous cycle and pituitary gonadotrophin release, *J. Endocrinol.*, 69, 67, 1976.
139. **Slama-Scemama, A.**, Effect of pinealectomy on gonadotrophins in immature female rats, *J. Neural Transmis.*, 39, 251, 1976.
140. **Takeo, Y., Anazawa, M., Shirama, K., Shimizy, K., and Maekawa, K.**, Pinealectomy and sexual rhythm in female rats, *Endocrinol. Jpn.*, 22, 219, 1975.
141. **Hus-Citharel, A., Roseau, S., and Zurburg, W.**, Effects of precocious pinealectomy and hemicastration on pituitary and plasma LH levels in immature male rats, *J. Neural Transm.*, 40, 33, 1977.
142. **Trentini, G. P., de Gaetani, C. F., Martini, L., and Mess, B.**, Effect of pinealectomy and of bilateral cervical ganglionectomy on serum LH levels in constant estrous-anovulatory rats, *Proc. Soc. Exp. Biol. Med.*, 153, 490, 1976.
143. **Reiter, R. J.**, *The Pineal, Vol. 3, 1978,* Eden Press Montreal, 1978, chap. 9.
144. **Collu, R. and Fraschini, F.**, The pineal gland — a neuroendocrine transducer, *Adv. Metab. Disord.*, 6, 161, 1972.
145. **Brownstein, M. J.**, The pineal gland, *Life Sci.*, 16, 1363, 1975.
146. **Orts, R. J., Kocan, K. M., and Wilson, I. B.**, Inhibitory action of melatonin on a pineal antigonadotropin, *Life Sci.*, 17, 845, 1975.
147. **Kappers, J. A.**, The mammalian pineal gland, a survey, *Acta Neurochir.*, 34, 109, 1976.
148. **Reiter, R. J. and Vaughan, M. K.**, Pineal antigonadotrophic substances: polypeptides and indoles, *Life Sci.*, 21, 159, 1977.
149. **Benson, B.**, Current studies on pineal peptides, *Neuroendocrinology*, 24, 241, 1977.
150. **Pavel, S.**, Arginine vasotocin as a pineal hormone, *J. Neural Transm.*, Suppl., 13, 135, 1978.
151. **Wilson, B. W., Lynch, H. J., and Ozaki, Y.**, 5-Methoxytryptophol in rat serum and pineal: detection, quantitation, and evidence of daily rhythmicity, *Life Sci.*, 23, 1019, 1978.
152. **Ebels, I. and Benson, B.**, A survey of evidence that unidentified pineal substances affect the reproductive systems in mammals, in *The Pineal and Reproduction*, Reiter, R. J., Ed., S. Karger, Basel, 1978, 51.
153. **Pévet, P. and Swaab, D. F.**, Immunocytochemical evidence for the presence of an α-MSH-like compound in the rat pineal gland, *J. Physiol.*, 75, 101, 1979.
154. **Negro-Vilar, A., Sanchez-Franco, F., Kwiathowski, M., and Samson, W. K.**, Failure to detect radioimmunoassayable arginine vasotocin in mammalian pineals, *Brain Res. Bull.*, 4, 789, 1979.
155. **Panke, E. S., Reiter, R. J., Rollag, M. D., and Panke, T. W.**, Pineal serotonin *N*-acetyltransferase activity and melatonin concentrations in prepubertal and adult Syrian hamsters exposed to short daily photoperiods, *Endocr. Res. Commun.*, 5, 311, 1978.
156. **Panke, E. S., Rollag, M. D., and Reiter, R. J.**, Pineal melatonin concentrations in the hamster, *Endocrinology*, 104, 194, 1979.
157. **Tamarkin, L., Reppert, S. M., and Klein, D. C.**, Regulation of pineal melatonin in the Syrian hamster, *Endocrinology*, 104, 385, 1979.
158. **Rollag, M. D., Panke, E. S., Trakulrungsi, W. K., Trakulrungsi, C., and Reiter, R. J.**, Quantitation of daily melatonin synthesis in the hamster pineal gland, *Endocrinology*, 106, 231, 1980.
159. **Lynch, H. J.**, Diurnal oscillations in pineal melatonin content, *Life Sci.*, 10, 791, 1971.
160. **Quay, W. B.**, *Pineal Chemistry*, Charles C Thomas, Springfield, 1974, chap. 7.
161. **Ralph, C. L., Mull, D., Lynch, H. J., and Hedlund, L.**, A melatonin rhythm persists in rat pineals in darkness, *Endocrinology*, 89, 1361, 1971.

162. **Wilkinson, M., Arendt, J., Bradtke, J., and de Ziegler, D.,** Determination of dark-induced elevation of pineal *N*-acetyltransferase activity with simultaneous radioimmunoassay of melatonin in pineal, serum, and pituitary of the male rat, *J. Endocrinol.*, 72, 243, 1977.
163. **Reiter, R. J.,** Anti- and counter antigonadotrophic effects of melatonin: an apparent paradox, in *Neural Hormones and Reproduction,* Scott, D. E., Kozlowski, G. P., and Weindl, A., Eds., S. Karger, Basel, 1978, 344.
164. **Tamarkin, L., Brown, S., and Goldman, B.,** Neuroendocrine Regulation of Seasonal Reproductive Cycles in the Hamster, Abstr. 4th Ann. Mtg. Soc. Neurosci., New York, 1975, 458.
165. **Reiter, R. J., Blask, D. E., Johnson, L. Y., Rudeen, P. K., Vaughan, M. K., and Waring, P. J.,** Melatonin inhibition of reproduction in the male hamster: its dependency on time of day of administration and on an intact and sympathetically innervated pineal gland, *Neuroendocrinology,* 22, 107, 1976.
166. **Sackman, J. W., Little, J. C., Rudeen, P. K., Waring, P. J., and Reiter, R. J.,** The effects of pineal indoles given late in the light period on reproductive organs and pituitary prolactin levels in male golden hamsters, *Hormone Res.*, 8, 84, 1977.
167. **Tamarkin, L., Hollister, C. W., Lefebvre, N. G., and Goldman, B. D.,** Melatonin induction of gonadal quiescence in pinealectomized Syrian hamsters, *Science,* 198, 935, 1977.
168. **Goldman, B., Hall, V., Hollister, C., Roychoudhury, P., Tamarkin, L., and Westrom, W.,** Effects of melatonin on the reproductive system in intact and pinealectomized male hamsters maintained under various photoperiods, *Endocrinology,* 104, 82, 1979.
169. **Brown, G. M., Basinska, J., Bubenik, G., Sibony, D., Grota, L. J., and Stancer, H. C.,** Gonadal effects of pinealectomy and immunization against *N*-acetylindolealkylamines in the hasmter, *Neuroendocrinology,* 22, 289, 1977.
170. **Knigge, K. M. and Sheridan, M. N.,** Pineal function in hamsters bearing melatonin antibodies, *Life Sci.*, 19, 1235, 1976.
171. **Joseph, S. A. and Knigge, K. M.,** Changes in hypothalamic content of LRF in animals immunized against melatonin, in *Current Studies on Hypothalamic Function 1978,* Vol. 1, Veale, W. L. and Lederis, K., Eds., S. Karger, Basel, 1978, 175.
172. **Reiter, R. J., Vaughan, M. K., Blask, D. E., and Johnson, L. Y.,** Melatonin: its inhibition of pineal antigonadotrophic activity in male hamsters, *Science,* 185, 1169, 1974.
173. **Reiter, R. J., Vaughan, M. K., Rudeen, P. K., Vaughan, G. M., and Waring, P. J.,** Melatonin-pineal relationships in female golden hamsters, *Proc. Soc. Exp. Biol. Med.*, 149, 290, 1975.
174. **Reiter, R. J., Rudeen, P. K., and Vaughan, M. K.,** Restoration of fertility in light-deprived female hamsters by chronic melatonin treatment, *J. Comp. Physiol.*, 111, 7, 1976.
175. **Reiter, R. J., Vaughan, M. K., Rudeen, P. K., and Philo, R. C.,** Melatonin induction of testicular recrudescence in hamsters and its subsequent inhibitory action on the antigonadotrophic influence of darkness on the pituitary-gonadal axis, *Am. J. Anat.*, 147, 235, 1976.
176. **Turek, F. W. and Losee, S. H.,** Melatonin-induced testicular growth in golden hamsters maintained on short days, *Biol. Reprod.*, 18, 299, 1978.
177. **Reiter, R. J. and Vaughan, M. K.,** A study of indoles which inhibit pineal antigonadotrophic activity in male hamsters, *Endocr. Res. Commun.*, 2, 299, 1975.
178. **Turek, F. W., Desjardins, C., and Menaker, M.,** Melatonin: antigonadal and progonadal effects in male golden hamsters, *Science,* 190, 280, 1975.
179. **Reiter, R. J., Vaughan, M. K., and Waring, P. J.,** Studies on the minimal dosage of melatonin required to inhibit pineal antigonadotrophic activity in male golden hamsters, *Hormone Res.*, 6, 258, 1975.
180. **Reiter, R. J., Rudeen, P. K., Sackman, J. W., Vaughan, M. K., Johnson, L. Y., and Little, J. C.,** Subcutaneous melatonin implants inhibit reproductive atrophy in male hamsters induced by daily melatonin injections, *Endocr. Res. Commun.*, 4, 35, 1977.
181. **Reiter, R. J.,** *The Pineal, Vol. 5,* Eden Press, Montreal, 1980, chap. 12.
182. **Chen, H. J., Brainard, G. C., III, and Reiter, R. J.,** Melatonin given in the morning prevents the suppressive action on the reproductive system of melatonin given in the afternoon, *Neuroendocrinology,* 31, 129, 1980.
183. **Reiter, R. J.,** The pineal gland: a regulator of regulators, *Prog. Psychobiol. Physiol. Psychol.*, 9, 323, 1980.
184. **Cohen, M., Roselle, D., Chabner, B., Schmidt, T. J., and Lippman, M.,** Evidence for a cytoplasmic melatonin receptor, *Nature (London),* 274, 894, 1978.
185. **Cardinali, D. P., Vacas, M. I., and Boyer, E. E.,** High affinity binding of melatonin in bovine medial basal hypothalamus, *IRCS Med. Sci.*, 6, 357, 1978.
186. **Cardinali, D. P., Vacas, M. I., and Boyer, E. E.,** Specific binding of melatonin in bovine brain, *Endocrinology,* 105, 437, 1979.
187. **Niles, L. P., Wong, Y. W., Mishra, R. K., and Brown, G. M.,** Melatonin receptors in brain, *Eur. J. Pharmacol.*, 55, 219, 1979.

188. **Vacas, M. I. and Cardinali, D. P.**, Diurnal changes in melatonin binding sites of hamster and rat brains. Correlation with neuroendocrine responsiveness to melatonin, *Neurosci. Lett.*, 15, 259, 1979.
189. **Turek, F. W., Desjardins, C., and Menaker, M.**, Melatonin-induced inhibition of testicular function in adult golden hamsters, *Proc. Soc. Exp. Biol. Med.*, 151, 502, 1976.
190. **Turek, F. W., Desjardins, C., and Menaker, M.**, Differential effects of melatonin on the testes of photoperiodic and nonphotoperiodic rodents, *Biol. Reprod.*, 15, 94, 1976.
191. **Fraschini, F., Collu, R., and Martini, L.**, Mechanisms of inhibitory action of pineal principles on gonadotrophin secretion, in *The Pineal Gland*, Wolstenholme, G. E. W. and Knight, J., Eds., Churchill Livingstone, London, 1971, 159.
192. **Kamberi, I. A., Mical, R. S., and Porter, J. C.**, Effect of anterior pituitary perfusion and intraventricular injection of catecholamines and indoleamines on LH release, *Endocrinology*, 87, 1, 1970.
193. **Kamberi, I. A., Mical, R. S., and Porter, J. C.**, Effects of melatonin and serotonin on FSH and prolactin, *Endocrinology*, 88, 1288, 1971.
194. **Kamberi, I. A.**, The role of brain monoamines and pineal indoles on the secretion of gonadotropins and gonadotropin releasing factors, *Prog. Brain Res.*, 39, 261, 1973.
195. **Martin, J. E. and Klein, D. C.**, Melatonin inhibition of the neonatal pituitary response to luteinizing hormone-releasing factor, *Science*, 191, 301, 1976.
196. **Martin, J. E., Engle, J. N., and Klein, D. C.**, Inhibition of the *in vitro* pituitary response to luteinizing hormone-releasing hormone by melatonin, serotonin, and 5-methoxytryptamine, *Endocrinology*, 100, 675, 1977.
197. **Martin, J. E. and Sattler, C.**, Developmental loss of the acute inhibitory effect of melatonin on the *in vitro* pituitary luteinizing hormone and follicle stimulating hormone responses to luteinizing hormone-releasing hormone, *Endocrinology*, 105, 1007, 1979.
198. **Longenecker, D. E. and Gallo, D. G.**, The inhibition of PMSG-induced ovulation in immature rats by melatonin, *Proc. Soc. Exp. Biol. Med.*, 137, 623, 1971.
199. **Reiter, R. J. and Sorrentino, S., Jr.**, Inhibition of luteinizing hormone release and ovulation in PMS-treated rats by peripherally administered melatonin, *Contraception*, 4, 385, 1971.
200. **Kao, L. W. L. and Weisz, J.**, Release of gonadotrophin-releasing hormone (Gn-RH) from isolated, perifused medial-basal hypothalamus by melatonin, *Endocrinology*, 100, 1723, 1977.
201. **Thieblot, L., Berthelay, J., et Blaise, S.**, Action de la melatonine sur la sécrétion gonadotrope du Rat, *C. R. Seances Soc. Biol. Paris*, 160, 2306, 1966.
202. **Johnson, L. Y. and Reiter, R. J.**, The pineal gland and its effects on mammalian reproduction, in *The Pineal and Reproduction*, Reiter, R. J., Ed., S. Karger, Basel, 1978, 116.
203. **Kinson, G. A.**, Pineal factors in the control of testicular function, *Ady. Sex Hormone Res.*, 2, 87, 1976.
204. **Alonso, R., Prieto, L., Hernandez, C., and Mas, M.**, Antiandrogenic effects of the pineal gland and melatonin in castrated and intact prepubertal male rats, *J. Endocrinol.*, 79, 77, 1978.
205. **Quay, W. B.**, Evidence for a pineal contribution in the regulation of vertebrate reproductive systems, *Gen. Comp. Endocrinol. Suppl.*, 2, 101, 1969.
206. **Frehn, J. L., Urry, R. L., and Ellis, L. C.**, Effect of melatonin and short photoperiod of Δ^4-reductase activity in liver and hypothalamus of the hamster and rat, *J. Endocrinol.*, 60, 507, 1974.
207. **Blask, D. E., Nodelman, J. L., Leadem, C. A., and Richardson, B. A.**, Influence of exogenously administered melatonin on the reproductive system and prolactin levels in underfed male rats, *Biol. Reprod.*, 22, 507, 1980.
208. **Blask, D. E. and Nodelman, J. L.**, Antigonadotrophic and prolactin-inhibitory effects of melatonin in anosmic male rats, *Neuroendocrinology*, 29, 406, 1979.
209. **Reiter, R. J., Petterborg, L. J., Trakulrungsi, C., and Trakulrungsi, W. K.**, Surgical removal of the olfactory bulbs increases sensitivity of the reproductive system of female rats to the inhibitory effects of late afternoon melatonin injections, *J. Exp. Zool.*, 212, 47, 1980.
210. **Banks, A. F. and Reiter, R. J.**, Melatonin inhibition of pineal antigonadotrophic activity in male rats, *Hormone Res.*, 6, 351, 1975.
211. **Klein, D. C.**, Circadian rhythms in indole metabolism in the rat pineal gland, in *The Neurosciences, Third Study Programme*, Schmidt, F. O., Ed., MIT Press, Cambridge, 1974, 509.
212. **Wilson, B. W., Lynch, H. J., and Ozaki, Y.**, 5-Methoxytryptophol in rat serum and pineal: detection, quantitation, and evidence for daily rhythmicity, *Life Sci.*, 23, 1019, 1978.
213. **McIsaac, W. M., Taborsky, R. G., and Farrell, G.**, 5-Methoxytryptophol: effect on estrus and ovarian weight, *Science*, 145, 63, 1964.
214. **Talbot, J. A. and Reiter, R. J.**, Influence of melatonin, 5-methoxytryptophol and pinealectomy on pituitary and plasma gonadotropin and prolactin levels in castrated adult male rats, *Neuroendocrinology*, 13, 164, 1973/74.
215. **Vilchez, J. A. and Debeljuk, L.**, Effect of 5-methoxytryptophol on the reproductive system of male rats, *J. Reprod. Fertil.*, 30, 305, 1972.

216. **Pomerantz, G. and Reiter, R. J.,** Influence of intraocularly-injected pineal indoles on PMS-induced ovulation in immature rats, *Int. J. Fertil.,* 17, 117, 1974.
217. **Vaughan, M. K. and Blask, D. E.,** Arginine vasotocin — a search for its function in mammals, in *The Pineal and Reproduction,* Reiter, R. J., Ed., S. Karger, Basel, 1978, 90.
218. **Benson, B. and Ebels, I.,** Pineal peptides, *J. Neural Trans.,* Suppl. 13, 157, 1978.
219. **Pavel, S.,** The mechanism of action of vasotocin in the mammalian brain, *Prog. Brain Res.,* 52, 445, 1979.
220. **Reiter, R. J.,** Comparative physiology: pineal gland, *Ann. Rev. Physiol.,* 35, 305, 1973.
221. **Reiter, R. J.,** Circannual reproductive rhythms in mammals related to photoperiod and pineal function: a review, *Chronobiologia,* 1, 365, 1974.
222. **Reiter, R. J.,** The pineal gland and seasonal reproductive adjustments, *Int. J. Biometeorol.,* 19, 282, 1975.
223. **Reiter, R. J.,** Interaction of photoperiod, pineal and seasonal reproduction as exemplified by findings in the hamster, in *The Pineal and Reproduction,* Reiter, R. J., Ed., S. Karger, Basel, 1978, 169.
224. **Reiter, R. J.,** Pineal-mediated regression of the reproductive organs of female hamsters exposed to natural photoperiods during the winter months, *Am. J. Obstet. Gynecol.,* 118, 878, 1974.
225. **Hoffman, R. A.,** Speculations on the regulation of hibernation, *Ann. Acad. Sci. Fenn., A, IV Biologica,* 71, 201, 1964.
226. **Frehn, J. L. and Chung-Ching, L.,** Effects of temperature, photoperiod and hibernation on the testes of golden hamsters, *J. Exp. Zool.,* 174, 317, 1970.
227. **Smit-Vis, J. H. and Smit, G. J.,** Hibernation and testes activity in the golden hamster, *Neth. J. Zool.,* 20, 502, 1970.
228. **Smit-Vis, J. H.,** The effect of pinealectomy and of testosterone administration on the occurrence of hibernation in adult male golden hamsters, *Acta Morphol. Neerl. Scand.,* 10, 269, 1972.
229. **Hall, V. and Goldman, B. D.,** Effects of gonadal steroid hormones on hibernation in the Turkish hamster *(Mesocricetus brandti), J. Comp. Physiol.,* 135, 107, 1980.
230. **Reiter, R. J.,** Influence of pinealectomy on the breeding capability of hamsters maintained under natural photoperiod and temperature conditions, *Neuroendocrinology,* 13, 366, 1973/74.
231. **Reiter, R. J.,** Evidence for refractoriness of the pineal-gonadal axis to the pineal gland in golden hamsters and its possible implications in annual reproductive rhythms, *Anat. Rec.,* 173, 365, 1975.
232. **Reiter, R. J., Vaughan, M. K., and Waring, P. J.,** Prevention by melatonin of short day induced atrophy of the reproductive systems of male and female hamsters, *Acta Endocrinol.,* 84, 410, 1977.
233. **Smit-Vis, J. H. and Akkerman-Bellaart, M. A.,** Spermiogenesis in hibernating golden hamsters, *Experientia,* 23, 844, 1967.
234. **Reiter, R. J.,** Exogenous and endogenous control of the annual reproductive cycle in the male golden hamster: participation of the pineal gland, *J. Exp. Zool.,* 191, 111, 1975.
235. **Turek, F. W., Elliott, J. A., Alvis, J. D., and Menaker, M.,** Effect of prolonged exposure to nonstimulatory photoperiods on the activity of the neuroendocrine-testicular axis of golden hamsters, *Biol. Reprod.,* 13, 475, 1975.
236. **Rollag, M. D., Panke, E. S., and Reiter, R. J.,** Pineal melatonin content in male hamsters throughout the seasonal reproductive cycle, *Proc. Soc. Exp. Biol. Med.,* 165, 330, 1980.
237. **Stetson, M. H., Matt, K. S., and Watson-Whitmyre, M.,** Photoperiodism and reproduction in golden hamsters: circadian organization and the termination of photorefractoriness, *Biol. Reprod.,* 14, 531, 1976.
238. **Turek, F. W.,** The interaction of photoperiod and testosterone in regulating serum gonadotropin levels in castrated male hamsters, *Endocrinology,* 101, 1210, 1977.
239. **Turek, F. W.,** Role of the pineal gland in photoperiod-induced changes in hypothalamic-pituitary sensitivity to testosterone feedback in castrated male hamsters, *Endocrinology,* 104, 636, 1979.
240. **Bittman, E. L.,** Hamster refractoriness: role of insensitivity of pineal target tissues, *Science,* 202, 648, 1978.
241. **Reiter, R. J., Petterborg, L. J., and Philo, R. C.,** Refractoriness to the antigonadotrophic effects of melatonin in male hamsters and its interruption by exposure of the animals to long daily photoperiods, *Life Sci.,* 25, 1571, 1979.

Chapter 4

PINEAL INVOLVEMENT IN THE PHOTOPERIODIC CONTROL OF REPRODUCTION AND OTHER FUNCTIONS IN THE DJUNGARIAN HAMSTER *PHODOPUS SUNGORUS**

Klaus Hoffmann

TABLE OF CONTENTS

* Experimental work was supported by the Deutsche Forschungsgemeinschaft, SPP "Mechanismen Biologischer Uhren".

I. INTRODUCTION

The Djungarian or hairy-footed hamster is a small representative of the Cricetinae or true hamsters. It inhabits the dry steppes of Mongolia and West Siberia and the eastern parts of the Lake Baikal region.[1,2] In general, two subspecies are mentioned. The western form, *Phodopus sungorus sungorus* Pallas 1770 exhibits in its northerly range, a marked seasonal dimorphism in pelage color, with a grayish-brown coat in summer and a whitish coat in winter.[3] *Phodopus* is easily maintained and bred in the laboratory if it is kept in mind that the animal comes from a dry habitat. It is much tamer than other hamster species, and breeding pairs can be kept together permanently. These features and its small size make *Phodopus* an ideal laboratory animal.

The Djungarian hamster shows strong reactions to photoperiod and its changes, and we have used the species extensively for studies of the photoperiodic mechanism, and of the involvement of the pineal gland in this mechanism. Though in general, *Phodopus* shows similarities to the golden hamster (*Mesocricetus auratus*), it differs in several respects from this often-used laboratory species. Similarities and dissimilarities in the photoperiodic phenomena and its mechanism of *Phodopus* and *Mesocricetus* will be discussed.

Our stock is derived from four animals that were captured in 1965 near Omsk in West Siberia (latitude 55°N).[3] To the best of my knowledge, all experiments on photoperiodic effects and pineal involvement in *Phodopus* have been performed with animals originating from our breeding colony.

II. PHOTOPERIODIC PHENOMENA

A. Natural Photoperiods

In our laboratory, breeding pairs and stock animals are maintained in constant temperature but under natural photoperiods (latitude 48°N). Under these conditions, the Djungarian hamsters exhibit a marked annual cycle in a number of easily assayable functions.[3] Figure 1 shows the changes occurring throughout the year in adult males. Testicular weight and tubular diameter decrease markedly in fall, and spermatogenesis completely ceases as evidenced by smears from the epididymal caudae and by the histological appearance of the seminiferous epithelium.[5] The annual cycle in weight of accessory glands and of the abdominal marking gland can be considered indirect evidence for an annual cycle of androgen levels. In general, the picture corresponds to that found in the golden hamster.[6,7] However, in the Djungarian hamster there is also a marked annual cycle in body weight, paralleling the gonadal cycle, with about a 30% decrease in winter. In addition, a seasonal dimorphism in pelage color is obvious. The annual cycle in body weight and pelage color is practically identical in females, though their average weight is lower.[3] *Phodopus* does not hibernate but may show daily torpor with body temperatures of about 20°C during the torpor phase.[3] Torpor is only observed in winter animals with partly or completely regressed testes.

B. Experimental Changes of Photoperiod

The dependence of the annual cycle in *Phodopus* upon photoperiod and its changes has been amply demonstrated.[3-5,8-14] Figure 2 gives an example. Long photoperiods in winter induce rapid recrudescence of testes and accessory glands, while short photoperiods in summer initiate regression. For plasma testosterone and plasma prolactin, markedly lower values in animals maintained in short days were reported.[15] However, it has to be kept in mind in such experiments that drastic circadian rhythms in plasma hormone levels have been found in this species (see Figure 3); thus the choice of the right time of day in such comparisons may be highly important.[16] In this respect, *Pho-*

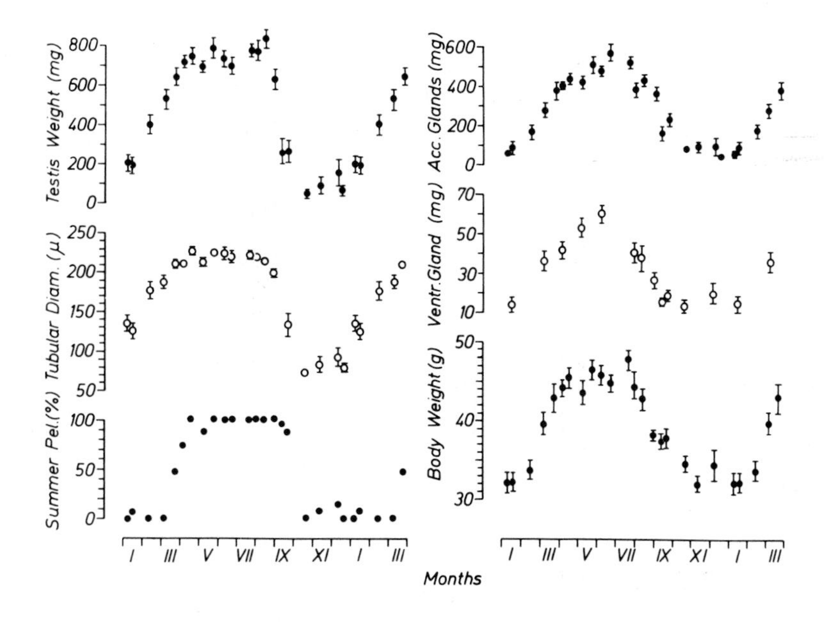

FIGURE 1. Annual cycle of (left) testis weight (both testes combined), tubular diameter, pelage color (% of animals in full summer pelt) and (right) weight of accessory glands, ventral marking gland, and body weight in adult male Djungarian hamsters. M ± SE of 15 animals for each point. The hamsters were exposed to the natural illumination and its changes throughout, but were kept at constant temperature (20 ± 3°C). Values for January to March have been repeated to facilitate inspection. (From Hoffmann, K., *Prog. Brain Res.*, 52, 397, 1979. With permission.)

dopus differs from *Mesocricetus* where no significant daily cycles in plasma gonadotrophins were found.[17,18] Differences in the uptake, binding, and metabolism of testosterone in different photoperiods were also reported in *Phodopus.*[15]

Not only gonads and gonad-dependent structures, but also other functions showing an annual cycle can be influenced by photoperiod in the Djungarian hamster. Testicular regression induced by short photoperiods is accompanied by a reduction in body weight, while stimulation by long photoperiods induces not only gonadal recrudescence, but also an increase in weight (compare Figure 8). These photoperiodically regulated changes in body weight are observed in both sexes.[3,5,9] They depend only partially on gonadal factors, and are also found in orchidectomized or ovariectomized hamsters.[19,20] Analysis of body composition in animals maintained in long and short days has shown that the lower weight in winter animals is mainly due to a reduction in fat (70%) while the decrease in protein (23%) is much smaller.[21] Though total fat is decreased in short photoperiods, brown adipose tissue is significantly increased.[22]

Pelage color is also controlled by photoperiod, moult into winter coat can be induced by short photoperiods in summer[11] (compare Figure 9); in winter, long photoperiods induce premature change into summer pelage.[9] Observations in castrated males have shown that these changes do not depend upon the presence of the gonads.[20] Recently an annual cycle of pituitary MSH content has been described in *Phodopus*[23] with high values in summer and low values in winter, and it has also been shown that MSH may stimulate melanogenesis of hair follicles in vitro.[24] These findings indicate that MSH may be involved in the control of the seasonal dimorphism of hair color in *Phodopus,* as has been suggested for the short-tailed weasel.[25]

While the change into the physiological winter state depends upon exposure to short photoperiods, gonadal recrudescence, moult into the summer coat, and weight gain

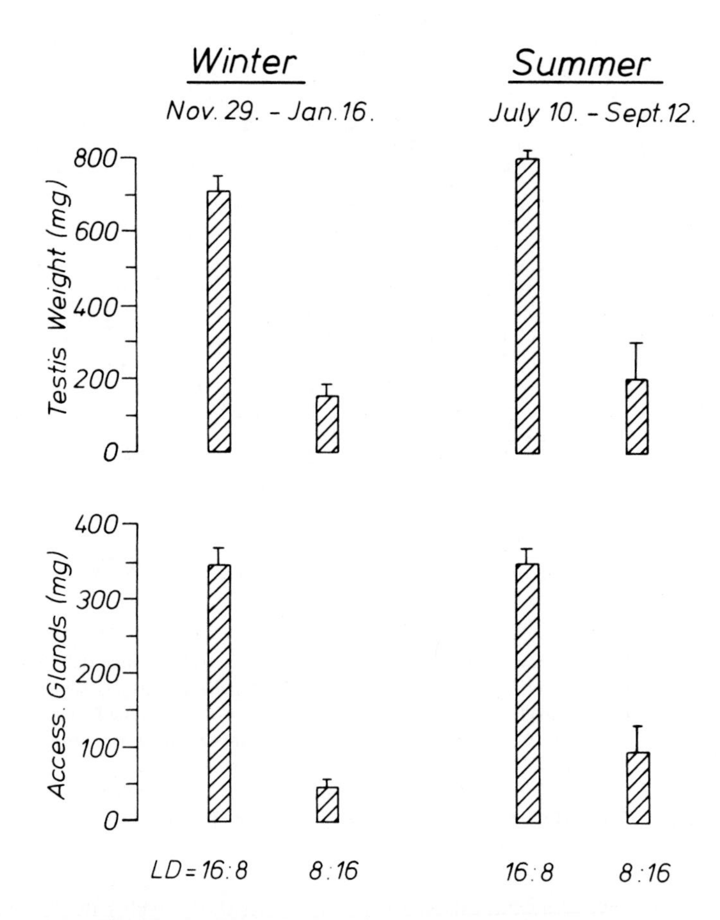

FIGURE 2. Weight of testes and accessory glands in male Djungarian hamsters after exposure to long (LD 16:8 = 16 hr of light and 8 hr of dark per day) and short (LD 8:16) photoperiods. In winter, long photoperiods induce rapid recrudescence, in summer short photoperiods lead to regression.[11]

may occur spontaneously if the animals are maintained in short photoperiods for a long time.[3,9,12,13] Recrudescence in spring may be largely due to this spontaneous process, as has also been described in the golden hamster.[6,7,26-28] In general, photoperiodic reactions in adult Djungarian hamsters are closely similar to those observed in the golden hamster, at least as far as gonads and gonad-dependent functions are concerned. However, in *Mesocricetus* there is no change in pelage colour, and marked changes in body weight have also not been described.

C. Photoperiod and Juvenile Development

In juveniles, a pronounced difference in photoperiodic sensitivity between the two hamster species exists. While in the golden hamster testicular development seems to be independent of photoperiod,[29] and does not differ from sighted controls in orbitally enucleated animals,[30] marked photoperiodic effects were found in young Djungarian hamsters.[31] Figure 4 shows testis weight and amount of spermatozoa in the epididymal caudae of males that were maintained, from birth to the age of 31 to 75 days, in either long or short photoperiods. While in long photoperiods nearly final testis weight was reached between 35 to 40 days, and masses of spermatozoa could be found in the epididymal caudae at that age, in short photoperiods, testes remained small and infantile in most animals, and in all but 16 of 196 animals, spermatozoa in the epididymal

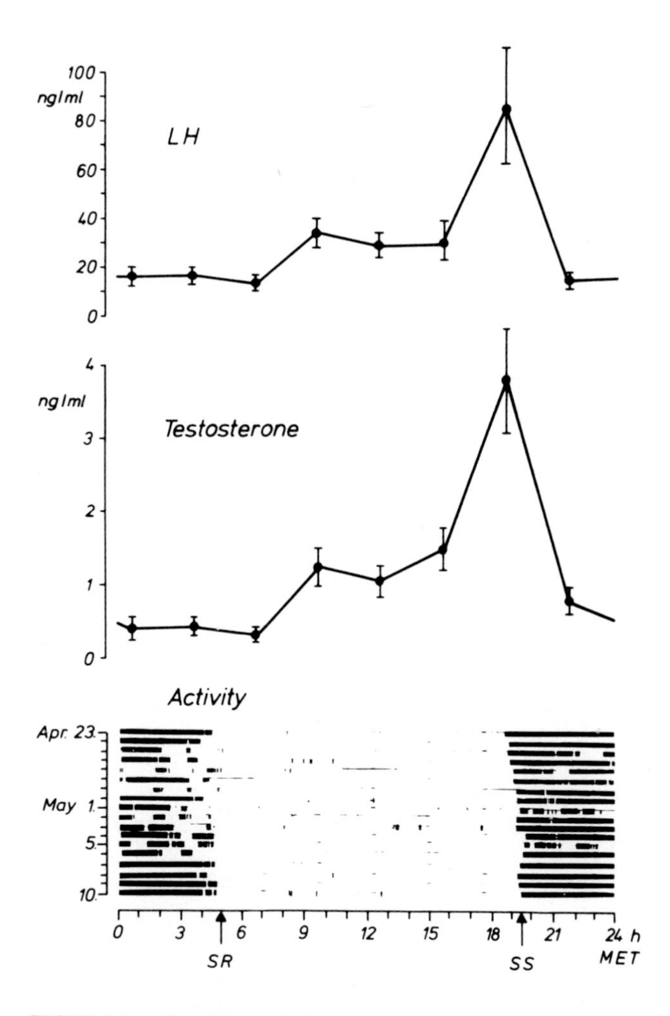

FIGURE 3. Circadian cycle in plasma concentrations of LH (top) and testosterone (middle) in adult male Djungarian hamsters maintained in natural photoperiods and sacrificed May 25 to 27, 1977 (LH) or April 29 to May 2, 1976 (testosterone) at the times indicated. At the bottom, the wheel-running activity of a representative animal for the time of April 23 to May 10 is given (SR = sunrise, SS = sunset). M ± SE of 15 to 20 determinations for each point. All animals had large active testes. Note the marked and parallel daily cycle in both hormones, with maximal values at the beginning of activity time and minimal values at its end. Testosterone and activity after Hoffmann and Nieschlag.[16] We are indebted to Professor W. Wuttke, Max-Planck-Institut für biophysikalische Chemie, Göttingen, for the determination of the LH concentrations.

caudae were conspicuously absent. The growth of the accessory glands was also very retarded in nearly all of the males reared in short days.[31] In female *Phodopus*, a similar inhibition of sexual maturation by short photoperiods was observed. Females kept from birth in LD 8:16 had significantly smaller ovaries and uteri than those reared in LD 16:8, at 45 days of age, and the number of antral follicles was much smaller in the short-day animals.[32]

Further experiments in which young male *Phodopus* were exposed first to long and then to short photoperiods, or vice versa, have shown that photoperiodic sensitivity starts between 7 and 14 days of age;[32] a careful quantitative histological examination

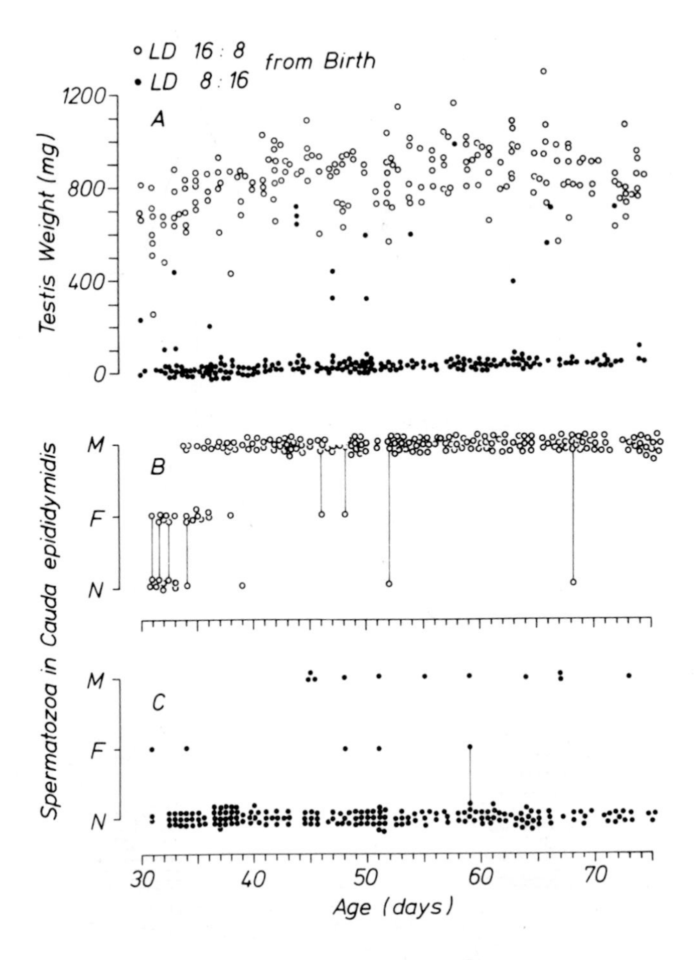

FIGURE 4. The effect of raising Djungarian hamsters from birth in long (LD 16:8) or short (LD 8:16) photoperiods on testis weight (A) and spermatogenesis (B and C) as evidenced by presence of spermatozoa in smears from the epididymal caudae. N = no spermatozoa; F = a few spermatozoa (up to 50); M = many spermatozoa (in most cases caudae packed with motile spermatozoa). If conditions in both caudae coincided, one point is indicated per animal, if different, two points connected by vertical line. (From Hoffmann, K., *J. Reprod. Fertil.*, 54, 29, 1978. With permission.)

of testes in young animals revealed differences first between long and short photoperiods at day 13.[33] The difference in photoperiodic reaction between juvenile golden and Djungarian hamsters seems not to depend upon differences in the maturation of the pineal gland and its innervation. A fluorescence microscopic and microspectrometric examination in 15-day-old golden and Djungarian hamsters did not reveal differences between the two species.[34]

Short photoperiods do not suppress sexual maturation permanently. When males were exposed to long or short photoperiods for a longer time, and testicular size was determined in regular intervals, full testicular size was reached at 40 days in all animals maintained in long photoperiods, while in short photoperiods, first signs of testicular development could be discerned between 116 and 164 days of age, and adult size was reached between 158 and 207 days (Figure 5). Pelage color showed similar spontaneous changes. Up to an age of 30 days all animals have summer pelage regardless of photoperiod. This was maintained in long photoperiods. In short photoperiods, however,

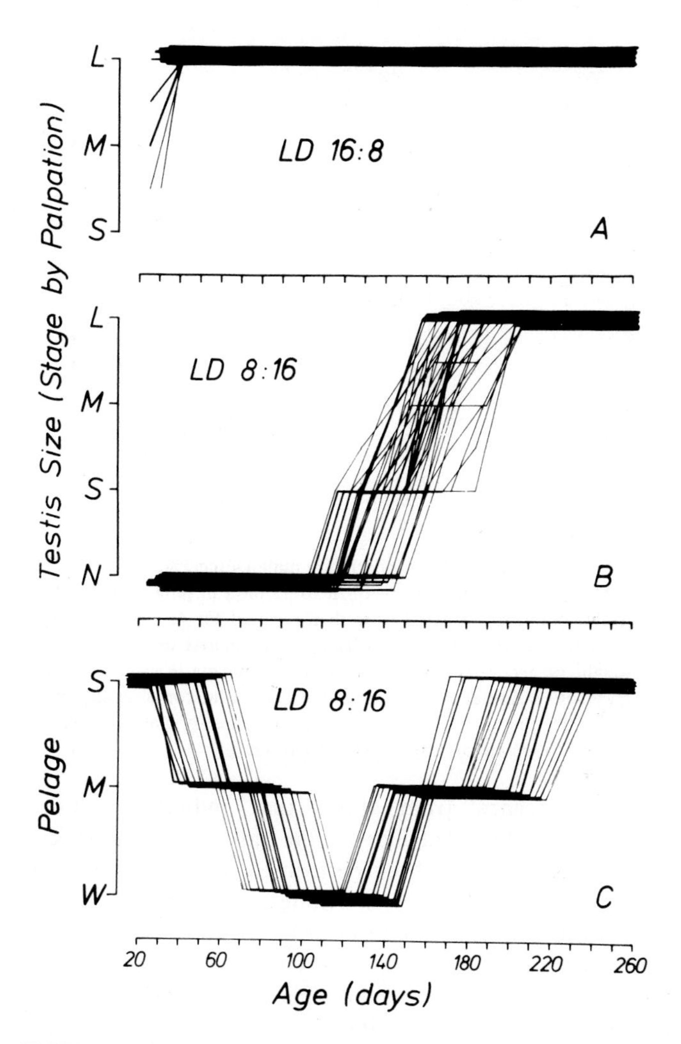

FIGURE 5. Testicular size (determined by palpation at intervals of about 14 days) of Djungarian hamsters maintained from birth in long (A) or short (B) photoperiods: N = nonpalpable; S = small; M = medium; L = large (corresponds to size at full spermatogenesis). C — pelage condition of the animals kept in short photoperiods (S = summer pelage; M = moult; W = winter pelage). (From Hoffmann, K., *J. Reprod. Fertil.*, 54, 29, 1978. With permission.)

animals started to moult into winter pelage, stayed in this condition until about 110-to- 150-days-old when they again moulted into summer coat (Figure 5, bottom). Body weight increased continuously in long days, and final weight was reached at about 160 days. In males maintained in short photoperiods from birth, body weight first rose until about 80 days, then plateaued, and started to increase again when gonads started to grow.[31] In general, the delayed development of testes and accessory glands, the spontaneous moult into summer coat, and the final weight increase in juveniles in short photoperiods closely paralleled the spontaneous recrudescence and accompanying changes that occur in adults.

D. Critical Photoperiod

In the golden hamster a sharply defined critical photoperiod of 12½ hr light per day has been described, dividing light times that maintain gonadal size and activity,

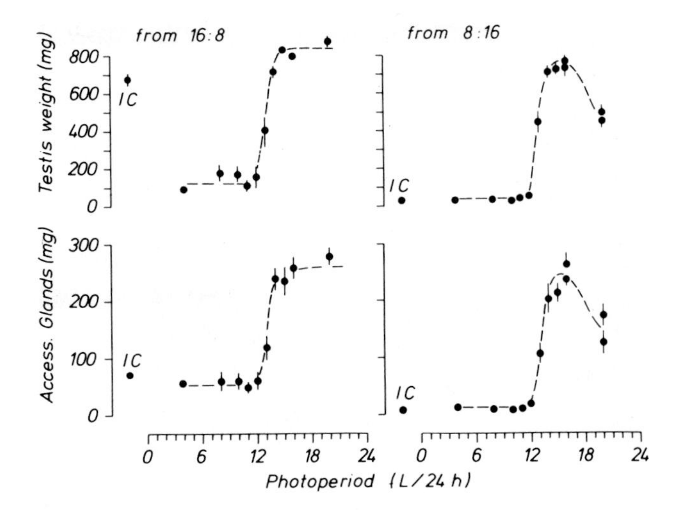

FIGURE 6. Critical photoperiod in male Djungarian hamsters. Animals were maintained from birth to 35 days in long (left) or short (right) photoperiods, and were then exposed for 45 further days to the photoperiods indicated. Mean (± SE) paired testis weight and weight of accessory glands for 12 to 19 animals in each group are given. IC = initial control (at 35 days). IC value for accessory gland in LD 16:8 is low since development of accessory glands is delayed relative to growth of testes.[31] After Hoffmann.[45]

or stimulate recrudescence from those that induce involution or are not stimulatory.[29,35,36] To my knowledge this is the only mammalian species in which systematic studies of this kind have been performed. For comparative reasons we have performed a similar study in *Phodopus.* Young males were kept from birth to 35 days of age in photoperiods of either 8 or 16 hr of light, and were then exposed for 45 days to different photoperiods ranging from 4 to 20 hr light per day. Figure 6 shows weights of testes and accessory glands at the end of this period. There is a fairly abrupt change from photoperiods that maintain gonadal size to those that induce regression at about 13 hr light per day (Figure 6, left). The critical photoperiod is slightly longer than in *Mesocricetus,* which corresponds to the fact that *Phodopus* lives in higher latitudes, and is exposed to more severe conditions in winter. A similar picture results for the animals that had first been exposed to short photoperiods, and whose testes were undeveloped at the start of the different photoperiods (Figure 6, right). Photoperiods of more than 13 hr stimulated testicular and accessory gland development, shorter photoperiods did not. However, in LD 20:4 (20 hr light per day) there is less stimulation than in LD 14:10, 15:9 or 16:8. This experiment has been performed twice, and in both cases a significantly smaller degree of stimulation was observed in LD 20:4 VS LD 16:8. Litter mates were used for both conditions. Similar results have been obtained in adult winter males in which gonadal recrudescence was stimulated by 16 or 20 hr light per day.[37] In animals coming from long photoperiods (maintenance) no such dip of the curve can be seen. It is interesting to note that in the golden hamster, Elliott[35] reported a similar indention of the curve at 18 hr of light per day for stimulation of recrudescence, while no such effect was observed for maintenance.

These findings indicate that, except for the slight shift in critical photoperiod, in both hamster species the photoperiodic mechanism is similar. The fact that in both species some very long photoperiods are less stimulatory than shorter ones indicates that the mechanism may be somewhat more complicated than the model proposed by Elliott[35] suggests, though most other evidence strongly supports this model.

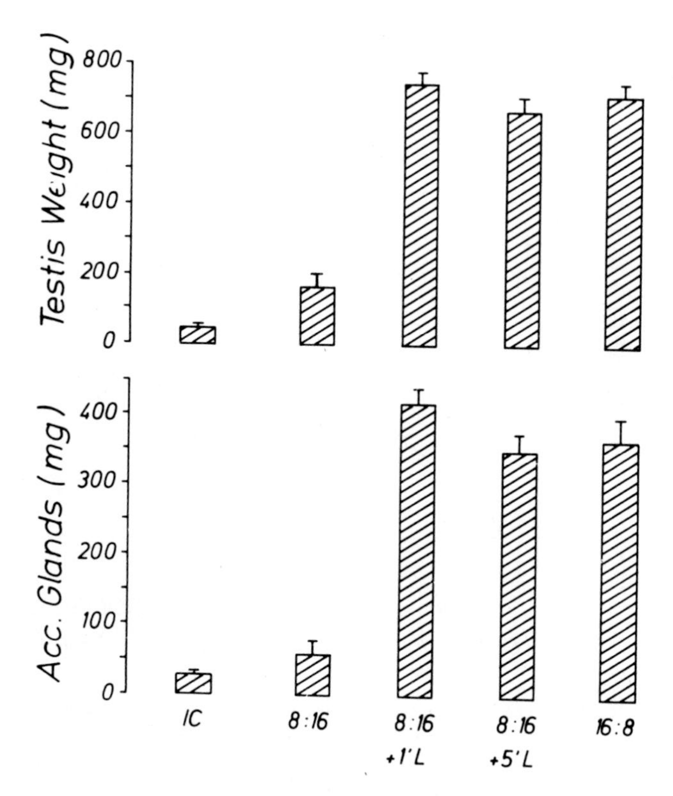

FIGURE 7. Recrudescence of testes (top) and accessory glands (below) after 46 days in the photoperiods indicated. At the beginning all animals had regressed testes and accessory glands (see IC = initial controls). In short photoperiods (8:16) there was little recrudescence, in long photoperiods (16:8) full recrudescence. When in short photoperiods the long dark time was interrupted by 1 (8:16 + 1 min L) or 5 (8:16 + 5 min L) min of light each night at midnight, recrudescence was indistinguishable from that in full long photoperiods (16:8). (From Hoffmann, K., *Experentia,* 35, 1529, 1979. With permission.)

The fairly sharp delineation of the critical photoperiod indicates that the hamsters are able to measure length of photoperiod rather precisely. In the golden hamster there is ample evidence that the photoperiodic mechanism is based on a circadian rhythm of photosensitivity. Elliott[35,36] has shown that 1 hr of light per circadian cycle may induce either short-day or long-day responses, depending upon the circadian phase into which the light pulse falls. In *Phodopus* no data on the circadian nature of the photoperiodic mechanism are available. We have shown recently, however, that interruption of the long dark time in short photoperiods by only 1 min of light each night (resulting in a cycle of 8 hr light, 7 hr 59 min dark, 1 min light, and 8 hr dark) can be as stimulatory as full photoperiods (16 hr light, 8 hr dark).[38] This holds for recrudescence of testes and accessory glands (Figure 7) as well as for increase in body weight (Figure 8), and induction of pelage change into summer coat. These findings may be important in connection with the pattern of pineal activity (see section V., "Pineal Rhythms"). In addition, these results show that even brief light interruptions during the normal dark phase, as may occur during maintenance of animals or during controls may seriously influence results of photoperiodic experiments.

III. PINEAL INVOLVEMENT

As in the golden hamster,[6,7] in *Phodopus* the pineal gland plays an important role

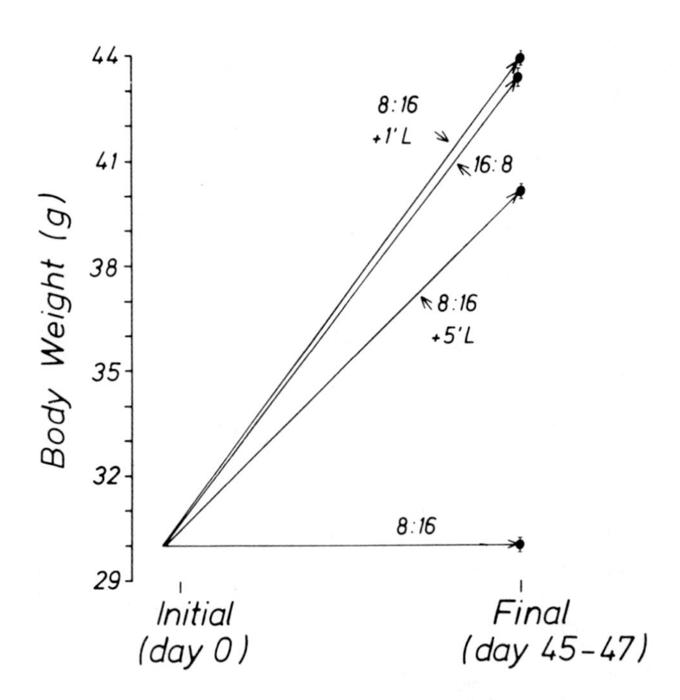

FIGURE 8. Change in body weight accompanying recrudescence of testes and accessory glands in different photoperiods. The same animals were used as in Figure 7. After values in Hoffmann.[38]

in the photoperiodic mechanism. If male hamsters are pinealectomized and then exposed to short photoperiods, involution of testes and accessory glands is inhibited (Figure 9, left).[10] Pinealectomy prevents the effect of short photoperiods, not only upon gonads and gonad-dependent functions, but also on pelage color (Figure 9, right), the normal moult into the winter pelt does not occur in these animals.[11] The decrease in body weight that normally follows exposure to short photoperiods is also abolished. While a commonly expressed opinion on pineal functions maintains that it is predominately concerned with regulation of reproductive functions, the results in *Phodopus* suggest that the pineal is involved in the transduction of photoperiodic effects, not only upon the neuroendocrine-gonadal axis, but also upon other or all functions that are regulated by photoperiod. Similar ideas were recently expressed for the golden hamster.[39] The preoccupation with gonadal effects of the pineal is probably due to the fact that in work on the influence of illumination on physiological processes, changes in reproduction are the most obvious in many animals, and have thus attracted more attention than variations in other functions.

Most workers in the field assume that the pineal gland conveys only inhibitory effects, and that this inhibition is overcome by long photoperiods or constant light. Based largely on observations in the golden hamster, Reiter[7,40] has repeatedly ventured that in long photoperiods the animals are "physiologically pinealectomized" and thus pinealectomy does not give measurable results. However, in the Djungarian hamster we obtained evidence that the pineal gland may also participate in the transduction of the effects of long photoperiods that stimulate gonadal recrudescence or hasten development. In a total of five experiments in adult *Phodopus*, we found that gonadal recrudescence was significantly retarded in males which, having experienced gonadal involution after exposure to short natural or artificial photoperiods, were pinealectomized and then exposed to long photoperiods.[5,11,13,14,41] Figure 10 shows the results of three such experiments. In all cases, in long photoperiods, development of testes and

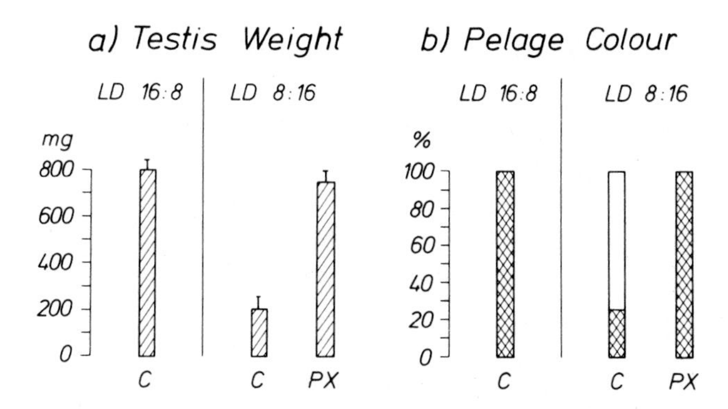

FIGURE 9. Effect of pinealectomy in Djungarian hamsters. Testicular regression (a) and moult into winter pelage (b) in short photoperiods are prevented by pinealectomy. After Hoffmann.[11]

accessory glands was significantly retarded in pinealectomized males, as compared to controls. There was also a significantly smaller increase in body weight in the pinealectomized hamsters, and change into summer pelage was also somewhat delayed.[40] Similarly, in juveniles raised from birth in long days, testicular development was significantly delayed in pinealectomized animals (Figure 11).[32,42]

Such findings are not confined to the Djungarian hamster. In ferrets, Herbert[43] has shown that the advance in onset of estrus which can be induced by long photoperiods is prevented in pinealectomized females. In general, these observations suggest that the pineal has not only inhibitory effects on sexual and other functions regulated by photoperiod, but is also partially involved in transferring the stimulatory action of long photoperiods. The latter is an active process rather than merely the elimination of an inhibition. The degree of activation via the pineal may differ between species, and in the golden hamster most of the findings support the assumption that the pineal is solely inhibitory. However, even in the latter species evidence was provided recently for a participation of the pineal gland in conveying effects of long photoperiods, in this case in its action to break photorefractoriness.[44]

Figure 11 also shows that pinealectomy partially prevents the inhibitory effects of short photoperiods. However, in short days there is still significantly less development than in long days in the pinealectomized hamsters. A similar trend can be seen in Figure 10 (bottom) which demonstrates that in pinealectomized adult males recrudescence of testes and accessory glands is significantly less in short days. Similar trends were found for body weight and pelage color.[32,41] If all photoperiodic influences were transferred via the pineal gland, no such differences should exist. These findings thus indicate that there may be photoperiodic influences which bypass the pineal gland. In the golden hamster, there are also some reports indicating that pinealectomy does not completely abolish all effects of short photoperiods or of blinding.[46-49]

IV. MELATONIN APPLICATION

It is commonly assumed that in the pineal gland neuronal information, modified by the external light cycle, is transduced into chemical messages which act at distant target sites and there modulate the function of the neuroendocrine axis.[12,40,50] While as yet no general consensus exists as to the compounds involved in conveying the photoperiodic message, melatonin is considered a likely candidate by many workers in the field.[12,45,51] We have assayed the effect of melatonin in *Phodopus* under several photoperiodic conditions, and at different seasons and thus in different physiological

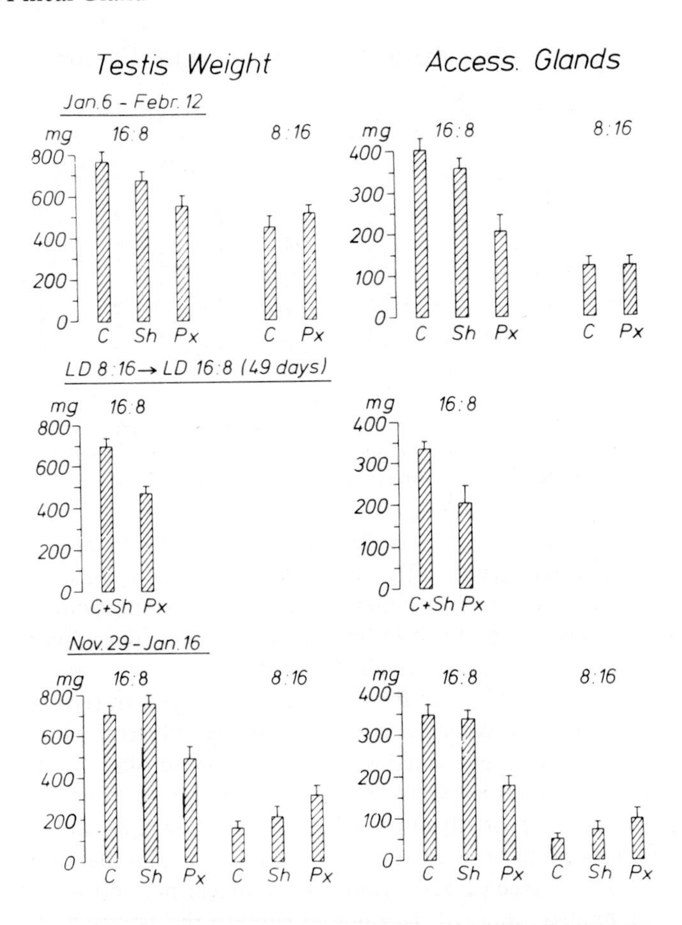

FIGURE 10. Weight of testes and accessory glands in pinealectomized (Px), sham-operated (Sh) and intact (C) Djungarian hamsters after exposure to long (16:8) or short (8:16) photoperiods. Prior to start of experiment, all animals had regressed testes and accessory glands, due to exposure to short natural (top and bottom) or artificial (middle) photoperiods. Operations were performed after regression, just prior to exposure to the photoperiods indicated. Note that in all three experiments, recrudescence of testes ($p < 0.05$ to < 0.002) and accessory glands ($p < 0.01$ to < 0.002) in long photoperiods was significantly retarded in pinealectomized animals as compared to sham-operated animals or unoperated controls. In lower diagram, pinealectomized hamsters, testicular ($p < 0.05$) and accessory gland ($p < 0.02$) development was significantly retarded in short VS long photoperiods. After Hoffmann.[5]

states of the animals. The halflife of melatonin in the organism is short and, moreover, early experiments with injected melatonin in golden hamsters had given no measurable results;[52] on the other hand, drastic effects of implanted melatonin on testis size and pelage color had been described in the short-tailed weasel.[53] We therefore implanted melatonin, either in beeswax or in silastic tubing, thus providing a continuous release. This mode of application may not be the optimal technique to reveal its physiological action (see below). Nevertheless, significant effects were observed in such experiments.

Figure 12 shows the effect of implanted melatonin on testicular development of juvenile *Phodopus* maintained in long photoperiods. There was a drastic and dose-dependent delay; with the highest dosage of melatonin, testis size was not significantly different from that of short-day controls.[32,54] A similar retardation was found in adult

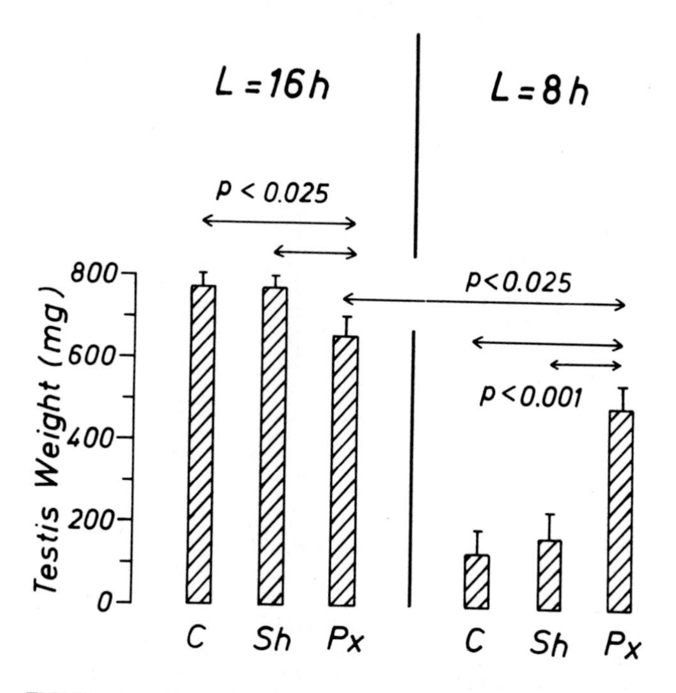

FIGURE 11. Testicular weight in Djungarian hamsters at 45 days of age. Animals lived from birth in long (L = 16hr) or short (L = 8hr) photoperiods. They were pinealectomized (Px) or sham-operated (sh) at 2 days of age or were left intact (C). (From Brackmann, K. and Hoffmann, K., *Naturwissenschaften,* 64, 341, 1977. With permission.)

animals that were exposed to long photoperiods, after having been implanted with melatonin (Figure 13, top).[8,9]

From such findings, one might conclude that melatonin inhibits gonadal development. However, when the same experiment was performed in summer animals with active testes (Figure 13, middle) melatonin not only failed to induce regression in long photoperiods, but also prevented testicular involution in short photoperiods and thus, in this case, mimicked pinealectomy.[10,11,55] On the other hand, when melatonin was implanted at the end of summer at a time when first indications of gonadal regression could be observed in some animals, implantation induced rapid testicular involution in long photoperiods (Figure 13, bottom).[11,14] In the golden hamster, implantation of melatonin was also found to have drastic effects in some cases. Thus in short photoperiods gonadal involution was inhibited,[47,48] and in long photoperiods melatonin could induce gonadal regression, though somewhat inconsistently.[56,57]

It is at present difficult to interpret these contradictory findings. Some points seem to be relevant, however. In all cases in which strong effects of melatonin implantation on testicular size and activity were found in *Phodopus,* similar effects were also observed, not only on size of accessory glands, but also on pelage color and body weight. Thus, in winter animals in which melatonin prevented rapid recrudescence brought about by long days (compare Figure 13, top), it also retarded gain in body weight and change into summer pelage.[9] In summer animals, in which melatonin prevented testicular involution in short photoperiods (compare Figure 13, middle), it also prevented moult into winter pelt and decrease in body weight.[55,58] Melatonin implants in late summer, which induced rapid involution of testes (compare Figure 13, bottom) and accessory glands in spite of long photoperiods, also led to decrease in body weight,

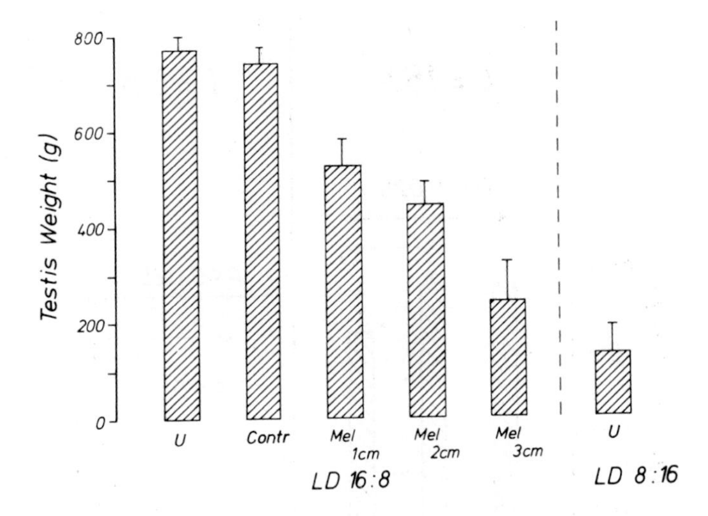

FIGURE 12. Effect of implanted melatonin on testicular development in juvenile Djungarian hamsters maintained in long photoperiods from birth. At the age of 6 days the animals were implanted with empty (Contr) or melatonin-filled (Mel) silastic tubes of different length. U = animals without implants. For comparison, testicular development in short photoperiods (LD 8:16) is also given. All animals were sacrificed at 45 days of age. Testicular development in animals with melatonin implants is significantly retarded VS controls in all cases ($p<0.01$ to <0.002); with the highest dose, there is no significant difference to the values obtained in animals maintained in short days. (From Brackmann, M., *Naturwissenschaften,* 64, 642, 1977. With permission.)

and to moult into winter pelage in some animals.[41,58] This parallel action of melatonin on gonads, and on other functions that do not depend upon gonadal activity, indicates that melatonin does not directly influence the neuroendocrine-gonadal axis, but rather the photoperiodic mechanism. This view is strengthened by other findings. Thus, in both hamster species, *Mesocricetus* and *Phodopus,* melatonin was unable to suppress spontaneous gonadal recrudescence which occurs after some time in spite of continuing short days.[9,41,59] Moreover, while in *Phodopus,* in which juvenile development can be influenced by photoperiod, melatonin can strongly inhibit sexual development (Figure 12); melatonin does not influence gonadal size and activity in *Mesocricetus,*[60] in which development up to puberty is independent of the photoperiodic treatment[29] and is normal even in blinded animals.[30] Such findings are not in accordance with the assumption that melatonin is an anti- or progonadotrophic hormone, but rather suggest that it is somehow involved in the conduction of the photoperiodic message upon all functions influenced by photoperiod.

The site of melatonin action in these experiments is unknown. It has been speculated that melatonin might act within the pineal gland itself where it might regulate the production and/or release of the "true antigonadotrophic" substance.[61,62] However, melatonin may induce gonadal regression in pinealectomized hamsters. Figure 14 shows the effects of melatonin implantation in late summer in *Phodopus* maintained in long photoperiods. In pinealectomized as well as in control animals, melatonin led to a significant regression of testes.[14] It should also be noted that in the pinealectomized hamsters that did not receive melatonin, there was a slight but significant involution as compared to the initial controls or to sham-operated hamsters not treated with melatonin. This indicates that the process of involution, which in the controls is stopped and reversed by the long photoperiods, continues in the pinealectomized ani-

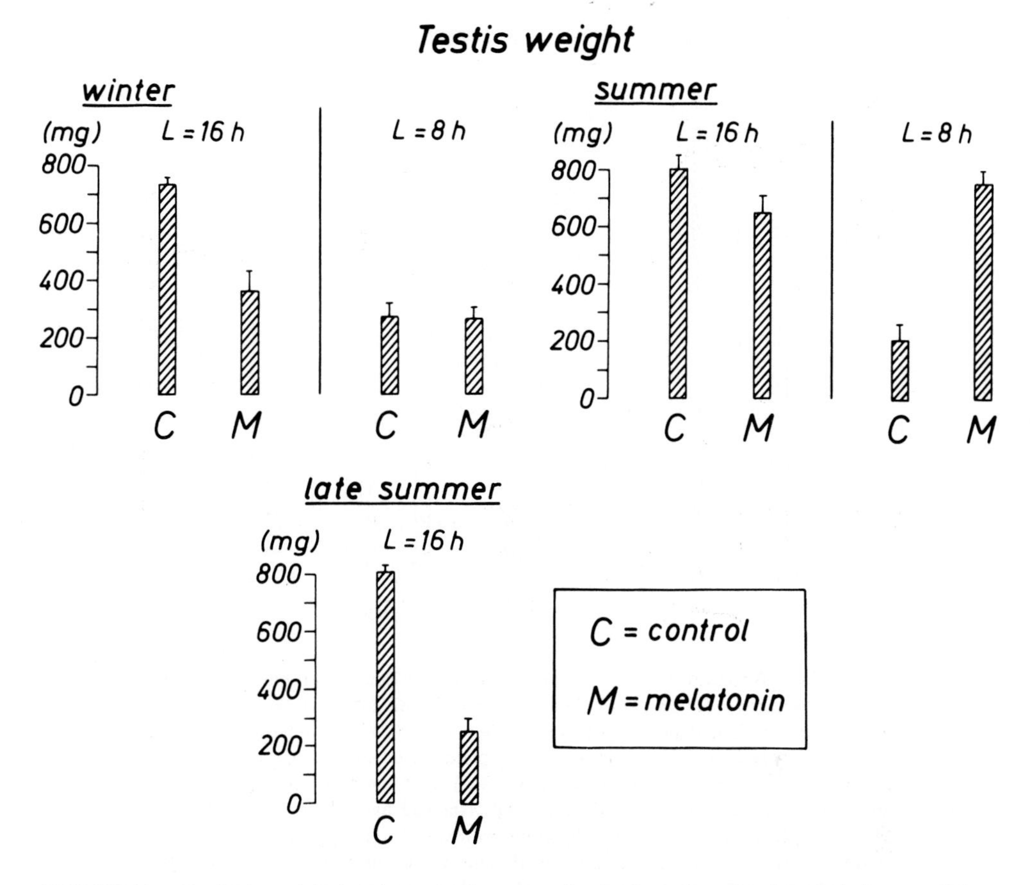

FIGURE 13. Testicular weight in Djungarian hamsters after implantation of melatonin in beeswax (M) or beeswax only (C). Prior to the experiment all animals were kept in natural illumination and their gonadal state differed correspondingly (see Figure 1). Top: testes were regressed at the beginning of the experiment (January 2); end of experiment after 37 days. Middle: testes large and active at the beginning of experiment (July 10); end of experiment after 2 months. Bottom: Testes still large at beginning of experiment (August 22) but some indications of regression noticable in controls sacrificed at this time. Experiment ended after 1 month. Note that the same treatment has different effects in different seasons. After Hoffmann.[11]

mals, which further confirms that the pineal is also involved in conveying the stimulatory effects of long photoperiods (compare section III., paragraph 3). Effects of melatonin in pinealectomized animals have been found also in the golden hamster.[12,57,63,64] Such observations are strong evidence against the hypothesis that the pineal gland is the major physiological site of melatonin action.

V. PINEAL RHYTHMS

In all mammalian species studied so far, a marked circadian cycle of pineal melatonin production and release has been found, and thus the continuous release of melatonin as from implants certainly differs from the physiological situation. This may account for the perplexing inconsistencies of the effect of melatonin implantation mentioned above. Recent experiments in golden hamsters indicate that pineal melatonin may be the major factor in conveying the photoperiodic message, and that not only the amount of melatonin produced and released but also its temporal pattern is important.[12,45,51,62-65]

The question that has to be asked is whether the pineal rhythm of melatonin production and release is different in stimulatory and nonstimulatory photoperiods. *N*-acetyl-

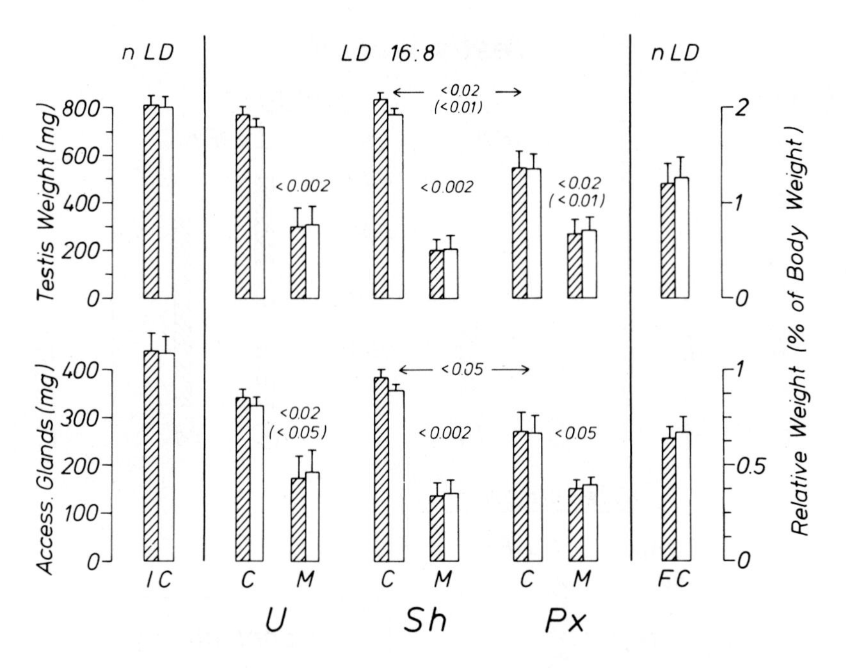

FIGURE 14. Weight of testes and accessory glands in Djungarian hamsters exposed to long photoperiods for 36 days in late summer (beginning August 21). Animals were pinealectomized (Px), sham-operated (Sh) or left intact (U) immediately prior to start of experiment. Half of each group received implants of melatonin (M) in beeswax, the other half beeswax only (C). Initial (IC) and final (FC) controls received no implants and were left in natural photoperiods (n LD). Hatched bars = absolute weight (left ordinate), open bars = relative weight (right ordinate). Note regression in animals that received melatonin. For significantly lower values in pinealectomized C-animals compared to sham-operated controls see text. (From Hoffmann, K. and Küderling, I., *Naturwissenschaften,* 64, 339, 1977. With permission.)

transferase (NAT) activity is considered the rate-limiting factor in the synthesis of melatonin. We have measured the rhythm of pineal NAT activity in *Phodopus* under different photoperiodic conditions, and have found marked differences in its pattern and thus presumably also in melatonin synthesis and release.[66,67] During the dark time there was a large increase in NAT activity (Figure 15), which was very high but brief in long photoperiods (a) and less high but more extended in short photoperiods (b). Moreover, interruption of the long dark time in short photoperiods by only 1 min of light each night completely suppressed NAT activity for the rest of the night (c), the pattern was indistinguishable from that of long photoperiods (compare with a). It should be recalled that the same photoperiodic regime, i.e., interruption of the dark time by 1 min of light at midnight, also mimicked the effects of long photoperiods on testicular recrudescence (Figure 7) and on increase in body weight (Figure 8).

The parallel action on photoperiod-dependent functions and on the pattern of pineal NAT activity and hence melatonin production and release, supports the hypothesis that the latter may be important in conveying the photoperiodic message. Peak values of NAT activity in long and short photoperiods differed, but the total amount of activity and thus, melatonin production, were probably not too dissimilar. However, the time during which NAT activity was above base level was much extended in short photoperiods, as compared to long photoperiods or to short photoperiods in which the dark time was interrupted by light. It could be that the time span during which melatonin levels are high is important.

In golden hamsters the rhythms of pineal NAT activity and melatonin content were

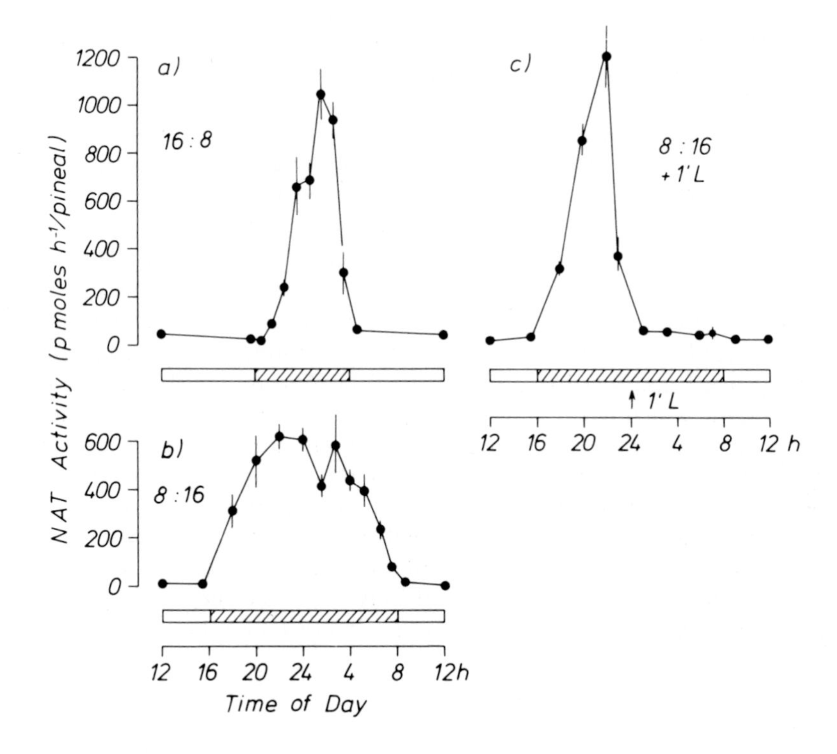

FIGURE 15. Pineal *N*-acetyltransferase activity in pineals of male Djungarian hamsters that had been maintained for 3 to 4 weeks in long (a) or short (b) photoperiods or in short photoperiods in which the dark time was interrupted by 1 min of light each night at midnight (c). Note similarity in pattern in a and c, and compare with effect of the same photoperiodic conditions on testicular recrudescence (Figure 7) and on change in body weight (Figure 8). After Hoffmann et al.[67]

also determined recently in two laboratories. Surprisingly, no nocturnal rise in NAT activity was found in one laboratory,[68] and only a threefold increase was found in the other.[69] Nevertheless, an approximate tenfold increase of pineal melatonin at nighttime was reported.[68,70,71] While peak melatonin concentrations were found similar in magnitude and duration in stimulatory and in nonstimulatory photoperiods, large differences in phase relative to the light-dark cycle were observed.[68,71] This might suggest that the phase of peak melatonin relative to a cycle of sensitivity is important in this species. However, this hypothesis has to be reconciled with reports that thrice-daily injections of melatonin in pinealectomized golden hamsters can induce gonadal regression, irrespective of the time of day of injections,[63,64] and that such injections are also effective after destruction of the suprachiasmatic nuclei which are considered essential for the persistence and expression of many circadian rhythms.[64] Due to the sampling interval, however, differences in length of time of elevated melatonin values might have been overlooked. In general, the hypothesis that the temporal pattern of melatonin release is important in conveying the photoperiodic message, seems to be well supported for both hamster species, though no definite statement can be made. To my knowledge, plasma melatonin levels have not been reported in hamsters so far.

VI. CONCLUDING REMARKS

In general, the studies in *Phodopus* reported here suggest that the pineal gland is not specifically concerned with the regulation of gonads and gonad-dependent functions, but rather plays an integral role in the transduction of photoperiodic stimuli upon the neuroendocrine axis, and thereby may influence many functions. The work in *Phodopus* also suggests that the pineal is not exclusively inhibitory, but may actively participate in the conveyance of the stimulatory effects of long photoperiods. At least some observations in the golden hamster and in the ferret indicate that these features are not limited to *Phodopus* but may be more general, and that there might be quantitative rather than qualitative differences between species. The drastically different pattern in pineal NAT activity in different photoperiods, and the parallelism between effects on photoperiodically influenced functions and the pineal rhythm in *Phodopus*, suggest that the pattern of melatonin production and release are important in conveying the photoperiodic message. Implanted and thus continuously available melatonin had inconsistent effects. Since this mode of application is rather different from the natural situation in which melatonin release is always rhythmical, these effects may be pharmacological rather than physiological. However, none of the experiments with melatonin application performed so far mimics the natural situation. Nevertheless, the parallel action of melatonin on reproduction and on other functions that depend upon photoperiod and its changes, and its ineffectiveness in processes that cannot be changed or halted by photoperiod, suggest that its action is not specifically upon gonads or gonadotrophins, but that melatonin may influence the whole photoperiodic mechanism which probably regulates a host of functions, and that it may normally be part of this mechanism.

REFERENCES

1. **Flint, W. E.,** *Die Zwerghamster der paläarktischen Fauna,* A. Ziemsen, Lutherstadt Wittenberg, 1966.
2. **Pogosianz, H. E. and Sokova, O. I.,** Maintaining and breeding of the Djungarian hamster under laboratory conditions, *Z. Versuchstierkunde.,* 9, 292, 1967.
3. **Figala, J., Hoffmann, K., and Goldau, G.,** Zur Jahresperiodik beim Dsungarischen Zwerghamster *Phodopus sungorus* Pallas, *Oecologia,* 12, 89, 1973.
4. **Hoffmann, K.,** Annuale Periodik bei Säugern und ihre photoperiodische Steuerung, *Arzneim. Forsch.,* 28, 1836, 1978.
5. **Hoffmann, K.,** Photoperiod, pineal, melatonin and reproduction in hamsters, *Prog. Brain Res.,* 52, 397, 1979.
6. **Reiter, R. J.,** Exogenous and endogenous control of the annual reproductive cycle in the male golden hamster: participation of the pineal gland, *J. Exp. Zool.,* 191, 111, 1975.
7. **Reiter, R. J.,** Interaction of photoperiod, pineal and seasonal reproduction as examplified by findings in the hamster, in *The Pineal Gland and Reproduction,* Reiter, R. J., Ed., S. Karger, Basel, 1978, 169.
8. **Hoffmann, K.,** Melatonin inhibits photoperiodically induced testis development in a dwarf hamster, *Naturwissenschaften,* 59, 218, 1972.
9. **Hoffmann, K.,** The influence of photoperiod and melatonin on testis size, body weight, and pelage colour in the Djungarian hamster *(Phodopus sungorus), J. Comp. Physiol.,* 85, 267, 1973.
10. **Hoffmann, K.,** Testicular involution in short photoperiods inhibited by melatonin, *Naturwissenschaften,* 61, 364, 1974.
11. **Hoffmann, K.,** Die Funktion des Pineals bei der Jahresperiodik der Säuger, *Nova Acta Leopold.,* 46, 217, 1977.
12. **Hoffmann, K.,** Photoperiodism in vertebrates, in *Handbook of Behavioral Neurobiology,* Vol. 4, Aschoff, J., Ed., Plenum Press, New York, 1981, 449.
13. **Hoffmann, K. and Küderling, I.,** Pinealectomy inhibits stimulation of testicular development by long photoperiods in a hamster *(Phodopus sungorus), Experientia,* 31, 122, 1975.
14. **Hoffmann, K. and Küderling, I.,** Antigonadal effects of melatonin in pinealectomized Djungarian hamsters, *Naturwissenschaften,* 64, 339, 1977.

15. **Horst, H. J.**, Photoperiodic control of androgen metabolism and binding in androgen target organs of hamsters *(Phodopus sungorus), J. Steroid Biochem.*, 11, 945, 1979.
16. **Hoffmann, K. and Nieschlag, E.**, Circadian rhythm of plasma testosterone in the male Djungarian hamster *(Phodopus sungorus), Acta Endocrinol.*, 86, 193, 1977.
17. **Turek, F. W., Alvis, J. D., Elliott, J. A., and Menaker, M.**, Temporal distribution of serum levels of LH and FSH in adult male golden hamsters exposed to long or short days, *Biol. Reprod.*, 14, 630, 1976.
18. **Tamarkin, L., Hutchinson, J. S., and Goldman, B. D.**, Regulation of serum gonadotropins by photoperiod and testicular hormone in the Syrian hamster, *Endocrinology*, 99, 1528, 1976.
19. **Hoffmann, K.**, Effect of castration on photoperiodically induced weight gain in the Djungarian hamster, *Naturwissenschaften*, 65, 494, 1978.
20. **Hoffmann, K.**, unpublished data, 1979.
21. **Brackmann, M.**, unpublished data, 1978.
22. **Heldmaier, G. and Hoffmann, K.**, Melatonin stimulates growth of brown adipose tissue, *Nature (London)*, 247, 224, 1974.
23. **Logan, A. and Weatherhead, B.**, Photoperiodic dependence of seasonal variations in melanocyte-stimulating hormone content of the pituitary gland in the Siberian hamster *(Phodopus sungorus), J. Endocrinol.*, 83, 41P, 1979.
24. **Weatherhead, B. and Logan, A.**, Seasonal variations in the response of hair follicle melanocytes to melanocyte-stimulating hormone, *J. Endocrinol.*, 81, 167P, 1979.
25. **Rust, C. C.**, Hormonal control of pelage cycles in the short-tailed weasel *(Mustela erminea bangsi), Gen. Comp. Endocrinol.*, 5, 222, 1965.
26. **Reiter, R. J.**, Evidence for the refractoriness of the pituitary-gonadal axis to the pineal gland in golden hamsters and its possible implications in annual reproductive rhythms, *Anat. Rec.*, 173, 365, 1972.
27. **Turek, F. W., Elliott, J. A., Alvis, J. D., and Menaker, M.**, Effect of prolonged exposure to non-stimulatory photoperiods on the activity of the neuroendocrine-testicular axis of golden hamsters, *Biol. Reprod.*, 13, 475, 1975.
28. **Stetson, M. H., Matt, K. S., and Watson-Whitmyre, M.**, Photoperiodism and reproduction in golden hamsters: circadian organization and the termination of photorefractoriness, *Biol. Reprod.*, 14, 531, 1976.
29. **Gaston, S. and Menaker, M.**, Photoperiodic control of hamster testis, *Science*, 167, 925, 1967.
30. **Reiter, R. J., Sorrentino, S., and Hoffman, R. A.**, Early photoperiodic conditions and pineal antigonadic function in male hamsters, *Int. J. Fertil.*, 15, 163, 1970.
31. **Hoffmann, K.**, Effects of short photoperiods on puberty, growth and moult in the Djungarian hamster *(Phodopus sungorus) J. Reprod. Fertil.*, 54, 29, 1978.
32. **Brackmann, M.**, Untersuchungen zur photoperiodischen Steuerung der Jugendentwicklung beim Dsungarischen Hamster, Ph.D. thesis, University of Munich, Fed. Rep. Germany, 1978.
33. **Frieling, R.**, Untersuchungen zur Spermatogenese und zum EinfluB der Photoperiode auf die Hodenentwicklung beim Dsungarischen Hamster (*Phodopus sungorus* Pallas), V.M.D. thesis, University of Munich, Fed. Rep. Germany, 1979.
34. **Van Veen, T., Brackmann, M., and Moghimzadeh, E.**, Postnatal development in the pineal organ in the hamsters *Phodopus sungorus* and *Mesocricetus auratus, Cell. Tissue Res.*, 189, 241, 1978.
35. **Elliott, J. A.**, Photoperiodic Regulation of Testis Function in the Golden Hamster: Relation to the Circadian System, Ph.D. thesis, University of Texas, Austin, 1974.
36. **Elliott, J. A.**, Circadian rhythms and photoperiodic time measurement in mammals, *Fed. Proc. Fed. Am. Soc. Exp. Biol.*, 35, 2339, 1976.
37. **Hoffmann, K.**, unpublished data, 1980.
38. **Hoffmann, K.**, Photoperiodic effects in the Djungarian hamster: one minute of light during darktime mimics influence of long photoperiods on testicular recrudescence, body weight and pelage colour, *Experientia*, 35, 1529, 1979.
39. **Vriend, J., Reiter, R. J., and Anderson, G. R.**, Effects of the pineal and melatonin on thyroid activity of male golden hamsters, *Gen. Comp. Endocrinol.*, 38, 189, 1979.
40. **Reiter, R. J.**, Comparative physiology: pineal gland, *Ann. Rev. Physiol.*, 35, 305, 1973.
41. **Küderling, I.**, Untersuchungen zum Einfluss der Epiphysis cerebri auf die Steuerung der Jahresperiodik beim Dsungarischen Zwerghamster (*Phodopus sungorus* PALLAS) in Abhängigkeit von der Photoperiode, Dipl. Biol. thesis, University of Munich, Fed. Rep. Germany, 1976.
42. **Brackmann, M. and Hoffmann, K.**, Pinealectomy and photoperiod influence testicular development in the Djungarian hamster, *Naturwissenschaften*, 64, 341, 1977.
43. **Herbert, J.**, The role of the pineal gland in the control by light of the reproductive cycle of the ferret, in *The Pineal Gland*, Wolstenholme, G. E. W. and Knight, J., Eds., Churchill Livingstone, Edinburgh, 1971, 303.

44. **Bittman, E. L.**, Discussion remark, in *Biological Clocks in Seasonal Reproductive Cycles,* Follett, B. K., Ed., John Wright & Sons, Bristol, 1981, in press.
45. **Hoffmann, K.**, The role of the pineal gland in the photoperiodic control of seasonal cycles in hamsters, in *Biological Clocks in Seasonal Reproductive Cycles,* Follett, B. K., Ed., John Wright & Sons, Bristol, 1981, in press.
46. **Reiter, R. J. and Johnson, L. Y.**, Depressant action of the pineal gland on pituitary luteinizing hormone and prolactin in male hamsters, *Hormone Res.,* 5, 311, 1974.
47. **Reiter, R. J., Vaughan, M. K., Blask, D. E., and Johnson, L. Y.**, Melatonin: its inhibition of pineal antigonadotrophic activity in male hamsters, *Science,* 185, 1169, 1974.
48. **Reiter, R. J.**, Regulation of pituitary gonadotrophins by the mammalian pineal gland, in *Neuroendocrine Regulation of Fertility,* Kumar, A., Ed., S. Karger, Basel, 1976, 169.
49. **Turek, F. W.**, Role of the pineal gland in photoperiod-induced changes in hypothalamic-pituitary sensitivity to testosterone feedback in castrated hamsters, *Endocrinology,* 104, 636, 1979.
50. **Klein, D. C.**, The pineal gland: a model of neuroendocrine regulation, in *The Hypothalamus,* Reichlin, S., Baldessarini, R. J. and Martin, J. B., Eds., Raven Press, New York, 1978, 303.
51. **Reiter, R. J.**, Reproductive effects of the pineal gland and pineal indoles in the Syrian hamster and the albino rat, *The Pineal Gland,* Vol. 2, Reiter, R. J., Ed., CRC Press, Boca Raton, 1981.
52. **Reiter, R. J.**, Pineal regulation of hypothalamicopituitary axis: gonadotrophins, in *Handbook of Physiology,* Endocrinology IV, Part 2, Knobil, E. and Sawyer, W. H., Eds. American Physiological Society, Washington 1974, 365.
53. **Rust, C. C. and Meyer, R. K.**, Hair color, molt, and testis size in male short-tailed weasels treated with melatonin, *Science,* 165, 921, 1969.
54. **Brackmann, M.**, Melatonin delays puberty in the Djungarian hamster, *Naturwissenschaften,* 64, 642, 1977.
55. **Hoffmann, K.**, The action of melatonin on testis size and pelage color varies with the season, *Int. J. Chronobiol.,* 1, 333, 1973.
56. **Turek, F. W., Desjardins, C., and Menaker, M.**, Melatonin: antigonadal and progonadal effects in male golden hamsters, *Science,* 190, 280, 1975.
57. **Turek, F. W.**, Antigonadal effect of melatonin in pinealectomized and intact male hamsters, *Proc. Soc. Biol. Med.,* 155, 31, 1977.
58. **Hoffmann, K.**, unpublished data, 1973.
59. **Turek, F. W. and Losee, S. H.**, Melatonin-induced testicular growth in golden hamsters maintained on short days, *Biol. Reprod.,* 18, 299, 1978.
60. **Turek, F. W.**, Effect of melatonin on photic-independent and photic dependent testicular growth in juvenile and adult male golden hamsters, *Biol. Reprod.,* 20, 1119, 1979.
61. **Quay, W. B.**, *Pineal Chemistry in Cellular and Physiological Mechanisms,* Charles C Thomas, Springfield, Ill., 1974, 193.
62. **Reiter, R. J., Blask, D. E., Johnson, L. Y., Rudeen, P. K., Vaughan, M. K., and Waring, P. J.**, Melatonin inhibition of reproduction in the male hamster: its dependency on time of day of administration and on an intact and sympathetically innervated pineal gland, *Neuroendocrinology,* 22, 107, 1976.
63. **Tamarkin, L., Hollister, C. W., Lefebvre, N. G., and Goldman, B. D.**, Melatonin induction of gonadal quiescence in pinealectomized Syrian hamsters, *Science,* 198, 953, 1977.
64. **Bittman, E. L., Goldman, B. D., and Zucker, I.**, Testicular responses to melatonin are altered by lesions of the suprachiasmatic nuclei in golden hamsters, *Biol. Reprod.,* 21, 647, 1979.
65. **Tamarkin, L., Westrom, W. K., Hamill, A. I., and Goldman, B. D.**, Effect of melatonin on the reproductive system of male and female Syrian hamsters: a diurnal rhythm in sensitivity to melatonin, *Endocrinology,* 99, 1534, 1976.
66. **Hoffmann, K., Illnerová, H., and Vaněček, J.**, Pineal *N*-acetyltransferase activity in the Djungarian hamster, *Naturwissenschaften,* 67, 408, 1980.
67. **Hoffmann, K., Illnerová, H., and Vaněček, J.**, The effect of photoperiod and of 1 min light at night time on the pineal rhythm in *N*-acetyltransferase activity in the Djungarian hamster, *Phodopus surgorus, Biol. Reprod.,* in press.
68. **Tamarkin, L., Reppert, S. M., and Klein, D. C.**, Regulation of pineal melatonin in the Syrian hamster, *Endocrinology,* 104, 385, 1979.
69. **Rudeen, P. K., Reiter, R. J., and Vaughan, M. K.**, Pineal serotonin *N*-acetyltransferase activity in four mammalian species, *Neurosci. Lett.,* 1, 225, 1975.
70. **Panke, E. S., Rollag, M. D., and Reiter, R. J.**, Pineal melatonin concentrations in the Syrian hamster, *Endocrinology,* 104, 194, 1979.
71. **Panke, E. S., Rollag, M. D., and Reiter, R. J.**, Effects of photoperiod on hamster pineal melatonin concentrations, *Comp. Biochem. Physiol.,* 66A, 691, 1980.

Chapter 5

MELATONIN LEVELS AS THEY RELATE TO REPRODUCTIVE PHYSIOLOGY

Harry J. Lynch and Richard J. Wurtman

TABLE OF CONTENTS

I. INTRODUCTION

Probably the first publication relating the pineal gland to reproductive function was Otto Heubner's[1] 1898 case report describing a young boy who had shown precocious puberty and was found at post-mortem to have a pineal tumor. Over the next half-century, many other children with pineal tumors and precocious sexual development were described, as well as a smaller number of patients whose pineal tumors were associated with delayed sexual development.

In 1954, Kitay and Altschule[2] reviewed the literature on human pineal tumors and proposed an explanation for their disparate reproductive consequences: the pineal secretes an antigonadal substance (or substances), whose absence (in children with tumors destroying the pineal parenchyma) caused sexual precocity and whose excess (in children with parenchymal, secreting tumors) could cause hypogonadism.

Some 40 years earlier, McCord and Allen,[3] zoologists interested in endocrine factors that might affect morphogenesis in tadpoles, had discovered that the animals became translucent 30 min after being exposed to extracts of pineal tissue, in dramatic contrast to tadpoles fed extracts of other tissues. The phenomenon was so striking and unfailingly reproducible that it was made the subject of a special study. Organic extracts of pineal tissue were tested, and quantities as small as one part of an acetone extract per 100,000 parts water effectively elicited the blanching response.

In the late 1950s, Lerner et al.[4] began to search for the skin-lightening constituent of pineal extracts, hoping to identify a neurogenic agent that might cause depigmentation in human vitiligo. After an heroic bioanalytical effort involving 200,000 cattle pineals and 4 years work, they succeeded in isolating and chemically characterizing the pineal constituent that elicits the amphibian melanophore response. This substance, named melatonin, has unusual chemical properties. The *N*-acetylation of its amine group and the *O*-methylation of its hydroxyl causes not physiologic *inactivation,* but *activation;* unlike its amine precursor, serotonin, it is highly lipid-soluble and crosses the blood-brain barrier easily.

The *O*-methylation of the indolic hydroxyl group of melatonin drew the attention of Axelrod who had identified another methoxylating enzyme, catechol-*O*-methyltransferase. Working with Weissbach, Axelrod went on to show that pineal tissue from all vertebrate animals examined has the enzymatic capacity, when provided with *N*-acetylserotonin and a suitable methyl donor, to synthesize melatonin.[5] An enzyme peculiar (but apparently not unique[6,7]) to the pineal, hydroxyindole-*O*-methyltransferase (HIOMT), catalyzes this transformation.

Meanwhile, Fiske et al. were studying the mechanisms by which long periods of exposure of rats to light accelerated the maturation of their gonads. In 1961,[8] they reported that the weights of pineals of rats kept in a continuously lit environment decrease, while the weights of their ovaries increase (and gonadal maturation, as manifested by the estrous cycle, accelerates). From this observation developed the notion that the function of the nonphotoreceptive mammalian pineal (which is homologous with the photoreceptive parietal structures in phylogenetically more primitive vertebrate classes) was still somehow related to environmental illumination. Wurtman et al.[9] then showed that exposure of female rats to continuous light, or removal of their pineals, had similar but nonadditive effects on the weights of their ovaries; they further showed that administration of crude pineal extracts[10] or of melatonin[11] could exert antigonadal effects and reverse the effects of pinealectomy.

Axelrod and Weissbach[12] developed a radiochemical assay that permitted in vitro assessment of HIOMT activity and allowed inferences to be made about relative rates of melatonin secretion; this assay provided the basis for a decade of research on the role of the pineal in regulating reproductive physiology. Wurtman and Axelrod,[13] and then many others, used treatment-induced changes in pineal enzyme activity as a basis for inferring parallel changes in melatonin secretion. The suppressive effect of light on pineal HIOMT activity was taken as evidence that light exposure also inhibited melatonin synthesis and secretion in vivo, and that by diminishing the amount of this gonad inhibitor in the animal, light stimulated gonadal maturation and reproductive function.[13] This strategy of inferring melatonin synthesis from in vitro enzyme activities has been amply justified by the data on melatonin levels discussed below.

II. BIOSYNTHESIS AND METABOLISM OF MELATONIN

Biosynthesis of melatonin and other pineal methoxyindoles is initiated by uptake of the amino acid tryptophan from circulating blood into the pineal. The first step in the biosynthetic pathway involves oxidation of tryptophan at the 5 position to form the amino acid 5-hydroxytryptophan. This step is catalyzed by the enzyme tryptophan hydroxylase and requires the presence of O_2, ferrous iron, and a reduced pteridine cofactor.[14]

The second step in melatonin biosynthesis, conversion of 5-hydroxytryptophan to serotonin (5-hydroxytryptamine), is catalyzed by the enzyme aromatic L-amino acid decarboxylase ("dopa decarboxylase"). This enzyme is ubiquitous and participates in catecholamine biosynthesis;[15,16] it requires pyridoxal phosphate as a cofactor.

Most of the serotonin produced outside the pineal organ (e.g., in brain or enterochromaffin cells) is metabolized by oxidative deamination, which is catalyzed by the enzyme monoamine oxidase (MAO). MAO is also found in the pineal and metabolizes a portion of the pineal serotonin to yield 5-hydroxyindole acetaldehyde, an unstable intermediate which is then either oxidized to 5-hydroxyindole acetic acid or reduced to 5-hydroxytryptophol. Both products are substrates for hydroxyindole-*O*-methyltransferase (HIOMT), yielding 5-methoxyindole acetic acid or 5-methoxytryptophol, respectively.[17]

The conversion of serotonin to melatonin is more important for studying pineal physiology. First, serotonin is *N*-acetylated by the enzyme serotonin-*N*-acetyltransferase;[18] a methyl group is then transferred from *S*-adenosylmethionine to the 5-hydroxy position of *N*-acetylserotonin, yielding melatonin. This latter reaction is catalyzed by HIOMT. Almost invariably, a treatment that increases the activity of one of these enzymes will affect the other similarly. Hence. for about a decade, debate raged over which of the two pineal enzymes was rate limiting in controlling melatonin synthesis in vivo. Now it appears that neither is: rather, melatonin synthesis rates may be controlled physiologically by alterations in the availability of the precursor of melatonin, serotonin. Almost all of the very large serotonin stores in pineal parenchymal cells are bound (probably to a cytoplasmic protein), and therefore are unavailable for conversion to melatonin (by *N*-acetylation) or for metabolism via other pathways (e.g., oxidative deamination, by MAO). Treatments that accelerate melatonin synthesis, such as the *absence* of light, apparently do so by increasing the firing of the postganglionic sympathetic nerves that innervate the pineal;[19] they release more norepinephrine, which, acting via beta-receptors and cyclic AMP, may diminish serotonin binding to this cytoplasmic protein, allowing "free" serotonin to be converted to *N*-acetylserotonin, and then to melatonin. (Such treatments concurrently accelerate synthesis of other serotonin-derived methoxyindoles, such as methoxytryptophol[20] — which are completely independent of serotonin-*N*-acetyltransferase).[21]

Studies of the physiological disposition of radioactively labeled melatonin, administered to rats as a bolus injection,[22,23] have shown that the hormone disappears rapidly from plasma and tissues by first-order decay and is excreted in urine primarily as glucuronic acid and sulfate conjugates of 6-hydroxymelatonin. Hirata and Hayaishi[24] recently showed that another melatonin metabolite, *N*-acetyl-5-methoxykynurenamine, is formed from melatonin in vitro by brain tissue. A two-step enzymatic reaction is involved: first, the indole ring structure is cleaved; then, the number-one carbon, adjacent to the indole nitrogen, is removed. These investigators suggest that melatonin may normally be metabolized in brain by this metabolic pathway.

III. MELATONIN ASSAYS

Only very recently have sufficiently sensitive and precise assay methods become available to permit the study of melatonin levels in relation to reproductive physiology. The idea that such a relationship might exist is based on observations, accumulated over the past 20 years, on the effects of pinealectomy and of exogenous melatonin, and quantitative inferences about in vivo changes in melatonin levels based on in vitro enzyme measurements. During this time, analytical methods for measuring melatonin have evolved. Melatonin was discovered and isolated by bioassays;[4] rhythmic changes in its pineal concentrations were demonstrated by fluorometric assay;[25] its absolute identification in body fluids was accomplished by gas chromatography-mass spectrometry;[26] and now, melatonin levels in physiologic states can be measured conveniently and reliably by radioimmunoassays.[27]

This review necessarily discusses rather cursorily the various analytical methods used to measure melatonin levels. At this time it does not seem possible to move from a simple tabulation of the observations made on melatonin levels to an all-embracing formulation of its role in reproductive physiology. However, the information now available can provide some perspective, and should aid design of future experimental work.

To underscore the general consistency in the information acquired over the years and to demonstrate how new approaches evolved in the study of the reproductive role of melatonin, we discuss this information in roughly chronological order.

A. Melatonin Estimates by Bioassay

Bioassay methods for measuring melatonin use its most striking physiological effect, the dermal melanophore response of amphibians — specifically, the melatonin-induced nucleocentric aggregation of melanin pigment granules within the living melanophore. The assay methods take either of two forms: (1) photometric measurement of changes in the transmission or reflectance of incident light by isolated frog skin exposed to melatonin solution,[4] or (2) microscopic assessment of the degree of melanin aggregation within the melanophore in intact amphibian larvae, as judged by an arbitrary scale of melanophore stages, the Hogben index.[28]

Both methods have documented merit. Lerner[4] used the first method to trace the isolation of melatonin from bovine pineal tissue. The second method, involving the microscopic evaluation of pigment dispersion in melanophores, exchanges the objectivity of the reflectometer for the capacity of the eye to discriminate. With this technique, attention is restricted to the unicellular effector organ that responds to melatonin stimulation, the melanophore. Most bioassay studies that have been done use this latter method.

The first quantitative bioassay used to measure physiologic changes in pineal melatonin content, that of Tomatis and Orias,[29] combined elements of both methods. Melanophores were scored relative to the Hogben index in pieces of isolated toad skin exposed to pineal tissue homogenate. Following an experimental design suggested by earlier work with HIOMT measurements, Tomatis and Orias[29] showed that pineal glands of rats exposed to constant darkness contain significantly more melatonin than those of rats exposed to constant light. They further showed that blinding or superior cervical ganglionectomy increases pineal melatonin content.

Table 1
MELATONIN LEVELS ESTIMATED BY BIOASSAY[a]

Date	Species	Sample	Melatonin level: Day	Night	Unit	Comment
			Using Isolated Frog Skin and Reflectometry			
1960	Cattle[4]	Pineal	0.2	—	ng/mg	Pooled tissue
1961	Rat[38]	Pineal	0.4	—	ng/mg	Pooled tissue
1964	Human[39]	Sciatic nerve	2	—	ng/100 g	Pooled tissue
		Urine	1	—	ng/48 hr	48-hr collection
			Using Isolated Toad Skin and Melanophore Stage			
1967	Rat[29]	Pineal	0.09-0.22	—	ng/mg	2-4 glands/assay
		After 3-80 days of constant light	0.05-0.07	—	ng/mg	2-4 glands/assay
		After 7-80 days of constant darkness	0.22-0.35	—	ng/mg	2-4 glands/assay
			Using Larval Anurans and Melanophore Stage			
1971	Rat[40]	Pineal	0.5	6.8	ng/gland	1 gland/assay
	Chicken[40]	Pineal	5.8	38	ng/gland	1 gland/assay
	Quail[40]	Pineal	0.6	3.2	ng/gland	1 gland/assay
1973	Human[31]	Plasma	ND-12[b]	100	pg/mℓ	20-mℓ sample
1974	Human[33]	Urine	ND-0.7	1.1-24	ng/8 hr	8-hr collections
1975	Chicken[41]	Serum	ND	190	pg/mℓ	8-10 mℓ sample
	Rat[42]	Serum	ND	20-50	pg/mℓ	Pooled from 2 rats
	Human[43]	Plasma	162	258	pg/mℓ	
1976	Human[32]	Plasma	ND	50-100	pg/mℓ	20-mℓ sample

[a] We apologize for any distortion of the facts that may have resulted from placing this information in a Procrustean table. Some of the numerical values listed here were estimated from published graphs, and we realize that such terms as "day" and "night" might be misleadingly vague as descriptions of environmental conditions or as time-points for describing the phenomenon under consideration. All entries were provided by well controlled, "state-of-the-art" studies. Before judging the merit of a particular datum, the reader is urged to consult the reference cited.

[b] ND = not detectable.

Most studies on animals' melatonin levels that have employed bioassays (Table 1) have used an adaptation of the quantitative bioassay for melatonin developed in Ralph's laboratory.[30] This method is based on the dermal melanophore response of larval *Rana pipiens* to melatonin in their bathing medium. The assay is sensitive to as little as 100 pg/mℓ melatonin in the final test solution and is remarkably specific in the presence of naturally occurring melatonin analogs. Thus, melatonin in pineal tissue homogenates can be measured by bioassay without further purification. Similarly, melatonin in extracts of body fluids can be assayed readily if the extracts are cleared of organic solvents which might modify responsiveness of the melanophores.

The assertion that the bioassay is sensitive to 100 pg/mℓ of melatonin requires further definition. While the dermal melanophores of tadpoles respond discernably to this concentration of melatonin, accuracy and precision in the bioassay of melatonin from natural sources require replicate sampling at various concentrations. Therefore, as the total melatonin content of a given sample approaches the melanophores' limit of sensitivity, the accuracy of the quantitative estimate diminishes. While pineal melatonin content can be measured readily, the meaningful measurement of melatonin in small samples of rat or hamster blood or human cerebrospinal fluid (CSF) is simply not feasible. Nevertheless, Pelham et al.,[31] using 20-mℓ blood samples, successfully demonstrated diurnal variations with nocturnal increases in the concentration of a "melatonin-like substance" in human plasma. Using the same assay method, Vaughan et al.[32] reaffirmed the rhythmic nocturnal rise in human plasma melatonin and demonstrated that this circadian rhythm persisted through 2½ days of constant light.

To measure urinary melatonin, the limited sensitivity of the bioassay method necessitated development of a column chromatographic technique using Amberlite® XAD-2[33] to concentrate melatonin from 4- to 8-hr human urine collections. Such studies demonstrated the relationship between diurnal lighting and rhythmic melatonin excretion with nocturnal increases, revealed that the rhythm persists through 28 hr of constant light,[34] and showed that 5 to 7 days were required to reentrain rhythmic melatonin excretion to a 12-hr phase shift in the lighting schedule.[35] Studies of human urinary melatonin also showed that, in a "timeless environment," the rhythmic pattern of melatonin excretion can be dissociated from a "self-selected" wake-sleep schedule,[36] and that blind human subjects exhibit a circadian rhythm in melatonin excretion, which is not phase-locked to their customary wake-sleep schedule.[37]

B. Melatonin Estimates by Spectrophotofluorometric Assay

A distinctive characteristic of all 5-hydroxy- and 5-methoxyindoles is their fluorescence in strong mineral acid at 540 to 550 nm, when activated at 295 nm. This phenomenon, discovered by Udenfriend, Bogdanski, and Weissbach,[44] facilitated development of a sensitive chemical method for measuring serotonin and studying its tissue distribution.[45] In 1963, Quay[46] described techniques for selective extraction (using different organic solvents and pH adjustments) and measurement of a variety of indoles, including melatonin; in 1964, he demonstrated, for the first time, circadian and estrous rhythms in pineal melatonin content.[47] Quay found that selective extraction of melatonin occurs with *p*-cymene and an alkaline solution. Of course, since all 5-hydroxy- and 5-methoxyindole derivatives have almost identical fluorescence characteristics, a truly selective extraction is necessary. While *p*-cymene affords such selectivity, Quay also found that significant technical difficulties remain in the practical application of this method. For example, impurities in commercial reagent-grade solvents appear to affect the percent recovery of the primary indole solutes much more than they do the specificity of the extraction.[46]

Maickel and Miller[48] discovered that when indoles containing HO- or CH_3O-substituents in the 5 position are heated with *O*-phthalaldehyde (OPT) in 3 *N* hydrochloric acid, they yield highly fluorescent products and thus enhance the sensitivity of the fluorometric assay. Again, the fluorescent spectra of the reaction products of OPT with serotonin, 5-methoxytryptamine, *N*-acetylserotonin, and melatonin (i.e., activation and emission wavelength maxima) are extremely similar. Therefore, a separation procedure is needed to measure the various compounds in each other's presence. The procedure developed by Maickel and Miller[48] uses multiple solvent extractions to separate the indoles, and the final step in each case is reaction with OPT. The estimated sensitivity of this method permits accurate measurement of 10 to 20 ng of each compound.

Although the few data in Table 2 represent nearly all the information on melatonin levels thus far obtained by fluorometric assays, they are of historical and practical significance. This assay provided the first demonstration of circadian rhythms in rat pineal melatonin content, and showed that the rhythms were correlated with light and dark periods and possibly were influenced to a lesser degree by the estrous cycle. Fluorescence assays also provided substantial data on the physical properties of melatonin and related compounds, and on methods for their physical separation; these data have proven useful in subsequent efforts to measure melatonin by other means.

Table 2
MELATONIN LEVELS ESTIMATED BY SPECTROPHOTOFLUOROMETRY

Date	Species	Sample	Melatonin level			Comment
			Day	Night	Unit	
With Selective Extraction						
1964	Rat[47]	Pineal	1	3	ng/gland	4-6 glands/assay
With Multiple Solvent Extraction and OPT Enhanced Fluorescence						
1968	Rat[48]	Pineal	0.4	—	ng/mg	
1970	Rat[49]	Pineal	0.48	—	ng/mg	4-5 mg tissue/assay
	Dog[49]	Pineal	0.17	—	ng/mg	4-5 mg tissue/assay

C. Melatonin Estimates by Radioimmunoassay

Various methods have been used to produce antimelatonin antisera for the radioimmunoassay (RIA) of melatonin. As the melatonin molecule is not itself antigenic, melatonin or one of its analogs must be coupled to a carrier molecule that will impart antigenicity to the complex. This product is then used to immunize (i.e., to stimulate antibody production in) an animal, usually a rabbit or goat. The specificity and potency of the resulting antisera are functions of the biochemical nature of the antigens used, the program of treatment, and above all, the individual immune response of the animal that is "immunized". Thus, a variety of relatively specific antisera have been produced for use in the radioimmunoassay of melatonin. In a recent review, Arendt[50] summarized the brief history of melatonin radioimmunoassays, outlined some of the methods used to produce anti-melatonin sera in rabbits, and enumerated a battery of criteria for defining the specificity of antisera.

In 1974, Grota and Brown[51] reported that antibodies to indoles can be raised after conjugation of serotonin or *N*-acetylserotonin to antigenic protein by means of the formaldehyde condensation reaction. The resulting antisera were described and proposed as the basis for radioimmunoassay and immunohistochemical localization of melatonin in brain structures.

The first report of a practical RIA for melatonin, validated by comparison with bioassay measurements, and accompanied by data on human plasma melatonin levels, was that of Arendt et al.[43] in 1975. At least five additional RIA methods were developed[52-56] within the next 2 years and melatonin measurements in pineal tissue, blood, CSF, and urine from humans and various other animal species began to inundate the literature. The advent of RIAs for melatonin also required tritiated melatonin with sufficiently high specific activity (e.g., 26 Ci/mmol), which is used as the tracer in most of the current RIAs. The method developed by Rollag and Niswender[54] uses a radioiodinated melatonin analog as a tracer; the high specific activity obtainable with radioiodine (1620 Ci/mmol) allows roughly tenfold greater sensitivity than that obtained with tritiated melatonin.

In 1977, an international cross-validation study was conducted by Wetterberg[27] using a reference serum containing 0.86 nmol/ℓ (200 pg/mℓ) melatonin. Results obtained in seven laboratories using six different antisera indicate that melatonin can be measured reasonably accurately by the various RIA methods. Estimates of the melatonin concentration in reference samples ranged from 128 ± 15 to 239 ± 35 pg/mℓ; the mean value obtained was 176 pg/mℓ.

All of the RIA methods currently in use involve initial extraction of melatonin from its biological source into a nonpolar organic solvent. It is then purified further by aqueous washes or by column or thin-layer chromatography, according to the specificity of the antiserum used and the challenge presented by the specific tissue or fluid under study. In our method,[57] measurement of melatonin extracted from pineal tissue homogenates, plasma, or CSF, using antiserum provided by Dr. Lawrence Levine of Brandeis University[52] presents very little challenge, and only simple aqueous washes of the chloroform extracts are required before assaying the samples. Urine extracts, in contrast, present a very serious challenge, and require isolation of melatonin by multiple[58] or continuous[59] unidimensional development on thin-layer chromatography plates.

Protocols for sample preparation and radioimmunoassay of melatonin are even more numerous than the 2 dozen or so laboratories studying melatonin levels: for example, a given laboratory may use entirely different methods for extracting melatonin from blood, urine, and tissues.[57] Also, different experimental questions impose different analytical constraints. Thus, studies designed to monitor entrainment of a melatonin secretion rhythm to an altered lighting schedule, or to demonstrate suppression of rhythmic melatonin secretion under constant light, are far less demanding than studies aimed at detecting very subtle changes in the absolute amounts of melatonin secreted or slight phase shifts in its rhythmic pattern.

Representative contributions from various laboratories using RIA methods to estimate melatonin levels include the following (Table 3):

1. Documentation of the universal relationship between environmental light and darkness and rhythmic changes in blood levels of melatonin in various wild and domesticated animals; also species, strain, and individual differences in the time course and amplitude of rhythmic changes in blood melatonin[56]
2. Evaluation of temporal and quantitative relationships between serum and CSF melatonin levels in the partially restrained rhesus monkey; using continuous withdrawal of CSF and measuring melatonin in 90- to 120-min fractions, developing a precise picture of the effects of various experimental manipulations on rhythmic melatonin secretion[60-62]
3. Development of procedures for urinary melatonin measurement as a noninvasive means of monitoring melatonin secretion in longitudinal studies on the entrainment of rhythmic melatonin secretion in individual human subjects[34,35,63] and individual experimental animals[64,65]
4. Correlation of rhythmic variations in hamster pineal melatonin content with the dramatic changes in the animal's reproductive status caused by seasonal or experimentally imposed changes in the daily photoperiod[66-69]

D. Melatonin Estimates by Gas Chromatography

Because of their specificity, sensitivity, and rapidity, gas chromatographic procedures show great promise for assaying melatonin. The exquisite resolving power of gas-liquid chromatography coupled with increasingly discriminating modes of detection, has made this method a most valuable analytical tool (Table 4).

1. Gas Chromatography with Electron-Capture Detection

For electron-capture detection, melatonin is extracted from its biological source and a volatile halogenated derivative is formed. In 1970, Degen and Barchas, using the extraction procedure of Miller and Maickel,[49] followed by the derivatization procedure, found that melatonin could be measured in pineal tissue with 100 pg as the lower limit of detectable material.[79] Greiner and Chan[80] used the method to measure melatonin contents of pineal glands collected at autopsy from 22 human subjects; melatonin

Table 3
MELATONIN LEVELS ESTIMATED BY RADIOIMMUNOASSAY

			Melatonin level			
Date	Species	Sample	Day	Night	Unit	Comment
1975	Human[43]	Plasma	132	188	pg/mℓ	Mean of 14 samples
1976	Rat[55]	Pineal	118	—	pg/gland	
1976	Sheep[54]	Serum	10-30	100-300	pg/mℓ	
1976	Rat[70]	Urine	0.18	1.4	ng/12 hr	
1977	Human[71]	Serum	20	78	pg/mℓ	Mean of 5 subjects
1977	Calves[72]	CSF	38	637	pg/mℓ	CSF and plasma were
		Plasma	19	121	pg/mℓ	sampled simultaneously
1977	Human[56]	Plasma	20	50-100	pg/mℓ	Range for 4 subjects
	Sheep[56]	Plasma	80	100-160	pg/mℓ	sampled at midlight
	Rat[56]	Plasma	6.3	75	pg/mℓ	and middark[a]
	Chicken[56]	Plasma	50	200	pg/mℓ	
	Pig[56]	Plasma	22	76	pg/mℓ	
	Donkey[56]	Plasma	24	128	pg/mℓ	
	Cow[56]	Plasma	20	320	pg/mℓ	
	Camel[56]	Plasma	29	221	pg/mℓ	
	Lizard[56]	Plasma	20	500	pg/mℓ	
1978	Human[73]	Serum	Spring & autumn	12-14	pg/mℓ	Sampled at monthly
			Winter & summer	20-24	pg/mℓ	intervals at 8 AM
1978	Human[74]	Serum	Premenstrual and menstrual	300	pg/mℓ	Mean values for 5 healthy women sampled
			During ovulation	100	pg/mℓ	before breakfast
1978	Sheep[75]	Serum	140	297	pg/mℓ	
1978	Hamster[67]	Pineal	95-232	760-1335	pg/gland	
1978	Human[76]	Plasma	19	74	pg/mℓ	
1978	Human[35]	Urine	1.3	4.3	ng/4 hr	
		Plasma	23	97	pg/mℓ	
1979	Human[77]	CSF, lumbar	1.7-68.7	—	pg/mℓ	Range of values from
		CSF, cisternal	1.6-41.8	—	pg/mℓ	26 patients aged 8-70
1979	Human[78]	Plasma	50-120	280-400	pg/mℓ	Range for 5 subjects
1979	Monkey[62]	CSF	2-6	9-32	pg/mℓ	Range for 2 animals
		Serum	7-18	8-23	pg/mℓ	Range for 2 animals
1979	Rat[65]	Urine	0.53	1.44	ng/12 hr	
1979	Hamster[68]	Pineal	14	443	pg/gland	Pooled samples
1980	Hamster[69]	Pineal	40-91	847	pg/gland	

[a] This comment applies to all data from Reference 56.

values ranged from 0 to 71.1 ng per gland. A diurnal variation in pineal melatonin content, based on the times of death, was observed; the melatonin concentration increased in the evening and through the night and started to fall in the morning.

2. Gas Chromatography with Mass Spectrometric Detection (GC-MS)

This technique involves the use of a gas chromatograph interfaced with a mass spectrometer. Specificity of the assay is based on both the gas chromatographic retention time of a given compound and the recording of the ion densities generated by specific fragments.

To form derivatives of indoles with appropriate vapor pressure for gas chromatography, a variety of reactions are employed.[26,56,79,82] The method has been used for the quantitation and absolute identification of melatonin in extracts of various tissues[56,83] and body fluids,[56,84,85] and to validate various RIAs for melatonin.[54,56,85] Because this method can identify and measure virtually any compound that can be isolated by gas-

Table 4
MELATONIN LEVELS ESTIMATED BY GAS CHROMATOGRAPHY

Date	Species	Sample	Melatonin level			Comment
			Day	Night	Unit	
			With Electron-Capture Detection			
1970	Rat[79]	Pineal	0.35	—	ng/gland	2 glands/sample
1978	Human[80]	Pineal	ND — 71.1		ng/gland	22 pineals at autopsy[a]
			With Mass Spectrometry			
1972	Rat[81]	Pineal	3.9	—	ng/gland	
1972	Chicken[82]	Serum	ND	71	pg/mℓ	
1973	Rat[83] (intact)	Hypothalamus	348	—	ng/g	
	Rat[83] (pinealectomized)	Hypothalamus	325	—	ng/g	
1977	Sheep[56]	Pineal	40-180		ng/pineal	Correlated with RIA estimates
	Human[84]	Plasma	25-40	125-440	pg/mℓ	Range for 3 subjects
		Plasma	10-55	—	pg/mℓ	Range for 5 subjects
		CSF	55-80	—	pg/mℓ	Range for 2 subjects
1978	Rat[85]	Pineal	0.5	3	ng/gland	
		Serum	20	80	pg/mℓ	
			With Negative-Chemical-Ionization Mass Spectrometry			
1978	Human[87]	Plasma	1.5	42.6	pg/mℓ	
1980	Rat[88]	Plasma	4	52	pg/mℓ	

[a] Diurnal variations in melatonin content were observed, with a minimum around noon.

liquid chromatography, it has produced some valuable results. For example, while using it to validate our RIA for measuring melatonin in rat serum, Wilson et al.[85] demonstrated a rhythmic variation in the concentration of 5-methoxytryptophol in rat serum. Sisak et al.[86] also used this method to quantitate 6-hydroxymelatonin (20 ng/mℓ) in a human urine sample. The general utility of the GC-MS technique for measuring melatonin in body fluids is, unfortunately, limited by the expense of the instrumentation and the need for a highly skilled operator.

3. Gas Chromatography with Negative-Chemical-Ionization Mass Spectrometry

Negative chemical ionization increases GC-MS sensitivity for electron-capturing compounds 150-fold. It permits routine measurement of melatonin in human plasma at concentrations as low as 1 pg/mℓ. Theoretically, GC-MS offers the greatest specificity since quantified substances are simultaneously identified (by retention time and molecular weight of fragment ions). When these methods are used to quantify endogenous compounds in a biological matrix, their applications are limited by problems of increased chemical background and irreversible adsorption phenomena. This latest technique, negative-chemical-ionization mass spectrometry, provides significantly increased sensitivity and specificity, and facilitates measuring melatonin at concentrations present in human plasma.

This analytical technique revealed daytime values as low as 1.5 g/mℓ and nighttime values as high as 42.6 pg/mℓ in plasma from two male and two female subjects.[87] In rats, daytime plasma melatonin levels were found to be 4 pg/mℓ; nighttime levels were 52 pg/mℓ; and, 2½ hr after treatment with 3 mg/kg isoproterenol in the daytime, 30 pg/mℓ.[88]

IV. EFFECTS OF EXOGENOUS MELATONIN ON MAMMALS

A complete review of the evidence that exogenous melatonin affects mammalian reproductive physiology is beyond the scope of this article; this topic has recently been reviewed by other authors.[89-91] Briefly, in 1963, melatonin administration (1 to 10 μg/day) to female rats was observed to diminish ovarian weight and suppress the vaginal estrous cycle;[11] its probable site of action is the neuroendocrine structures in the brain[92,93] (even though some evidence exists that melatonin affects the gonads directly).[94]

The efficacy of a melatonin dose in modifying reproductive function varies markedly depending on the time of its administration (relative to the prevailing light-dark schedule) and on the species studied. The first evidence of this dependence was the demonstration that melatonin administration blocks the pineal serotonin rhythm when given in the 8th hr, but not during the 14th hr of the daily 14-hr photoperiod.[95] It was subsequently shown that large doses of the methoxyindole (1.25 to 5 mg) completely inhibited ovulation if administered to rats during the critical period of proestrus (2 to 4 PM). (This inhibition could be overcome by administration of luteinizing hormone, which suggests that melatonin does not act directly on the ovaries).[96] As described in detail elsewhere in this volume, the reproductive effects of exogenous melatonin are quantitatively far greater in hamsters than in the albino rat. Recent studies have shown that if exogenous melatonin is administered late in the daily photoperiod to intact hamsters (kept under a lighting regimen that maintains their reproductive competence), the gonads regress;[97-99] moreover, if the melatonin is administered three times daily (25-μg injections at 3-hr intervals) to pinealectomized hamsters, the same gonadal regression is seen regardless of the lighting schedule (even constant light).[100,101]

Recent development of a new technique for programmed intermittent or cyclic subcutaneous infusion of melatonin[102] should facilitate further study of the relationship between the time of melatonin administration and its physiologic effects.

The utility of pinealectomy for studying the physiologic effects of melatonin requires that the operation either completely deprive animals of the methoxyindole or at least terminate diurnal and gonadal-dependent rhythms of melatonin secretion. It was believed initially that, among mammalian organs, only the pineal contained HIOMT activity and thus could synthesize melatonin;[12] however, subsequent investigations demonstrated that there is HIOMT activity in the retina[6] and Harderian gland,[7] and the melatonin is synthesized from tryptophan in mammalian retina.[6] The extent to which these organs contribute to circulating melatonin remains controversial; it is well-established that melatonin continues to circulate after pinealectomy, but that its levels do not exhibit normal circadian or diurnal rhythms.[56,83,103] Melatonin has been detected in urines of pinealectomized rats both by bioassay and RIA,[103] in the blood of pinealectomized sheep by RIA,[56] and in hypothalamic tissues of pinealectomized rats by GC-MS.[83] (Other investigators failed to detect melatonin by bioassay[42] or GC-MS[88] in blood of pinealectomized rats; this may reflect differences in the sensitivity of assay methods used). In any event, there is abundant evidence[104,105] that pinealectomy accelerates gonadal maturation and subsequently modifies reproductive functions, and that most of the reproductive effects of pinealectomy can be reversed by exogenous melatonin.

V. LOCUS OF MELATONIN SECRETION

There are two potential routes for melatonin secretion in mammals. Since the pineal gland is an intracranial organ, it could secrete melatonin into the surrounding subarachnoid space; in mammals, such as humans, where it forms part of the roof of the third ventricle, it could also secrete melatonin directly into ventricular CSF. The alternative route of melatonin secretion, of course, would be into the circulation via the rich vascularization of the pineal. Melatonin has been identified and measured in CSF obtained from calves,[72] sheep,[106] rhesus monkeys,[62] and humans.[77,107,108] Its reported CSF concentrations vary substantially among and within species (for a review, see Reppert[62]). In two studies, melatonin concentrations in CSF exceeded those in blood sampled simultaneously: the first involved calves;[72] the other, children undergoing treatment for leukemia.[107] Other studies have consistently shown blood melatonin levels to be greater than those of the CSF. Melatonin was measured in human daytime CSF samples presumably originating in the lumbar sac and basal cisterns. Since no gradient in melatonin concentration was detected, it was inferred that melatonin is not released directly from the pineal into the third ventricle.[77] The most meticulous studies investigating the route of melatonin secretion have been done with sheep[106] and rhesus monkeys.[60-62] The time courses of the appearance and disappearance of endogenous melatonin were monitored, as was the rate at which radioactively tagged melatonin was transferred from one compartment to the other. High-frequency sampling revealed a lag time between increases and decreases in blood melatonin levels and corresponding changes in CSF melatonin levels. It was concluded from these studies that the primary route of secretion from the pineal gland was into the circulation. The tendency of circulating melatonin concentrations to be higher than those in the CSF could, of course, be due partly to the ability of melatonin to bind to plasma albumin.[109]

VI. CLINICAL STUDIES

A. Age and Pubescence

Three studies have addressed the question of changes in melatonin levels with puberty. In the first, plasmas from five male and four female children, six identified as prepubertal and three as pubertal, were sampled at noon and at midnight. Noontime values were all less than 14 pg/mℓ plasma. At midnight, pubertal children had mean melatonin concentrations of 128 pg/mℓ, and prepubertal children had 162 pg/mℓ. In this small sample, no significant differences in melatonin levels were seen that could be related to sex or to stage of genital development.[50]

In another study, luteinizing hormone and melatonin were measured in blood taken from four pubertal boys at 20-min intervals for 24 hr. All four subjects showed a significant augmentation of both luteinizing hormone and melatonin levels during nocturnal sleep; significant correlations were also observed between the luteinizing hormone and melatonin levels. These data were interpreted as indicating that the plasma melatonin concentrations during sleep are insufficient to prevent spontaneous luteinizing-hormone secretion during puberty.[110]

The third study involved 51 healthy boys and girls between the ages of 11½ and 14 years. Melatonin and methoxytryptophol were assayed by GC-MS in blood taken between 1100 and 1300 hr. The methoxytryptophol concentration showed neither a sex distribution nor any change with development. The melatonin concentrations in the girls ranged from less than 5 to 280 pg/mℓ and did not change with development; in the boys it ranged from less than 10 to 2300 pg/mℓ. At genitalia stage 1, the mean value for melatonin was 218 pg/mℓ; at genitalia stages 2, 3, and 4, the mean values were 30, 17, and 18 pg/mℓ, respectively. Serum luteinizing hormone, follicle-stimulating hormone, testosterone, and estradiol, measured by RIA, correlated inversely with melatonin levels.[111]

B. Estrus and Menstrual Rhythms

Several enzymes required for melatonin biosynthesis are influenced by circulating hormones and exhibit rhythmic changes that parallel the estrous cycle in experimental animals.[112,113] Quay,[47] monitoring pineal melatonin content fluorometrically in rats, found a similar parallelism. Ozaki et al.[114] showed that the rate of melatonin excretion among female rats during the 12-hr daily dark period varied synchronously with the vaginal estrous cycle, reaching its nadir in proestrus. Wetterberg and associates[74] measured serum melatonin levels in five healthy women at 2- and 3-day intervals in the early morning, before breakfast: melatonin levels were elevated during menstruation and lowest at the time of ovulation.

VII. PATTERNS OF MELATONIN LEVELS IN MAMMALS

A number of photoperiodic events are controlled somewhat by the pineal gland. Winter regression of the gonads in response to shorter days,[115] the normal initiation and timing of estrus in seasonally breeding animals,[116] and change in pelage, which occurs in some animals,[117] all seem to depend on a functional pineal gland. It is becoming increasingly clear that many of these regulatory functions are mediated by melatonin secretion. Further, it is evident that it is the temporal pattern of circulating melatonin that is critical, rather than the mere presence or absence of melatonin or its absolute levels.

Investigators have sought to document the precise temporal pattern of melatonin levels in various mammalian species (Table 5). The larger mammals lend themselves to this kind of study because longitudinal studies can be made on individual animals by high-frequency sampling. Thus, the most detailed investigations have measured blood and CSF melatonin levels in cattle,[72] sheep,[106] humans,[50,107,108] and rhesus monkeys;[60-62] blood melatonin levels in sheep;[54,75] and blood and urinary melatonin levels in humans.[33-37] Because of the important contributions that experiments with hamsters have already made in the study of the relationships between environmental lighting, pineal function, and reproductive physiology, several groups[66-69,97] have documented the time course of melatonin secretion in hamsters indirectly, by measuring pineal me-

Table 5
CHARACTERISTICS OF DAILY MELATONIN RHYTHM IN VARIOUS MAMMALIAN SPECIES

	Artiodactyla		Primates		Rodentia	
	Cattle	**Sheep**	**Monkeys**	**Humans**	**Hamsters**	**Rats**
Tissues monitored	Blood[56,72] CSF[72]	Blood[54,56,75] CSF[106]	Blood[60-62] CSF[60-62]	CSF[50,77] Pineal[80] Blood Urine[33,64]	Pineal[66-69]	Pineal[64,70,120] Blood[70,64,120] Urine[64,70]
Under diurnal lighting	Elevated throughout dark phase[56,72]	Elevated throughout dark phase[54,56,75]	Elevated throughout dark phase[60-62]	Elevated throughout dark phase[33,63,121]	Elevated at one point in dark phase[66-68]	Elevated at one point in dark phase[70,120]
Under constant darkness		Rhythmic variation persists[54]	Rhythmic variation persists[61]	Rhythmic variation persists[126]		Rhythmic variation persists[64,65,125]
Under constant light		Rhythmic variation suppressed[54]	Rhythmic variation suppressed[61]	Rhythmic variation persists[34,122]	Rhythmic variation suppressed[68]	Rhythmic variation suppressed[65,125]
Light pulse during dark phase		Transiently suppressed[54]	Abruptly suppressed[62,127]	Not suppressed[63]	Abruptly suppressed[68]	Abruptly suppressed[70,119]

latonin content at frequent intervals throughout the 24-hr day. Similar studies have been conducted with the laboratory rat to establish the extent to which daily increases or decreases in pineal melatonin content parallel changes in plasma and urinary melatonin levels,[64,118] and in the activity of melatonin-forming enzymes in the pineal.[119,120]

Unless otherwise indicated, all tissues and fluids examined in the following species exhibit characteristic and similar daily rhythms in melatonin concentrations, with melatonin levels peaking during the daily dark period. The rhythms themselves persist when animals are kept in constant darkness, but disappear in constant light (Table 5).

A. Cattle

Melatonin levels in the blood[56,72] and CSF[72] show a daily rhythm, with a marked rise with the onset of darkness which was sustained throughout the dark period. The average melatonin concentration in CSF was five times higher than that in plasma during the night, but only twice as high during the day.[72] The effect of exposure to constant light or constant darkness was not tested, nor was the effect of imposing light during the normal dark period.

B. Sheep

Plasma melatonin levels are abruptly elevated with the onset of darkness, remain high throughout the dark period, and promptly decrease with the onset of light.[54,56] Melatonin concentrations in blood collected at 1-min intervals increase 5 to 10 min after the abrupt imposition of darkness and reach peak values over a 2- to 10-min interval. When light was resumed, melatonin fell to baseline levels in 5 to 10 min.[54] When sheep were exposed to constant darkness, a circadian rhythm persisted in blood melatonin; however, no rhythm was detectable in sheep exposed to constant light. When light was turned on at a time of high melatonin concentrations, serum melatonin levels dropped precipitously; they returned to peak values when the lights were turned off.[54] In a study done over a period of a year with ewes exposed to seasonal alterations in length of the photoperiod, nocturnal melatonin elevations were longest during estrus (long nights) and shortest during anestrus (short nights). There were no significant differences in mean concentrations of melatonin during the light or dark periods associated with these different photoperiods and reproductive states.[75] In another study, CSF melatonin levels paralleled blood levels during darkness and the quantity of melatonin normally secreted into the blood was estimated to be more than 100 times greater than that secreted in the CSF. The increase in CSF melatonin following the onset of darkness lagged behind that in the blood.[106]

C. Monkey

Parallel increases in plasma and CSF melatonin levels were observed in rhesus monkeys during darkness. More extensive studies with CSF samples showed that the increase in melatonin levels occurs shortly after lights are turned off, that these levels remain high during darkness, and that they decrease soon after the lights are turned on. The magnitude of the rhythm remains relatively stable from day to day in each animal, but shows substantial variation among animals.[60] Constant light suppresses the melatonin rhythms, although sporadic increases of melatonin can still be detected. During exposure to constant darkness, melatonin rhythms persist, but the duration of melatonin elevations increases from 12 hr to 16 to 18 hr.[61]

D. Human

The human species has been studied most extensively. Earliest observations showed that blood melatonin levels are higher during the dark portion of the day than during the light period.[31,32] Melatonin rhythms persist through 2½ days of constant light.

Further, on the basis of occasionally detectable melatonin levels in daytime samples and of irregularities in the pattern of nighttime melatonin levels, the tendency of melatonin to be secreted in bursts was noted.[32] The advent of more sensitive assays (RIA, GC-MS) has permitted short time-interval sampling, and thus a more detailed picture of the diurnal melatonin rhythm in human subjects has been obtained. In the first of such studies, blood melatonin levels rose before the onset of darkness and dropped before the onset of light.[108] This general pattern has been observed in other laboratories, although one group reported that, in some individuals, plasma levels rise following a considerable delay after the onset of darkness.[121] Episodic melatonin secretion was confirmed by high-frequency sampling (2.5- and 20-min), superimposed on a "maintained" baseline concentration;[36,121] secretory episodes also occur during the waking day. These episodic variations are not apparently related to sleep stages.[36]

Urinary melatonin has also been measured in the human. There is a diurnal rhythm in melatonin excretion,[33] which persists in constant light and during sleep deprivation.[34,122] In a timeless environment, when behavioral rhythms are free-running urine melatonin rhthyms free-run and even become dissociated from the free-running sleep-darkness rhythm.[36] After a phase shift in the lighting regimen (experimentally imposed, or after travel over a number of time zones), urinary melatonin rhythms reentrain to the new lighting regimens within a few days.[35,63] In addition to these diurnal variations, monthly variations correlated with the menstrual cycle,[74] and yearly variations have been reported in the morning levels of serum melatonin.[73]

E. Hamster

Melatonin has been measured only in the pineal of the hamster. In Syrian prepubertal and adult hamsters, melatonin contents increase in the latter portion of the dark period, and return to daytime values before or with the onset of light.[67-69] The peak concentrations reached during darkness, and the duration of these high levels (approximately 4 hr) apparently are not altered by the photoperiod; instead, the phase difference between the onset of darkness and the peak in pineal melatonin increases as the dark period is lengthened. Melatonin concentrations remained at daytime values when the light period is extended into the night,[68] while a pulse of light given during the dark period rapidly decreases dark-time levels.[68,69] The halflife of plasma melatonin is approximately 9 min and the total quantity of melatonin synthesized throughout the 24-hr day is estimated to be approximately 18.6 ng.[69]

F. Rat

Melatonin levels in the pineal, plasma, and urine change in parallel throughout the 24-hr period.[64,70] Frequent measurements of blood and pineal melatonin show significant elevations in melatonin levels 3 hr into the dark phase, and abrupt decreases about 3 hr before the onset of light.[120] Experimental imposition of light during the normal dark period causes rapid declines in plasma and pineal melatonin levels, with a halflife of approximately 5 min.[119] Urinary melatonin levels are suppressed in animals exposed to constant light of sufficient intensity, but persist in animals exposed to constant darkness,[65,123] in blinded animals,[64,124] or in animals exposed to constant light of sufficiently low intensity.[125] The urinary melatonin rhythm shifts after a phase shift in the lighting regimen;[64] the rate of shifting appears to depend on the intensity of the light presented during the light portion of the light-dark cycle.[125]

The existing opportunities to determine how much melatonin actually is present in body fluids, and to relate such concentrations to variations in reproductive function, should help to reveal, at long last, the purpose of the pineal.

ACKNOWLEDGMENTS

These studies were supported in part by grants from the National Science Foundation (PCM 77-15700) and the National Institutes of Health (HD 11722).

REFERENCES

1. **Heubner, O.,** Tumor der glandula pinealis, *Dtsch. Med. Wochenschr.*, 24(2), 214, 1898.
2. **Kitay, J. I. and Altschule, M. D.,** *The Pineal Gland, A Review of the Physiologic Literature,* Harvard University Press, Cambridge, 1954.
3. **McCord, C. P. and Allen, F. P.,** Evidences associating pineal gland function with alterations in pigmentation, *J. Exp. Zool.*, 23, 207, 1917.
4. **Lerner, A. B., Case, J. D., and Takahashi, Y.,** Isolation of melatonin and 5-methoxyindole-3-acetic acid from bovine pineal glands, *J. Biol. Chem.*, 235, 1992, 1960.
5. **Axelrod, J. and Weissbach, H.,** Purification and properties of hydroxyindole-*O*-methyltransferase, *J. Biol. Chem.*, 236, 211, 1961.
6. **Cardinali, D. P. and Rosner, J. M.,** Serotonin metabolism by the rat retina in vitro, *J. Neurochem.*, 18, 1769, 1971.
7. **Vlahakes, G. and Wurtman, R. J.,** A Mg_2^+-dependent hydroxyindole-*O*-methyltransferase in rat Harderian gland, *Biochim. Biophys. Acta,* 261, 194, 1972.
8. **Fiske, V. M., Bryant, G. K., and Putnam, J.,** Effect of light on the weight of the pineal in the rat, *Endocrinology,* 66, 489, 1960.
9. **Wurtman, R. J., Roth, W., Altschule, M. D., and Wurtman, R. J.,** Interactions of the pineal and exposure to continuous light on organ weights of female rats, *Acta Endocrinol.*, 36, 617, 1961.
10. **Meyer, C. J., Wurtman, R. J., Altschule, M. D., and Lazo-Wasem, E. A.,** The arrest of prolonged estrus in "middle-aged" rats by pineal gland extract, *Endocrinology,* 68, 795, 1961.
11. **Wurtman, R. J., Axelrod, J., and Chu, E. W.,** Melatonin, a pineal substance: effect on rat ovary, *Science,* 141, 277, 1963.
12. **Axelrod, J. and Weissbach, H.,** Purification and properties of hydroxyindole-*O*-methyltransferase, *J. Biol. Chem.*, 236, 211, 1961.
13. **Wurtman, R. J. and Axelrod, J.,** The pineal gland, *Sci. Am.*, 213, 50, 1965.
14. **Lovenberg, W., Jequier, E., and Sjoerdsma, A.,** Tryptophan hydroxylation in mammalian systems, *Adv. Pharmocol.*, 6A, 21, 1968.
15. **Lovenberg, W., Weissbach, H., and Udenfriend, S.,** Aromatic L-amino acid decarboxylase, *J. Biol. Chem.*, 237, 89, 1962.
16. **Snyder, S. H., Axelrod, J., Wurtman, R. J., and Fisher, J. E.,** Control of 5-hydroxytryptophan decarboxylase activity in the rat pineal gland by sympathetic nerves, *J. Pharmacol. Exp. Ther.*, 147, 371, 1965.
17. **Axelrod, J. and Weissbach, H.,** Enzymatic *O*-methylation of *N*-acetylserotonin to melatonin, *Science,* 131, 1312, 1960.
18. **Klein, D. C. and Weller, J. L.,** Indole metabolism in the pineal gland: a circadian rhythm in *N*-acetyltransferase, *Science,* 169, 1093, 1970.
19. **Wurtman, R. J., Axelrod, J., and Fisher, J. E.,** Melatonin synthesis in the pineal gland: effect of light mediated by the sympathetic nervous system, *Science,* 143, 1329, 1964.
20. **Wilson, B. W.,** The application of mass spectrometry to the study of the pineal gland, *J. Neural Transm.*, Suppl. 13, 279, 1978.
21. **Wurtman, R. J. and Ozaki, Y.,** Physiological control of melatonin synthesis and secretion: mechanisms generating rhythms in melatonin, methoxytryptophol, and arginine vasotocin levels and effects on the pineal of endogenous catecholamines, the estrous cycle, and environmental lighting, *J. Neural Transm.*, Suppl. 13, 59, 1978.
22. **Kopin, I. J., Pare, C. M. B., Axelrod, J., and Weissbach, H.,** The fate of melatonin in animals, *J. Biol. Chem.*, 236, 3072, 1961.
23. **Kveder, S. and McIsaac, W. M.,** The metabolism of melatonin (*N*-acetyl-5-methoxytryptamine and 5-methoxytryptamine), *J. Biol. Chem.*, 236, 3214, 1961.
24. **Hirata, F. and Hayaishi, O.,** In vitro and in vivo formation of two new metabolites of melatonin, *J. Biol. Chem.*, 249, 1311, 1974.

25. **Quay, W. B.** Circadian rhythm in rat pineal serotonin and its modification by estrous cycle and photoperiod, *Gen. Comp. Endocrinol.*, 3, 473, 1963.
26. **Smith, I., Mullen, P. E., Silman, R. E., Snedded, W., and Wilson, B. W.**, Absolute identification of melatonin in human plasma and cerebrospinal fluid, *Nature (London)*, 260, 718, 1976.
27. **Wetterberg, L.**, Melatonin in serum, *Nature (London)*, 269, 646, 1977.
28. **Waring, H.**, *Color Change Mechanisms of Cold-Blooded Vertebrate*, Academic Press, New York, 1963.
29. **Tomatis, M. E. and Orias, R.**, Changes in melatonin concentration in pineal gland in rats exposed to continuous light or darkness, *Acta Physiol. Lat. Am.*, 17, 227, 1967.
30. **Ralph, C. L. and Lynch, H. J.**, A quantitative melatonin bioassay, *Gen. Comp. Endocrinol.*, 15, 334, 1970.
31. **Pelham, R. W., Vaughan, G. M., Sandock, K. L., and Vaughan, M. K.**, Twenty-four-hour cycle of a melatonin-like substance in the plasma of human males, *J. Clin. Endocrinol. Metab.*, 37, 341, 1973.
32. **Vaughan, G. M., Pelham, R. W., Pang, S. F., Loughlin, L. L., Wilson, K. M., Sandock, K. L., Vaughan, M. K., and Koslow, S. H.**, Nocturnal elevation of plasma melatonin and urinary 5-hydroxyindoleacetic acid in young men: attempts at modification by brief changes in environmental lighting and sleep and by autonomic drugs, *J. Clin. Endocrinol. Metab.*, 42, 752, 1976.
33. **Lynch, H. J., Wurtman, R. J., Moskowitz, M. A., Archer, M. C., and Ho, M. H.**, Daily rhythm in human urinary melatonin, *Science*, 17, 169, 1975.
34. **Jimerson, D. C., Lynch, H. J., Post, R. M., Wurtman, R. J., and Bunney, W. E.**, Urinary melatonin rhythms during sleep deprivation in depressed patients and normals, *Life Sci.*, 23, 1501, 1977.
35. **Lynch, H. J., Jimerson, D. C., Ozaki, Y., Post, R. M., Bunney, W. E., and Wurtman, R. J.**, Entrainment of rhythmic melatonin secretion in man to a 12-hour phase shift in the light/dark cycle, *Life Sci.*, 23, 1557, 1978.
36. **Weitzman, E. D., Weinberg, U., D'eletto, R., Lynch, H., Wurtman, R. J., Czeisler, C., and Erlich, S.**, Studies of the 24-hour rhythm of melatonin in man, *J. Neural Transm.*, Suppl. 13, 325, 1978.
37. **Lynch, H. J., Ozaki, Y., Shakal, D., and Wurtman, R. J.**, Melatonin excretion of man and rats: effect of time of day, sleep, pinealectomy and food consumption, *Int. J. Biometerorol.*, 19, 267, 1975.
38. **Prop, N. and Kappers, J. A.**, Demonstration of some compounds present in the pineal organ of the rat by histochemical methods and paper chromatography, *Acta Anat.*, 45, 90, 1961.
39. **Barchas, J. D. and Lerner, A. B.**, Localization of melatonin in the nervous system, *J. Neurochem.*, 11, 489, 1964.
40. **Lynch, H. J.**, Diurnal oscillations in pineal melatonin content, *Life Sci.*, 10, 791, 1971.
41. **Pelham, R. W.**, A serum melatonin rhythm in chickens and its abolition by pinealectomy, *Endocrinology*, 96, 543, 1975.
42. **Pang, S. F. and Ralph, C. L.**, Pineal and serum melatonin at midday and midnight following pinealectomy or castration in male rats, *J. Exp. Zool.*, 193, 275, 1975.
43. **Arendt, J., Paunier, L., and Sizonenko, P. C.**, Melatonin radioimmunoassay, *J. Clin. Endocrinol. Metab.*, 40, 347, 1975.
44. **Udenfriend, S., Bogdanski, D. F., and Weissbach, H.**, Fluorescence characteristics of 5-hydroxytryptamine (serotonin), *Science*, 122, 972, 1955.
45. **Giarman, N. J. and Day, M.**, Presence of biogenic amines in the bovine pineal body, *Biochem. Pharmacol.*, 1, 235, 1959.
46. **Quay, W. B.**, Differential extraction for the spectrophotofluorometric measurement of diverse 5-hydroxy- and 5-methoxy-indoles, *Anal. Biochem.*, 5, 51, 1963.
47. **Quay, W. B.**, Circadian and estrous rhythms in pineal melatonin and 5-hydroxyindole-3-acetic acid, *Proc. Soc. Exp. Biol. Med.*, 115, 710, 1964.
48. **Maickel, R. P. and Miller, F. P.**, The fluorometric determination of indole-alkylamines in brain and pineal gland, *Adv. Pharmacol.*, 6, 71, 1968.
49. **Miller, F. P. and Maickel, R. P.**, Fluorometric determination of indole derivatives, *Life Sci.*, 9, 747, 1970.
50. **Arendt, J.**, Melatonin assays in body fluids, *J. Neural Transm.*, Suppl. 13, 265, 1978.
51. **Grota, L. J. and Brown, G. M.**, Antibodies to indolealkylamines: serotonin and melatonin, *Can. J. Biochem.*, 52, 196, 1974.
52. **Levine, L. and Riceberg, L. J.**, Radioimmunoassay for melatonin, *Res. Commun. Chem. Pathol. Pharmacol.*, 10, 693, 1975.
53. **Pang, S. F., Brown, G. M., Grota, L. J., and Rodman, R. L.**, Radioimmunoassay of melatonin in pineal glands, Harderian glands, retinas, and sera of rats or chickens, *Fed. Proc. Fed. Am. Soc. Exp. Biol.*, 35, 691, 1976.
54. **Rollag, M. D. and Niswender, G. D.**, Radioimmunoassay of serum concentrations of melatonin in sheep exposed to different lighting regimens, *Endocrinology*, 98, 482, 1976.

55. **Wurtzburger, R. J., Kawashima, K., Miller, R. L., and Spector, S.,** Determination of rat pineal gland melatonin content by radioimmunoassay, *Life Sci.,* 18, 867, 1976.
56. **Kennaway, D. J., Frith, R. G., Phillipou, G., Matthews, C. D., and Seamark, R. F.,** A specific radioimmunoassay for melatonin in biological tissue and fluids and its validation by gas chromatography-mass spectrometry, *Endocrinology,* 101, 119, 1977.
57. **Lynch, H. J., Ozaki, Y., and Wurtman, R. J.,** The measurement of melatonin in mammalian tissues, *J. Neural Transm.,* Suppl. 13, 251, 1978.
58. **Thoma, J. A.,** A quantitative approach to unidimensional multiple chromatography, *J. Chromatog.* 12, 441, 1963.
59. **Brenner, M. and Niederwieser, A.,** Durchlaufende dunnschich-chromatographie, *Experientia,* 17, 237, 1961.
60. **Reppert, S. M., Perlow, M. J., Tamarkin, L., and Klein, D. C.,** A diurnal melatonin rhythm in primate cerebrospinal fluid, *Endocrinology,* 104, 295, 1979.
61. **Perlow, M. J., Reppert, S. M., Tamarkin, L., Wyatt, R. J., and Klein, D. C.,** Photic regulation of the melatonin rhythm: monkey and man are not the same, *Brain Res.,* 182, 211, 1980.
62. **Reppert, S. M., Perlow, J. J., and Klein, D. C.,** CSF melatonin, in *Neurobiology of Cerebrospinal Fluid,* Wood, J. H., Ed., Plenum Press, New York, 1980, 579.
63. **Wetterberg, L.,** Melatonin in humans: physiological and clinical studies, *J. Neural Transm.,* Suppl. 13, 289, 1978.
64. **Adler, J., Lynch, H. J., and Wurtman, R. J.,** Effect of cyclic changes in environmental lighting and ambient temperature on the daily rhythm in melatonin excretion by rats, *Brain Res.,* 163, 111, 1979.
65. **Lynch, H. J. and Wurtman, R. J.,** Control of rhythms in the secretion of pineal hormones in humans and experimental animals, in *Biological Rhythms and Their Central Mechanism,* Suda, M., Hayaisha, O., and Nakagawa, H., Eds., Elsevier/North Holland, Amsterdam, 1979, 117.
66. **Panke, E. S., Reiter, R. J., Rollag, M. D., and Panke, T. W.,** Pineal serotonin *N*-acetyltransferase activity and melatonin concentrations in prepubertal and adult Syrian hamsters exposed to short daily photo-periods, *Endoc. Res. Commun.,* 5, 311, 1978.
67. **Panke, E. S., Rollag, M. D., and Reiter, R. J.,** Pineal melatonin concentrations in the Syrian hamster, *Endocrinology,* 104, 194, 1979.
68. **Tamarkin, L., Reppert, S. M., and Klein, D. C.,** Regulation of pineal melatonin in the Syrian hamster, *Endocrinology,* 104, 385, 1979.
69. **Rollag, M. D., Panke, E. S., Trakulrungsi, W., Trakulrungsi, C., and Reiter, R. J.,** Quantification of daily melatonin synthesis in the hamster pineal gland, *Endocrinology,* 106, 231, 1980.
70. **Ozaki, Y., Lynch, H. J., and Wurtman, R. J.,** Melatonin in rat pineal, plasma, and urine: 24-hour rhythmicity and effect of chlorpromazine, *Endocrinology,* 98, 1418, 1976.
71. **Smith, J. A., Padwick, D., Mee, T. J. X., Minneman, K. P., and Bird, E. D.,** Synchronous nyctochemeral rhythms in human blood melatonin and in human post-mortem pineal enzyme, *Clin. Endocrinol.,* 6, 219, 1977.
72. **Hedlund, L., Lischko, M. M., Rollag, M. D., and Niswender, G. D.,** Melatonin: daily cycle in plasma and cerebrospinal fluid of calves, *Science,* 195, 686, 1977.
73. **Arendt, J., Wirz-Justice, A., and Bradtke, J.,** Annual rhythm of serum melatonin in man, *Neurosci. Lett.,* 7, 327, 1977.
74. **Wetterberg, L., Arendt, J., Paunier, L., Sizonenko, P. C., van Donselaar, W., and Heyden, T.,** Human serum melatonin changes during the menstrual cycle, *J. Clin. Endocrinol. Metab.,* 42, 185, 1976.
75. **Rollag, M. D., O'Gallaghan, P. L., and Niswender, G. D.,** Serum melatonin concentrations during different stages of the reproductive cycle in ewes, *Biol. Reprod.,* 18, 279, 1978.
76. **Smith, J. A., Barnes, J. L., and Mee, T. J.,** The effect of neuroleptic drugs on serum and cerebrospinal fluid melatonin concentrations in psychiatric subjects, *J. Pharm. Pharmacol.,* 31, 246, 1978.
77. **Brown, G. M., Young, S. N., Gauthier, S., Tsui, H., and Grota, L. J.,** Melatonin in human cerebrospinal fluid in daytime: its origin and variation with age, *Life Sci.,* 25, 929, 1979.
78. **Mendlewicz, J., Linkowski, P., Branchey, L., Weinberg, U., Weitzman, E. D., and Branchey, M.,** Abnormal 24 hour pattern of melatonin secretion in depression, *Lancet,* 2, 1362, 1979.
79. **Degen, P. H. and Barchas, J. D.,** Gas chromatographic assay for melatonin, *Proc. West. Pharmacol. Soc.,* 13, 34, 1970.
80. **Greiner, A. C. and Chan, S. C.,** Melatonin content of the human pineal gland, *Science,* 199, 83, 1978.
81. **Catabeni, F., Koslow, S. H., and Costa, E.,** Gas chromatographic-mass spectrometric assay of four indole alkylamines, *Science,* 178, 166, 1972.
82. **Pelham, R. W., Ralph, C. L., and Campbell, I. M.,** Mass spectral identification of melatonin in blood, *Biochem. Biophys. Res. Commun.,* 46, 1236, 1972.
83. **Koslow, S. H. and Green A. R.,** Analysis of pineal and brain indole alkylamines by gas chromatography-mass spectrometry, *Adv. Biochem. Psychopharmacol.,* 7, 33, 1973.

84. **Wilson, B. W., Snedden, W., Silman, R. E., Smith, I., and Mullen, P.**, A gas chromatography-mass spectrometry method for the quantitative analysis of melatonin in plasma and cerebrospinal fluid, *Anal. Biochem.*, 81, 283, 1977.
85. **Wilson, B. W., Lynch, H. J., and Ozaki, Y.**, 5-Methoxytryptophol in rat serum and pineal: detection, quantitation and evidence for daily rhythmicity, *Life Sci.*, 23, 1019, 1978.
86. **Sisak, M. E., Markey, S. P., Colburn, R. W., Zavadil, A. P., and Kopin, I. J.**, Identification of 6-hydroxymelatonin in normal human urine by gas chromatography-mass spectrometry, *Life Sci.*, 25, 803, 1979.
87. **Lewy, A. J. and Markey, S. P.**, Analysis of melatonin in human plasma by gas chromatography: negative chemical ionization mass spectrometry, *Science*, 201, 741, 1978.
88. **Lewy, A. J., Tetsuo, M., Markey, S. P., Goodwin, J. K., and Kopin, I. J.**, Pinealectomy abolishes plasma melatonin in the rat, *J. Clin. Endocrinol. Metab.*, 50, 204, 1980.
89. **Minneman, K. P. and Wurtman, R. J.**, Effects of pineal compounds on mammals, *Life Sci.*, 17, 1189, 1975.
90. **Minneman, K. P. and Wurtman, R. J.**, The pharmacology of the pineal gland, *Ann. Rev. Pharmacol.*, 16, 33, 1976.
91. **Reiter, R. J., Rollag, M. D., Panke, E. S., and Bank, A. F.**, Melatonin: reproductive effects, *J. Neural Transm.*, Suppl. 13, 209, 1978.
92. **Fraschini, F., Mess, B., Piva, F., and Martini, L.**, Brain receptors sensitive to indole compounds: function in control of LH secretion, *Science*, 159, 1104, 1968.
93. **Anton-Tay, F., Chou, C., Anton, S., and Wurtman, R. J.**, Brain serotonin concentration: elevation following intraperitoneal administration of melatonin, *Science*, 162, 277, 1968.
94. **Ellis, L. C.**, Inhibition of rat testicular androgen synthesis in vitro by melatonin and serotonin, *Endocrinology*, 90, 17, 1972.
95. **Fiske, V. M. and Huppert, L. C.**, Melatonin action on pineal varies with photoperiod, *Science*, 162, 279, 1968.
96. **Ying, S. Y. and Greep, R. O.**, Inhibition of ovulation by melatonin in the cyclic rat, *Endocrinology*, 92, 333, 1973.
97. **Tamarkin, L., Brown, S., and Goldman, B.**, Neuroendocrine regulation of seasonal reproductive cycles in the hamster, *Neurosci. Abstr.*, 1, 458, 1975.
98. **Tamarkin, L. T. and Goldman, B.**, Effects of melatonin on the reproductive system in intact and pinealectomized hamsters, *J. Neural Transm.*, Suppl. 13, 398, 1978.
99. **Tamarkin, L., Westrom, W. K., Hamill, A. I., and Goldman, B. D.**, Effect of melatonin on the reproductive systems of male and female Syrian hamsters: a diurnal rhythm in sensitivity to melatonin, *Endocrinology*, 99, 1534, 1976.
100. **Tamarkin, L. and Goldman, B.**, Effects of melatonin on the reproductive system in intact and pinealectomized hamsters, *J. Neural Transm.*, Suppl. 13, 398, 1978.
101. **Goldman, B., Hall, V., Hollister, C., Roychoudhury, P., Tamarkin, L., and Westrom, W.**, Effects of melatonin on the reproductive system in intact and pinealectomized male hamsters maintained under various photoperiods, *Endocrinology*, 104, 82, 1979.
102. **Lynch, H. J., Rivest, R. W., and Wurtman, R. J.**, Artificial induction of melatonin rhythms by programmed microinfusion, *Neuroendocrinology*, 31, 106, 1980.
103. **Ozaki, Y. and Lynch, H. J.**, Presence of melatonin in plasma and urine of pinealectomized rats, *Endocrinology*, 99, 641, 1976.
104. **Wurtman, R. J., Axelrod, J., and Kelly, D. E.**, *The Pineal*, Academic Press, New York, 1968.
105. **Reiter, R. J., Vaughan, M. K., Vaughan, G. M., Sorrentino, S., and Donofrio, R. J.**, The pineal gland as an organ of internal secretion, in *Frontiers of Pineal Physiology*, Altschule, M. D., Ed., MIT Press, Cambridge, 1975, 54.
106. **Rollag, M. D., Morgan, R. J., and Niswender, G. D.**, Route of melatonin secretion in sheep, *Endocrinology*, 102, 1, 1978.
107. **Smith, J. A., Mee, T. J. X., Barnes, N. D., Thornburn, R. J., and Barnes, J. L. C.**, Melatonin in serum cerebrospinal fluid, *Lancet*, 2, 425, 1976.
108. **Arendt, J., Wetterberg, L., Heyden, T., Sizonenko, P. C., and Paunier, L.**, Radioimmunoassay of melatonin: human serum and cerebrospinal fluid, *Hormone Res.*, 8, 65, 1977.
109. **Cardinali, D. P., Lynch, H. J., and Wurtman, R. J.**, Binding of melatonin to human and rat plasma proteins, *Endocrinology*, 91, 1213, 1972.
110. **Fevre, M., Segel, T., Marks, J. F., and Boyar, R. M.**, LH and melatonin secretion patterns in pubertal boys, *J. Clin. Endocrinol. Metab.*, 47, 1383, 1978.
111. **Silman, R. E., Leone, R. M., Hooper, R. J. L., and Preece, M. A.**, Melatonin, the pineal gland and human puberty, *Nature (London)*, 282, 301, 1979.
112. **Wurtman, R. J., Axelrod, J., Snyder, S. H., and Chu, E. W.**, Change in the enzymatic synthesis of melatonin in the pineal during the estrous cycle, *Endocrinology*, 76, 798, 1965.

113. **Wallen, E. P. and Yochim, J. M.,** Pineal hydroxyindole-*O*-methyltransferase (HIOMT) and reproductive cyclicity in the rat, *Fed. Proc. Fed. Am. Soc. Exp. Biol.,* 30, 610, 1971.
114. **Ozaki, Y., Wurtman, R. J., Alonso, R., and Lynch, H. J.,** Melatonin secretion decreases during the proestrous stage of the rat estrous cycle, *Proc. Natl. Acad. Sci. U.S.A.,* 75, 531, 1978.
115. **Reiter, R. J. and Hester, R. J.,** Interrelationships of the pineal gland, the superior cervical ganglia and the photoperiod in the regulation of the endocrine systems of hamsters, *Endocrinology,* 79, 1168, 1966.
116. **Herbert, J.,** The pineal gland and light-induced oestrus in ferret, *J. Endocrinol.,* 43, 625, 1969.
117. **Rust, C. C. and Meyer, R. K.,** Hair color, molt, and testis size in male, short-tailed weasels treated with melatonin, *Science,* 165, 921, 1961.
118. **Lynch, H. J. and Wurtman, R. J.,** Control of rhythms in the secretion of pineal hormones in humans and experimental animals, in *Biological Rhythms and Their Central Mechanism,* Elsevier/North Holland, Amsterdam, 1979, 117.
119. **Illnerova, H., Vanecek, J., Krecek, J., Wetterberg, L., and Saaf, J.,** Effect of one minute exposure to light at night on rat pineal serotonin *N*-acetyltransferase and melatonin, *J. Neurochem.,* 32, 673, 1978.
120. **Wilkinson, M., Arendt, J., Brodtke, J., and de Ziegler, D.,** Determination of a dark-induced increase of pineal *N*-acetyltransferase activity and simultaneous radioimmoassay of melatonin in pineal, serum, and pituitary tissue of the male rat, *J. Endocrinol.,* 72, 243, 1977.
121. **Weinberg, U., D'eletto, R. D., Weitzman, E. D., Erligh, S., and Hollander, C.,** Circulating melatonin in man: episodic secretion throughout the light-dark cycle, *J. Clin. Endocrinol. Metab.,* 48, 114, 1979.
122. **Akerstedt, T., Froberg, J. E., Friberg, Y., and Wetterberg, L.,** Melatonin excretion, body temperature and subjective arousal during 64 hours of sleep deprivation, *Psychoneuroendocrinology,* 4, 219, 1979.
123. **Ralph, C. L., Mull, D., and Lynch, H. J.,** Locomotor activity rhythms of rats under constant conditions as predictors of melatonin content of their pineals (abstr.), *Am. Zool.,* 10, 302, 1970.
124. **Reiter, R. J., Sorrentino, S., Jr., Ralph, C. L., Lynch, H. J., Mull, D., and Jarrow, E.,** Some endocrine effects of blinding and anosmia in adult male rats with observations on pineal melatonin, *Endocrinology,* 88, 895, 1971.
125. **Rivest, R. W., Lynch, H. J., Ronsheim, P. M., and Wurtman, R. J.,** Effect of Light Intensity on Regulation of Melatonin Secretion and Drinking Behavior in the Albino Rat, *Int. Symp. on Melatonin, Bremen, Fed. Rep. Germany, 1980.*
126. **Vaughan, G. M., McDonald, S. D., Jordan, R. M., Allen, J. P., Bell, R., and Stevens, E. A.,** Melatonin, pituitary function and stress in humans, *Psychoneuroendocrinology,* 4, 351, 1979.
127. **Reppert, S. M., Perlow, M. J., Tamarkin, L., and Klein, D. C.,** Photic regulation of the melatonin rhythm: a distinct difference between man and monkey, *Pediatr. Res.,* 13, 362, 1979.

Chapter 6

ARGININE VASOTOCIN AND VERTEBRATE REPRODUCTION

Mary K. Vaughan

Table of Contents

I. INTRODUCTION

Adaptation to the unmitigating perils of an ever-changing climatic and predatory environment appears to have been the ultimate key to survival of extant vertebrate species. A singularly potent mechanism of coping with these pressures would include the haphazard mutation of the species-specific genomic lifethread with an accompanying natural selection of the fittest reproductively capable members. Alternatively, another tack has been the modification of the organismal response to previously existing endogenous stimuli. A noteworthy example which combines the classical elements of both approaches are the hormones of the neurohypophysis which perform dissimilar homeostatic functions throughout vertebrate phylogeny but yet are surmised to be derived from one ancestral template, arginine vasotocin (AVT).

Hans Heller[1] first recognized that pituitary glands from birds, reptiles, amphibians, and bony fishes contained a unique neurohypophyseal water balance principle. This finding provided the scientific community with an excellent opportunity to view functional aspects of adaptive radiation in the preservation of vertebrate homeostasis. As mammals evolved with their complicated antidiuretic, galactobolic, and uterotonic requirements, neurohypophyseal hormones with single amino acid substitutions (vasopressin and oxytocin) were found to be more effective than the all-purpose posterior pituitary messenger of older times, AVT (Table 1). For many years, comparative neuroendocrinologists considered AVT to be entirely banished from the brain of mammals. Recent development of highly specific and selective techniques have brought a resurgence of interest in this peptide with the discovery of AVT or an AVT-like peptide in the pineal gland of mammals. Is it possible that a phylogenetically ancient hormone, displaced from its normal site of secretion in mammals, has always encamped in certain circumventricular organs (pineal and subcommissural organ) unbeknownst to neuroendocrinologists? The ensuing discourse will elaborate briefly upon the changing roles that this peptide serves in reproductive competence of lower vertebrates and will concentrate on the evidence of its possible presence and evanescent role in mammals.

II. ARGININE VASOTOCIN AND REPRODUCTION IN LOWER VERTEBRATES

A. Class Agnatha

The simple features of the hagfish (Myxinformes) neurohypophysis ranks it as the most primitive which exists today;[2] the posterior pituitary of the lamprey (Petromyzoniformes) is a close second. Numerous investigators have demonstrated antidiuretic, oxytocic, and frog-water balance activities consistent with the presence of AVT in the Pacific hagfish (*Polistrotrema stoutii*)[3] and in several lampreys including the sea lamprey (*Petromyzon marinus*),[4-7] the river lamprey (*Lampetra fluviatilis*),[8-10] and the Western Brook lamprey (*Lampetra richardsoni*).[11] Immunocytochemical evidence suggests that AVT is the only neuropeptide found in the preoptico-hypophyseal neurosecretory system of adult migrating *L. fluviatilis*.[10]

The river lamprey undergoes an annual spawning migration from the sea to fresh water.[11,12] While a direct action of AVT on reproductive competence in this species has not been investigated, this nonapeptide may play an integral role in the mobilization of fat during the prolonged voluntary fast from the sea to the spawning grounds. During migration, the animals abstain from food; only sequestered fat is mobilized for energy.[12] This voluntary abstinence is similar to that observed in other migrating vertebrates such as birds and whales. Vasotocin injected into migrating river lampreys has an adipokinetic effect of increasing the concentration of blood glucose and muscle glycogen.[12] Bentley and Follett[12] speculated that the glucose and glycogen-sparing ac-

Table 1
STRUCTURES OF NATURALLY OCCURING NEUROHYPOPHYSEAL PEPTIDES AND THEIR PHYLETIC DISTRIBUTION

Peptide	Basic Structure: Cys-Tyr-X-Y-Asn-Cys-Pro 1 2 3 4 5 6 7	Z - Gly (NH_2) 8 9	Phyletic distribution
Arginine vasotocin	Ile Gln	Arg	Cyclostomes, actinopterygians, lungfishes, rays, sharks, holocephalans, amphibians, reptiles, birds, & mammals
Isotocin	Ile Ser	Ile	Actinopterygians
Mesotocin	Ile Gln	Ile	Lungfishes, amphibians, reptiles, & birds
Glumitocin	Ile Ser	Gln	Rays
Aspartocin	Ile Asn	Leu	Sharks
Valitocin	Ile Gln	Val	Sharks
Oxytocin	Ile Gln	Leu	Mammals
Arginine vasopressin	Phe Gln	Arg	Mammals
Lysine vasopressin	Phe Gln	Lys	Mammals (pigs)

tions of AVT might be a result of peripheral vasoconstriction or of mobilizing free fatty acids which are metabolized instead of glucose. Support for this latter hypothesis has been observed in the migrating anadromous sea lamprey,[13] the Coho salmon,[14] and the pigeon.[15] In these species, AVT elevated plasma-free fatty acids possibly through actions on other lipolytic hormones such as glucagon, adrenocorticotrophic hormone (both of which lampreys apparently lack), and/or growth hormone. Vasotocin injections elevate growth hormone levels in Coho salmon[14] as well as in rats;[16] thus, the glucogenic and lipolytic effect of vasotocin may be mediated via growth hormone in some species. Clearly these studies are still in the neonatal stage and await additional evidence before a definite role for AVT in the migratory response of these species can be affirmed or denied.

B. Class Chondrichthyes

Among the cartilaginous fishes, a variety of neurohypophyseal principles appear to be present. Sawyer[17] listed AVT as a possible hormone in two Selachian elasmobranchs (*Mustelus asterias* and *Squalus acanthias*), in one holocephalan (*Hydrolagus colliei*), and in two batoids (*Raia calvata* and *Raia batis*); the neurohormone was absent from several other species.

In the chondricthyian species tested to date, AVT was not a potent inducer of oviduct contraction. Isolated oviducts of sexually immature,[18] pregnant[18-20] and nonpregnant spiny dogfish[19,20] (*S. acanthias*), gravid gray smoothhounds (*Mustelus californicus*),[18] and sexually immature and gravid leopard sharks (*Trakis semifasicata*)[18] failed to respond to AVT. Only isolated oviducts from postpartum leopard sharks clearly responded to low concentrations of neuropeptides. While these results are not overwhelming, LaPoint[18] reasons that the oviduct does appear to manifest a greater sensitivity at parturition.

C. Class Osteichthyes

The initial identification of AVT as a natural neurohypophyseal principle in ray-finned fishes was made by Heller and Pickering[21,22] from large amounts of pollack (*Gadus virens*) pituitary powder collected by Dr. Grace Pickford. This neuropeptide has been chemically identified in pituitary extracts from other actinopterygians including the carp (*Cyprinus carpio*), whiting-pout (*Gadus luscus*), Boston hake[17] (*Urophycis tenuis*), and in the three living genera of sarcopterygian lungfishes.[23] An immunocytochemical study of the neurohypophyseal hormone-producing system of the lungfish (*Protopterus aethiopicus*)[24] and several other teleosts[25] demonstrated two types of neurosecretory cells which give rise to separate vasotocinergic and either mesotocinergic or isotocinergic axons. An immunoperoxidase technique has been employed for the localization and identification of AVT and its neurophysin in a stenohaline freshwater teleost, the white sucker (*Catostomus commersoni*),[26] and in the platyfish (*Xiphorphorus maculatus*).[27]

The caudal neurosecretory system of fishes and its neurohemal organ in teleosts, the urophysis, contain four neurosecretory factors identified as urotensin I-IV. The corresponding binding proteins (urophysins) have also been demonstrated.[28,29] Urotensin IV has hydrosmotic activity and chromatographic properties that are indistinguishable from AVT in the mudsucker (*Gillichthys mirabilis*),[30,31] milkfish (*Chanos chanos*), and rainbow trout (*Salmo gairdneri*).[32] The urophysis of silver eels (*Anguilla anguilla*) and several salt-water fishes[33] also contains radioimmunoassayable levels of AVT. Separate laboratories, however, have failed to visualize this peptide in the trout and white sucker urophysis using immunocytochemical techniques.[33,34] As will become increasingly apparent throughout this discourse, this ability to localize the peptide by histochemical methods is a recurring theme in all classes in which it has been investigated.

Besides the neurohypophysis and urophysis already mentioned, radioimmunoassay and bioassay evidence argue for the presence of AVT in the pineal gland of several fresh-water fish (Tables 2 and 3). AVT could not be demonstrated via immunofluorescent techniques.[33] Several possibilities were suggested by Holder[33] and his group to account for this curious discrepancy: (1) the fixation procedure used may have blocked the immunoreactive sites, or (2) the total amount of AVT present in the pineal (30 to 60 pg per pineal) was too weak to give any immunological reaction; (3) the lack of immunofluorescence reaction product might be related to a lack of storage granular vesicles. Finally, this group could not completely discount the possibility that the pineal and/or urophyseal substance was not AVT but rather a closely related peptide with the same biological activity and affinity for antibodies directed against AVT or arginine vasopressin (AVP).

Injections of neurohypophyseal hormone preparations evoke vigorous spawning reflex responses in individual killifish *Fundulus heteroclitus* regardless of sex or hypophysectomy.[35,36] Curiously, however, there is no blatant attempt to coordinate sexual responses when males and females were placed together unless the normal spawning season was emminent.[35,36] The normal spawning attitude of cyprinodonts includes an unmistakable S-shaped flexure of the body with quivering and flattening of anal and dorsal fins; this behavior locks the two fish together during a period of rapid mutual vibration during which the ova and sperm are emitted simultaneously.[36]

Synthetic AVT (47 to 62.5 ng/g) was extremely effective in eliciting spawning behavior in the fish; however, the effective dose is probably pharmacological and there is a long lag period of 15 to 30 min before the dose of neuropeptide becomes manifest. A tenfold larger dose of isotocin (480 to 490 ng/g) was required to elicit equivalent results; AVP was as effective as AVT. This latter observation prompted Pickford and Strecker[37] to speculate that vasopressor activity might be the possible common denominator which precipitated mating activity. One potent argument against the pressor hypothesis is that other teleostean vasoactive agents such as catecholamines and angiotensin II are completely ineffective in eliciting spawning behavior. A direct action of AVT at the level of the oviduct seems unlikely since its walls contain mostly connective tissue and elastic fibers with few, if any, smooth muscle fibers; oviposition mainly results from contractions of the striated muscular wall of the abdomen.[38]

The nucleus preopticus (NPO) has been proposed as the principal neural site of action of AVT in generating the spawning reflex. Electrolytic lesions that destroyed a major part of this nucleus impaired or abolished this reflex after intraperitoneal injection of synthetic AVT or hog pituitary suspension.[39] One difficulty with this schema was posed by Macey and coworkers;[39] the NPO neurosecretory cells would fulfill dual, but, seemingly, mutually exclusive functions. (1) The NPO cells serve as the source of the hormone that elicits spawning behavior and, incongruously, (2) as the center which detects and triggers the behavioral effect. If, however, three morphologically and electrophysiologically distinct populations of cells coexist within the magnocellular part of the nucleus (as has been suggested by Hayward[40] in the NPO of the goldfish), then one might postulate that AVT acts as an intranuclear messenger (neurotransmitter?) to stimulate spawning. Alternatively, AVT may be released at the normal spawning time of the year from another site of synthesis and secretion (such as the urophysis and/or pineal) to act at the NPO.

D. Class Amphibia

All amphibian species tested appear to possess a neurohypophyseal peptide with biological and chromatographic characteristics consistent with those of AVT. Sawyer[17] listed two urodeles (*Necturus maculosus* and *Triturus alpestris*) as well as several anurans (*Rana catesbiana, Rana temporaria, Rana esculenta, Rana pipiens, Bufo americanus, Bufo bufo,* and *Xenopus laevis*) which had the amphibian water balance principle.

Oviducts from many oviparous anuran (*R. catebiana, B. bufo, B. marinus,* and *X. laevis*) and urodele (*N. maculosus, Triturus cristatus,* and *Salamandra atra*) species respond vigorously and rapidly to all neurohypophyseal hormones tested; oviducts, however, were usually most sensitive to AVT.[19,41,42] The oviducts of several species revealed a marked seasonal variation in sensitivity to AVT stimulation.[18]

The fortuitous acquisition of limbs and the subsequent migration to land did not totally divorce the members of the new class Amphibia from the exigencies of reproduction in an aquatic environment. In an interesting set of experiments, Diakow[43] demonstrated that breeding behavior in the leopard frog (*R. pipiens*) may be induced by the amphibian water balance principle, AVT. Unreceptive female frogs repetitively emit a release call when mounted by fertile males. Inhibition of the release call in gravid frogs facilitates maintenance of the male's clasp so that oviposition and spawning can occur. Diakow[43] clearly showed that injections of AVT inhibit the release call in otherwise unreceptive females (Figure 1) presumably by causing an accumulation of water and abdominal distention.

E. Class Reptilia

Chromatographic and pharmacological evidence indicates that AVT is a neurohypophyseal hormone in the three major groups of living reptiles: *Chelonia, Squamata,* and *Crocodilia.*[17] Additionally, radioimmunoassay and bioassay data suggest the presence of this neurohormone in the pineal gland of two reptiles (Tables 2 and 3), a snake (*Natrix natrix* L.) and a tortoise (*Testudo hermanni* Gmel.). Interestingly, in this latter species, a marked seasonal difference in pineal content of AVT was noted (May values — 2100 pg per gland; January values — 322 pg per gland.

Neurohypophyseal lobe extracts and hormones can induce parturition and oviposition in gravid viviparous colubrid snakes,[45] ovoviparous lacertid lizards,[46] oviparous iquanid lizards[47] and in 36 oviparous turtle species.[48] In an in vitro assay system, AVT was ten times more potent than oxytocin in causing oviductal contractions in the turtle (*Chrysemys picta*) and the viviparous xantusiid lizard (*Klauberina riversiana*).[49,50] Guillette[51] examined the effects of various doses of AVT on parturition in a viviparous

Table 2
RADIOIMMUNOASSAY LEVELS OF ARGININE VASOTOCIN (AVT), ARGININE VASOPRESSIN (AVP), AND OXYTOCIN (OT) IN THE PINEAL GLAND OF VERTEBRATE SPECIES

Species	AVT pg/gland[a]	AVP pg/gland	OT pg/gland	Time of day	Ref.
Freshwater fish					
Salmo gairdneri irideus Gibbons	65.6				33
Salmo gairdneri Richardson	23.9				33
Salmo trutta morpha fario	43.5				33
Salvelinus fontinalis Mitchill	36.1				33
Anguilla anguilla (Linnaeus)	56.1				33
Reptiles					
Testudo hermanni (Gmelin)	2100 (May)				33
Testudo hermanni (Gmelin)	322 (January)				33
Birds					
Chicken	312 ± 39	ND[b]	ND	Morning	53
Chicken[d]	0.148[c]	0.042			54
Mammals					
Mouse	7.1 ± 0.9			Morning	54
Hamster	7.1 + 0.7			Morning	54
Guinea pig	7.6 ± 0.7			Morning	54
Gerbil	6.5 + 0.5			Morning	54
Cow (pg/gland)	258 ± 29	407 ± 66	238 ± 41	1200-1600	54
Cow (μU/mg)	8.9	12.1			111
Rat	ND	3.7 ± 0.7	3.7 ± 1.7		100
Sprague-Dawley (infant)[d]	ND	0.42			54
Sprague-Dawley (adult)[d]	ND	0.186			54
Sprague-Dawley (adult)	9.0 ± 0.45	ND	ND	Morning	53
Wistar (adult)	5.1 ± 0.3			Morning	53
Wistar (adult)	ND	2.7 ± 0.8	1.3 ± 0.8		100

Brattleboro (adult)	ND	ND	8.7 ± 3.7	100
Brattleboro (adult) (μU/gland)	<2.5	<3.0		111
Long-Evans (adult)	129 ± 32	<19		111
Human (fetal, 126-350 days) (ng/mg)	1.1 − 15.8			92
Rabbit (New Zealand, adult)[d]	ND	0.032		54

[a] except where noted
[b] ND = Nondectable levels
[c] value obtained by subtraction: AVP − (AVT + AVP) = AVT
[d] pg/μg protein

Table 3
BIOASSAY LEVELS OF ARGININE VASOTOCIN (AVT), ARGININE VASOPRESSIN (AVP), AND OXYTOCIN (OT) IN THE PINEAL GLAND OF VERTEBRATE SPECIES

Species	Rat ADH	Eel ventral aorta pg/ gland	Hydroosmotic	Ref.
Freshwater fish				
Salmo gairdneri irideus Gibbons		61.4		33
Salmo gairdneri Richardson		27.0		33
Salmo trutta morpha fario		44.7		33
Salvelinus fontinalis Mitchill		36.1		33
Anguilla anguilla (Linnaeus)		56.1		33
Reptiles				
Testudo hermanni (Gmelin)		2248 (May)		33
Natrix natrix		0.001-0.01		33
Mammals				
Rat				
Rat Fetal Wistar rat	160 ± 23		26,820 ± 1,530	86
Fetal Brattleboro rat	2 ±0.5		413 ± 87	109
Fetal Long-Evans rat	19 ± 3		3,591 ± 936	109
Newborn Wistar rat	35 ± 4		6,156 ± 260	86
Adult Long-Evans rat	3.5 ± 1		748 ± 153	109
Adult Brattleboro rat	0.3 ± 0.1		71 ± 13	109
Adult Wistar rat	5 ± 2		890 ± 120	86
Adult Wistar rat (noon sample)	6 ± 1		996 ± 130	119
Adult Wistar rat (midnight sample)	25 ± 3		2,621 ± 598	119
Adult hypox Wistar (noon)	7 ± 1		1,176 ± 165	119
Adult hypox Wistar (midnight)	3 ± 0.5		510 ± 80	119
Cat				
Cat (urethane anesthetized)	55 ± 5		8,415 ± 1077	126,130
Cat (male, urethane anesthetized)	55 ± 5		8,635 ± 1120	127

lizard, *Sceloporous jarrovi.* Both doses of AVT (0.02 μg and 2 μg) induced parturition in all experimental gravid lizards. No differences were observed between the two AVT treatment groups on several dependent parameters relating to parturition including initial contraction latency, parturition latency, and interbirth duration. Maximal sensitivity of the isolated oviduct occurred near the preferred body temperature (34°) in this species.[18] One interesting observation made by LaPoint[18] was that the evolution of viviparity per se was not coincident with the evolution of oxytocin.

Heller[19] suggested that the physiological state of the reptilian oviduct determines its sensitivity to neurohypophyseal hormones. In the turtle (*C. picta*), the amplitude of AVT-induced oviductal contractions was enhanced by estradiol whereas progesterone reduced both the duration and rest period between contractions.[52] Steroids had no effect, however, on AVT-induced oviductal contractions in the viviparous lizard, *K. riversiana.*[18]

F. Class Aves

Pharmacological evidence indicates a rather wide phyletic distribution of AVT in the neurohypophyses of birds although relatively few species have actually been examined. Using various techniques, the posterior lobes of the duck, domestic varieties of fowl, turkey and pigeon, parakeet, and the white-crowned sparrow have been shown

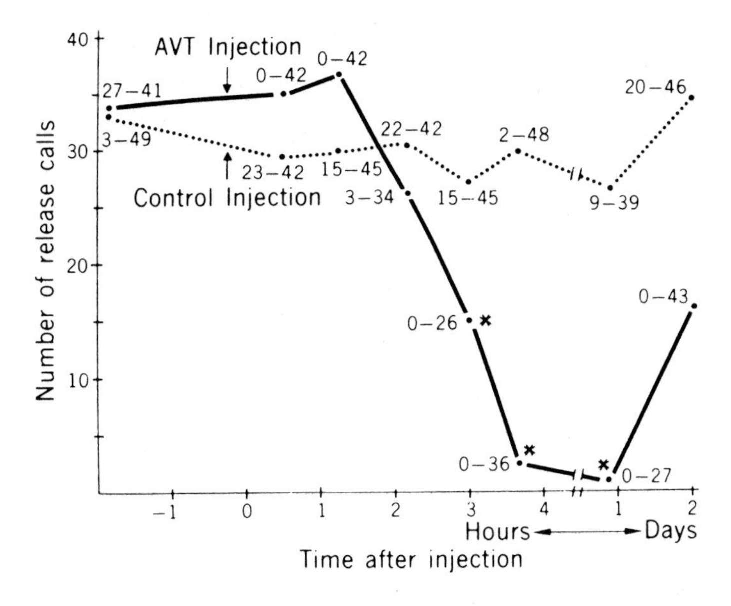

FIGURE 1. Median number of release calls in female *Rana pipiens* after 30 sec of manual stimulation. The numbers are ranges. X indicates that the value for the female injected with arginine vasotocin (AVT) is significantly lower than that for the controls (Mann-Whitney U test, two-tailed; $p < 0.05$). (From Diakow, C., *Science*, 199, 1456, 1978. Copyright 1978 by the American Association for the Advancement of Science.)

to contain AVT. Recently, Fernstrom and co-workers[53] reported that the chicken pineal gland contained no oxytocin or AVP but did have 300 pg per gland of AVT-like immunoreactivity. Since plasma AVT in the hydrated chicken is approximately 5 pg/mℓ, these authors concluded that the AVT present in the epiphysis was not due to blood trapped in the pineal vascular space. Not clearly supporting the above data, Negro-Vilar[54] and co-workers were only able to detect very low levels of AVT in the chicken pineal.

Injections of mammalian posterior lobe extracts[54] or AVT[49,55] stimulates uterine smooth muscle contractions and induces expulsion of the egg to the exterior in the act of oviposition. Sensitivity of the oviduct to AVT rises towards the time of normal oviposition. Of the four parts of the hen oviduct (magnum, isthmus, uterus, vagina), the uterus showed the greatest sensitivity to inactivate this neuropeptide.[55,56] In the hen, Munsick[49] showed that the injection of 120 ng AVT resulted in premature oviposition in 90 sec. In anesthetized hens, intravenous injections of vasotocin and oxytocin also increased intrauterine pressure.[57] Tanaka and Nakajo[58,59] observed a depletion in neurohypophyseal vasotocin-like activity coincident with egg-laying; a concomitant rise in the amphibian water-balance principle was seen in the blood.[60]

Both prostaglandin E_1 (PGE_1) and AVT induce contractions of the uterine portion of the hen oviduct in vitro and provoke premature oviposition in vivo.[61] Indomethacin, an inhibitor of prostaglandin biosynthesis, inhibited AVT-induced contractions of hen uterine strips in vitro. This observation prompted Rzasa[61] to speculate that AVT induces uterine contractions via a prostaglandin mediated mechanism.

Hypophysectomy in the hen delays oviposition; eventually, though, birds once again become capable of laying. Opel[62] demonstrated that AVT concentrations in the blood increase 30 to 50 times in neural lobectomized hens at the time of laying; simultaneously, the stalk median eminence content of AVT decreases. Thus, sufficient quantities of AVT necessary for oviposition may be secreted from the stalk of the severed neurohypophysis, from the hypothalamus, or perhaps from an alternate glandular source such as the pineal.

Sexual behavioral activity had been elicited following neurohypophyseal hormone administration in fish, amphibians and birds. Injections of AVT or oxytocin (in doses that induce oviposition in hens) arouse sexual behavior in cocks (*Gallus domesticus*) and the male pigeon (*Columba livia* Gmel.).[63] The male sexual activities registered during the 30-min observation period included: waltzing, comb-grabbing, mounting, and copulation of sexually receptive hens.

Both cholinergic and adrenergic neurotransmitters may be involved in the release of AVT from the neurohypophysis of birds.[64,65] Norepinephrine, regitine (an α-adrenergic blocking agent), and atropine (an anticholinergic agent) decrease blood levels of AVT in hens subjected to an intravenous injection of hypertonic NaCl. Conversely, acetylcholine increased blood titers of AVT. Niezgoda[64] suggested that acetylcholine may be involved in the release mechanism for AVT and a monoaminergic mediator may inhibit release.

III. ARGININE VASOTOCIN AND REPRODUCTION IN MAMMALS

A. Arguments For and Against its Presence in the Pineal and Fetal Neurohypophysis

The evolutionary diversity of neurohypophyseal hormones was not manifested immediately upon discovery of vasopressin and oxytocin. A foreshadowing presage to the disparity in posterior pituitary neurohormones was revealed in 1941 when Heller demonstrated that neurohypophyseal extracts from nonmammalian vertebrates contained a unique "frog water balance principle" not found in mammalian extracts. Munsick[49] later observed that a chicken posterior lobe extract affected the hen uterus and rat oviduct assay as would be predicted if the active substance contained the ring of oxytocin and the side chain of vasopressin. Curiously, such a compound had been fortuitously synthesized as an analogue for the mammalian neurohormones two years earlier by Katsoyannis and Du Vigneaud;[66,67] the hybrid peptide was named arginine vasotocin.

The comprehensive review of the pineal literature prior to 1954 by Kitay and Altschule[68] revealed that in many cases administration of crude pineal extract was associated with changes in blood pressure. These authors concluded, however, that many of the cardiovascular and respiratory effects were nonspecific and consistent with the results of injection of many different types of crude biological preparations. In 1963, Milcu and co-workers[69] employed a procedure for the extraction of protopituitarin from neurohypophyses to 250 g of an acetone — dried powder of bovine pineal glands. These investigators discovered a pineal peptide with biological and chromatographic characteristics similar to AVT using five bioassay systems: rat pressor, rat oxytocic (0 Mg), rat oxytocic (0.5m*M*/L*Mg*), hen oxytocic, and frog bladder (*Rana esculenta*) tests. Two years later, Pavel[70] detected a peptide with the ring of oxytocin and a substitution in the 8-position other than arginine in the pineal gland of the pig. The peptide was tentatively identified as lysine vasotocin (LVT) based on its rat pressor and rat oxytocic activities; it, too, had been synthesized in the early 1960s.[71,72]

Confirmatory evidence for the presence of AVT in the pineal from other laboratories did not appear until 1970 when Cheesman and Fariss[73] reported the isolation and characterization of a gonadotrophin-inhibiting substance from the bovine pineal gland. Structural elucidation of the active peptide was determined by degrading the molecule into two tetrapeptides which were then examined by mass spectroscopy.[74] Besides the mammalian pineal gland, AVT has been tentatively identified by bioassay in the neurohypophyses of fetal sheep and seals,[75] in human CSF,[76-80] and in the bovine subcommissural organ.[81] Radioimmunoassay evidence argues for its presence in the neural lobe of fetal sheep and humans,[82,83] and in the subcommissural organ of rabbits in addition to the epiphysis.[84]

Some discrepancy concerning absolute levels of AVT in the human fetal neurohypophysis exists in literature. Pavel[85] has reported biologic activity for AVT and AVP in 20 to 22-week human fetal gland over 50-fold higher than that published by Skowsky and Fisher.[82] Methodological variables may have contributed to the reported differences; these include (1) acetic acid extraction[85] vs acetone drying[82] and (2) bioassay[85] vs radioimmunoassay[82] detection techniques. Skowsky and Fisher[82] speculated that fetal neurohypophyseal vasotocin which is present in early fetal life decreases with gestational maturation. This hypothesis was seemingly borne out by observations of Pavel[86] and co-workers in the pineal gland of the rat. The AVT-biological activities of rat pineal extracts markedly decreased from fetus to adult (Table 3). Similarly, the subcommissural organ of young rabbits (3500 μU/mg) had significantly greater quantities of AVT than did mature rabbits (100 μU/mg).[81]

Neurohypophyseal hormones are noncovalently bound to specific low molecular weight cysteine-rich carrier proteins called neurophysins. Neurohypophyseal neurophysins are located in the paraventricular and supraoptic nuclei where AVP and oxytocin are synthesized and are stored in the same subcellular organelles in the posterior pituitary gland. Neurophysin-like proteins have been extracted and purified from the bovine[87,88] and human[89,90] pineal gland and are released into the culture media from incubated pineal ependymal cells from rat fetuses.[91] Legros[92] and co-workers confirmed the presence of both neurophysins and AVT in the fetal human epiphysis; AVT, however, was unbound to intact neurophysins. The pineal tissue utilized was obtained 30 min or more after death; thus, rapid enzymatic degradation of bound peptide may have already occurred.

Histological evidence for the presence of neurosecretory material in the pineal was first advanced by Bargmann[93] who described beaded stained fibers in the pineal of the hedgehog; these findings were independently substantiated by other investigators in that species and in the horseshoe bat.[94-96] The identity of the neuropeptide in the pineal and subcommissural organ has sparked an unabated controversy. In 1976, Bowie and Herbert[97] presented immunocytochemical evidence for the presence of AVT in the rat pineal gland. Antiserum to AVT stained a distinct population of polygonal-shaped cells. The presumptive AVT-containing cells had extensive perivascular and intercellular processes whose cytoplasmic endings were frequently cupped around neighboring cells. Pévet and co-workers[98] dispute the abovementioned radioimmunoassay and immunocytochemical evidence for the presence of AVT on two premises: (1) antibodies to AVT do not stain the nonapeptide but rather some similar AVT-like peptide in the pineal and (2) vasopressin and oxytocin-containing fibers present in the pineal and subcommissural organ probably crossreact with antisera of other investigators to give false-positive results for AVT. To substantiate their hypothesis, this European group has shown that immunocytochemistry of the rat pineal gland gave positive results with only one of five AVT antisera;[98] the same cells which reacted to the AVT-antiserum also stained with anti-MSH,[99] anti-LRH,[98] and antibody raised against the sheep antigonadotrophic pineal UMO5R fraction.[98] Using a radioimmunoassay system, this group was unable to detect AVT in the bovine, hamster, mouse, ovine, rabbit, and rat pineal or subcommissural organ[98] nor in the midterm fetal sheep or near-term fetal seal neurohypophysis.[100] In the rabbit subcommissural organ, an apparent AVT-value of 3.2 ± 0.9 pg/mg wet tissue was attributed to a crossreaction with AVP (3346.3 ± 522.0 pg/mg wet tissue) and oxytocin (1077.4 ± 236.8 pg/mg wet tissue) which were present in high concentration.[101] Further, Buijs and Pévet[102] have described AVP and oxytocin-containing extrahypothalamic fibers in the pineal (especially in the pineal stalk and in the anterior portion of the gland); vasopressin, oxytocin, and two neurophysins were subsequently detected and quantitated in the bovine epiphysis.[103] In addition to the European group, a recent paper by Negro-Vilar[54] failed to detect AVT under various physiological conditions in the infantile and adult rat and adult rabbit pineals. Unlike the assay of Fernstrom which directly measured AVT by a specific antibody, neither of the antibodies used by Negro-Vilar specifically measured only AVT.

Albeit, the cogent arguments of the European group as well as Negro-Vilar et al.[54] against the radioimmunoassay and immunocytochemical evidence for AVT seriously militate against the presence of the neuropeptide in the pineal, they have yet to thoroughly discount the equally potent bioassay evidence of Pavel and his group or the mass spectroscopic data. Recently, another Roumanian group announced the tentative identification of AVT from a bovine pineal extract prepared by the Milcu-Nanu method. The active fraction was tested in three bioassay systems (frog bladder, rat antidiuretic, and rat uterine assays) and the ratio of these activities were characteristic for AVT as was the electrophoretic mobility of the preparation. A further complicating factor in the controversy is that the histological techniques may not be as sensitive as the radioimmunoassay or bioassay methods as evidenced by the paper of Holder and co-workers[103] in lower vertebrates. In those species tested, radioimmunoassay and bioassay (eel ventral aorta) were in excellent agreement concerning the presence of AVT in the pineal; however, no immunoreactive response could be detected in the pineal gland of fish nor of reptiles even when pineal glands were embedded, sliced, and treated together with neurointermediate lobe or preoptic nuclei which reacted positively and showed a strong fluorescence. Application of new detection and quantification techniques, reanalysis by mass spectrometry, and/or sequencing of AVT or the AVT-like peptide may eventually reconcile the controversial question of the presence in the mammalian epiphysis of nonapeptide.

One final hypothesis with some precedence in the posterior lobe concerns a storage and/or prohormone form of the peptide which might differ by a few amino acids from the humoral form. There is now strong evidence that in the bovine posterior lobe there are two neurohypophyseal intermediates (prohormones?) of arginine vasopressin (Ala-Gly and Val-Asp-[Arg^8]-vasopressin.)[105] Current radioimmunoassay techniques are at present incapable of distinguishing between synthetic arginine vasopressin and either of these two compounds. Depending upon the antigenic and biological properties of the compound, a comparable situation, if it exists in the pineal, may allow false positive results with one or more of the assays or histochemical methods used to detect "AVT". Very preliminary evidence that such a "prohormone" might exist is inferred from the results of Neacsu.[106] He isolated a peptide (E_5) from bovine pineal glands which had 14 amino acids: the 9 amino acids of AVT plus methionine, valine, phenylalanine, alanine, and thyroxine (sic); the peptide was not sequenced unfortunately. E_5 had biological properties of AVT;[106] its antigenic properties are unknown. Whether this peptide is a prohormone form of AVT or is itself the hormone reacting in the bioassay system of Pavel or in the current radioimmunoassays remains to be elucidated.

B. Hypotheses Concerning Synthesis and Secretion from the Pineal

The question of synthesis and secretion of AVT from the pineal becomes a moot point if, indeed, AVT is not present in the mammalian pineal gland as has been discussed. However, several publications by Pavel document the biosynthesis and secretion of a basic peptide in the rat[107-109] and human fetal[110] pineal gland; the peptide had bioassay and chromatographic characteristics indistinguishable from synthetic AVT. Further, the susceptibility of the peptide to tryptic digestion, to specific oxidative inactivation by tyrosinase, and to reductive inactivation by sodium thioglycollate indicated that the AVT-like peptide had at least one basic amino acid, a tyrosine residue and a disulfide bond; these characteristics are consistent with the specific pharmacological profile of AVT.[110] The total amount of peptide released into the culture media during 38 days of incubation from human fetal pineal glands contained about ten times more bioassayable activity than did nonincubated fetal glands of comparably aged fetuses.[110] Since the pineal gland at this age is composed almost entirely of ependymal cells, Pavel and co-workers concluded that this was the cell of origin of the peptide. Cultured pineal ependymal cells from normal rat fetuses release 40 times more AVT over a 43-day period than did nonincubated rat pineal glands.[107] Pineal cells from fetal homozygous Brattleboro rats, which have a genetic AVP insufficiency, are capable of synthesizing AVT; as adults, however, only a very small amount (0.3μU per gland)[107] or no AVT[109] is detectable in the epiphysis.

A recent publication by Pavel and co-workers[108] pursued the question of vasotocin biosynthesis by the adult rat pineal. In these experiments, the pineal glands were removed immediately following decapitation and dissected into two halves: an apical half containing the pineal recess (ependymal cells) and a basal half containing most of the pinealocytes. Both halves of the pineal contained and released into the culture medium a basic peptide not differentiable from synthetic AVT. However, the amount of AVT contained in the nonincubated basal pineal halves was about seven times greater than contained in the nonincubated apical halves; similar results were seen in the bovine and pig (where lysine vasotocin was measured) pineal glands. This observation reflects a reversal in theory by Pavel. His original supposition contended that most of the AVT was found in the apical portion of the gland; no AVT was found in parenchymal portion of the gland. Since degradative changes in the basal portion occur at death, his new theory advocates that the apical pineal halves (ependymal portion) synthesize and release an amount of peptide AVT which is 14 times greater than that found in the nonincubated halves; the basal halves store AVT (Figure 2, #8).

The confusing question of neurotransmitter regulation of AVT release has been approached by scientists working in both the lower vertebrate and mammalian fields. As discussed earlier in Section II. F there is some evidence that acetylcholine may be intimately involved in the release of AVT from the neurohypophysis of hens;[64] a monoaminergic mediator may inhibit release. Sartin[112] and co-workers surveyed the effects of neurotransmitters on AVT release from rat pineal glands in vitro. Rat pineals were incubated with $10^{-8}M$ or $10^{-6}M$ norepinephrine, dopamine, serotonin, and acetylcholine. Only acetylcholine significantly elevated radioimmunoassayable levels of AVT in vitro (Figure 3); AVP was not detected in the incubation media. Corroborating the results of Niezgoda[64] at the mammalian level, Sartin concluded that a cholinergic transmitter may play a role in the regulation of AVT secretion (Figure 2, #9). A few reports have suggested that the pineal may possess some mechanism for cholinergic stimulation. Both the rabbit and the rat[114] pineal glands may have some parasympathetic innervation. Romijn[113] described two types of postganglionic nerve endings in the rabbit which were characterized by the presence of small dense-cored vesicles or small clear vesicles. Pharmacological and cytochemical experiments showed them to be noradrenergic and cholinergic, respectively; both types were often seen in the same nerve bundle. An indirect indicator of acetylcholine activity is the presence of significant acetylcholinesterase localized in small granular vesicles of sympathetic axons.[115,116]

In an in vivo test model, Cusack[117] and co-workers administered norepinephrine (2.5 mg/kg) or isoproterenol (2.0 mg/kg) subcutaneously to intact adult male rats; AVT secretion was significantly elevated by both compounds. These investigators conjectured that the increased sympathetic release of norepinephrine, which normally stimulates melatonin secretion, also elicited the release of AVT. Unfortunately, acetylcholine was not tried. Interestingly, separate synapses on the vasopressin neurosecretory

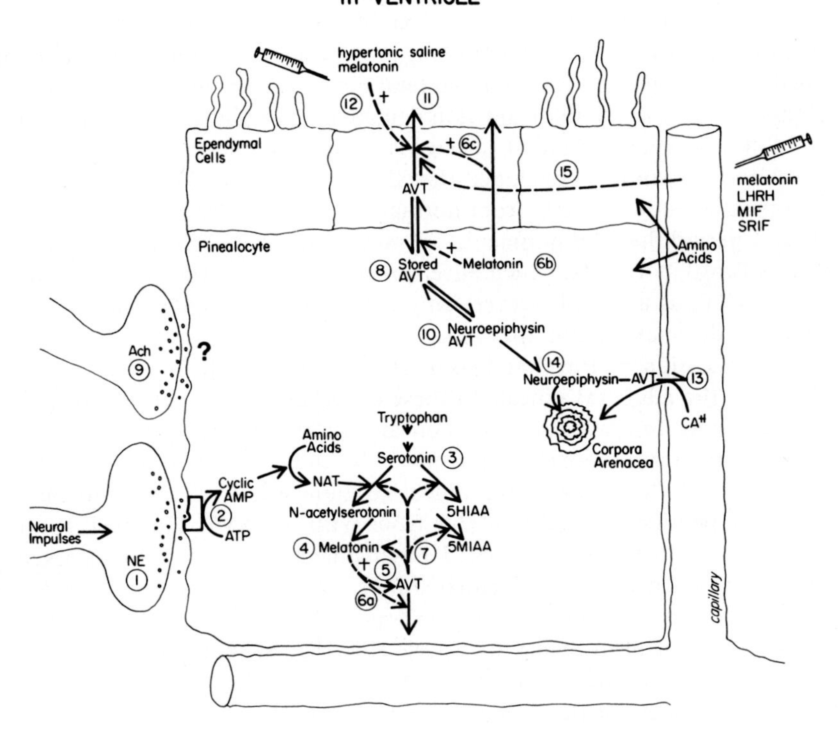

FIGURE 2. Hypotheses concerning the synthesis and regulation of secretion of arginine vasotocin (AVT) from the pineal gland. (1) Norepinephrine (NE) release from sympathetic nerve endings stimulates β-adrenergic receptors on pinealocytes and (2) stimulates the production of intracellular cyclic AMP.[134-136] The amino acid tryptophan is converted to serotonin in several steps. Serotonin (3) in turn, is converted to a family of indolic compounds including *N*-acetylserotonin, melatonin (4), 5-hydroxyindole acetic acid (5HIAA), and 5-methoxyindole acetic acid (5MIAA). Melatonin purportedly releases AVT (6a,b,c);[130,131] in turn, AVT inhibits various aspects of indole biosynthesis (5,7).[137] Melatonin may directly promote release of AVT in the pineal vascular system (6a) or, alternatively, may act intracellularly (6b) to affect conversion of stored AVT (8)[108] to a readily releasable form. The releasable pool of AVT may then be liberated into the CSF (11)[130,131] or possibly released into the blood in exchange for Ca + + (10,13,14).[133] Some evidence exists[112] that a cholinergic stimulus (9) may also affect AVT release. Intracarotid injections (15) of LHRH, TRH, somatostatin (SRIF), MIF, and melatonin (12)[130] or hypertonic saline (12)[126] release AVT. The NE and isoproterenol stimulation of AVT secretion[117] in vivo may be a result of stimulation of melatonin biosynthesis or a direct action of NE on AVT release (not shown).

cell of both cholinergic muscarinic and α-adrenergic origin have been shown to exist;[118] thus, the data of Cusak and Sartin may not necessarily be in conflict.

A diurnal rhythm of vasotocin in the rat pineal has been demonstrated by Calb and co-workers.[119] Pineal content of AVT was high at noon and low at midnight (LD 12:12; lights on 0600 hr) in both normal and hypophysectomized rats. Rats exposed to constant light had low levels of AVT while constant darkness for 24 hr increased pineal nonapeptide levels. These results of Calb et al.[119] however, may not reflect long term changes in peptide concentration following prolonged alteration in photoperiod as illustrated by Sartin and collaborators.[120] In that experiment, blinded rats were sacrificed every 2 weeks following surgical enucleation; pineal AVT content was measured by radioimmunoassay. While gonadotrophins increased, pineal AVT content decreased ($p < 0.05$) to its lowest level at week 8. The authors conjectured that blinding

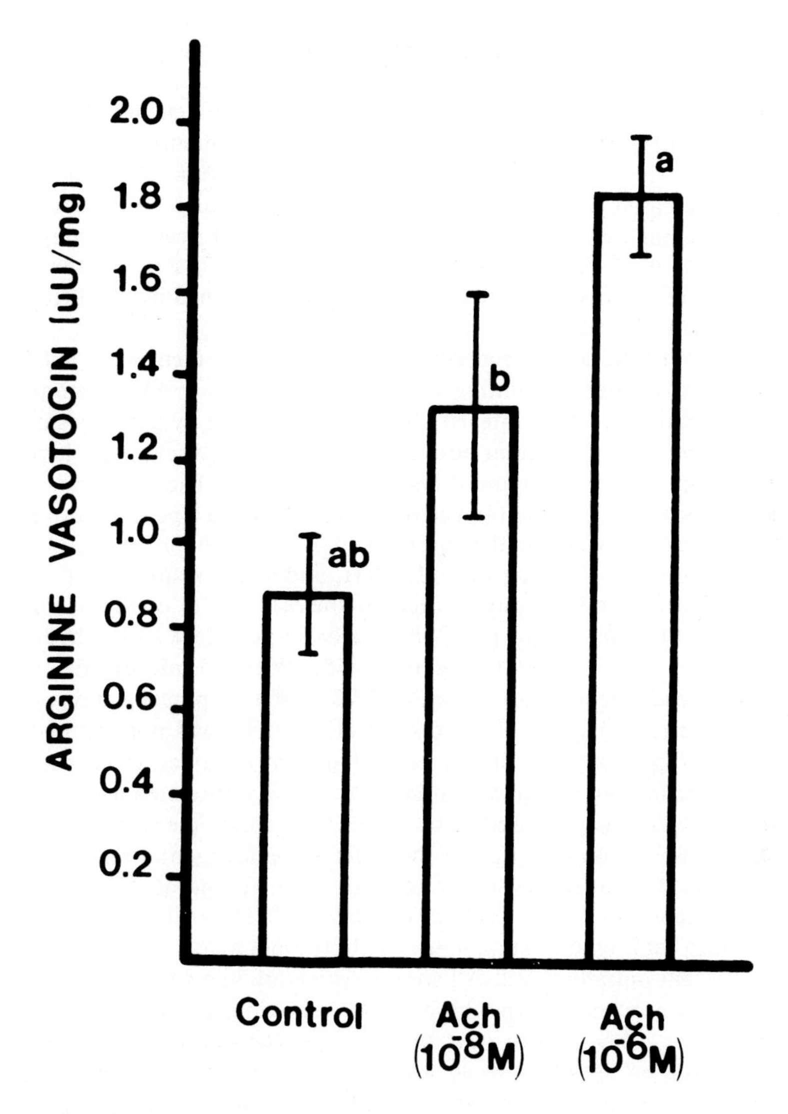

FIGURE 3. The effect of acetylcholine (Ach) on arginine vasotocin secretion from rat pineal glands in vitro. Like letters indicate statistical differences: a = $p < 0.05$; b = $p < 0.08$; n = 6 pineals for controls and for each treatment group. Incubation time was 8 hr. (From Sartin, J. L., Brout, B. C., and Orts, R. J., *Acta Endocrinol.*, 91, 571, 1979. With permission.)

reduced pineal AVT probably because of increased secretion; however, turnover studies are necessary to verify this hypothesis. One further note of caution concerns many rhythms of blinded animals which tend to free-run. Thus, if the AVT rhythm free-runs in the blinded rat, then Sartin may have fortuitously sampled a group of six rats with "normally" low levels of AVT.

Recent studies by Popoviciu[78] and Pavel[79] maintain that the release of AVT into the CSF of man is uniquivocably dependent upon rapid eye movement (REM) sleep. The release is not induced by darkness since CSF removed after a REM episode during daylight hours had high levels of AVT. Narcoleptic subjects had CSF-AVT levels during REM sleep about twice as high as healthy subjects. Subcutaneous injections of synthetic AVT in human subjects prolonged the REM sleep period[121] but showed no indication of altering latency to slow wave sleep or REM sleep as previously reported in cats.[118] Oneiric activity in humans after peptide injection was not obviously altered. In the cat, 600 molecules of AVT is an extremely potent sleep inducing agent presumably through its selective action on serotonin-containing neurons. In the rat, the administration of 50 μg/kg AVT markedly reduced REM latency without affecting sleep latency, total sleep, or total amount of REM in the first 2 hrs. However, the observation of decreased percentage of REM sleep in the second 2 hr recording period in the rat[123] was reminiscent of the decreased period of REM sleep following intraperitoneal injection of AVT during a 5-hr recording in the cat.[122] Perhaps the difference in potency of the administered peptide can be partially explained by the different route of administration (intraventricular[122] vs intraperitoneal).[123]

Besides the cholinergic and adrenergic neurotransmitters mentioned above, Pavel has shown that many other stimulating agents deplete or release AVT from the pineal. Maintenance of osmoregulatory integrity following challenge by a hyperosmotic stimulus is under hormonal influence in both lower vertebrates (where AVT is one of the principle neurohypophyseal hormones) and mammals.[124,125] Thus, it is not surprising that an intraventricular injection of hypertonic saline depleted a peptide with frog bladder activity identical to AVT from the pineal gland of the cat (Figure 2, #12).[126]

Intracarotid injections of 0.1 μg LHRH, TRH, and somatostatin (SRIF),[127] melanocyte inhibiting factor (MIF-I), and purified bovine MIF[128] into urethane-anesthetized cats also induced the release into the CSF of a basic peptide with the hydroosmotic and antidiuretic properties of AVT (Figure 2, #15). Within 5 min of injection of the above mentioned neuropeptides, the store of AVT in the cat pineal gland was reduced by 40%; 60 min after injection, the hydroosmotic activity was not restored whereas the antidiuretic activity was three times greater than control values. Pavel and co-workers[127,128] speculated that an additional antidiuretic activity unrelated to AVT was present; the other peptide was proposed to be AVP. To substantiate this hypothesis, they cite a previous observation that during the night the pineal gland releases AVT and stores AVP released from the posterior pituitary; the storage of AVP by the pineal was completely abolished by hypophysectomy indicating its neurohypophyseal origin.[119] The idea that hypothalamic-releasing hormones affect AVT release is not as remote as one might initially think. All stimuli which are known to release AVT have a concomitant effect on neurohypophyseal AVP secretion. LHRH, TRH, and SRIF have been shown by other investigators to release AVP from the neurohypophysis in vitro.[129] Pulse doses of SRIF produced AVP rises above baseline levels within 2 min from superfused rat pituitaries. The rises observed after a pulse dose of LHRH, and TRH were phenomenal: 10 μg LHRH — 704.2 ± 82.6%, 50 μg LHRH — 953.6 ± 118.5% rise, and 100 μg — 1897.2 ± 17.2% rise. Skowsky and Swan[129] aver that SRIF, TRH, and LHRH stimulate both the magnitude and frequency of in vitro AVP secretion and may augment physiologic in vivo neurohypophyseal AVP release.

Despite the sizable number of hormonal and exogenous stimuli which purportedly release AVT from the pineal, Pavel and colleagues[130,131] espouse the view that the physiologic-releasing factor for this neuropeptide is the pineal indole melatonin. Intraventricular[130] (Figure 2, #12) or intracarotid[131] (Figure 2, #15) injections of melatonin into urethane-anesthetized cats significantly decreased pineal AVT content and increased its level in CSF (Figure 2, #11) between 5 and 60 min after injection; plasma levels did not become detectable until 60 min. Intraventricular injection was more than 50 times as effective as intravenous injection.[130] A diurnal sensitivity to melatonin was also observed; melatonin was 500 times more potent in inducing AVT release during the night in the dark than during daylight hours. Since melatonin synthesis occurs predominantly during the scotophase, AVT release would necessarily be tied to melatonin synthesis. As with LHRH, SRIF, and TRH, melatonin does not selectively release AVT. Recently, Lemay and co-workers[132] demonstrated that melatonin stimulated vasopressin release in a dose-dependent manner (1×10^{-8} to $1 \times 10^{-3} M$ from perifused pooled rat neural lobes. Omission of Ca^{++} combined to an excess of Mg^{++} prevented the stimulatory effect suggesting a calcium dependent mode of secretion.

Extrapolating back to Pavel's previous papers, several hypotheses can be envisioned concerning the mechanisms involved in melatonin-AVT interaction at the pineal level. Endogenous melatonin within the pinealocyte (Figure 2, #6b) could diffuse across the ependymal cell to stimulate AVT release into the CSF (Figure 2, #11). Alternatively, melatonin might act intracellularly (Figure 2, #6b) to activate conversion of the stored AVT in the basal half of the gland to free peptide (Figure 2, #8) which then exits the epiphysis either through the CSF or into the pineal vascular system (Figure 2, #6a). If, indeed, AVT is bound to a pineal-specific neurophysin as previously discussed, a mechanism involving the exchange of the peptide-neuroepiphysin complex could be proposed (Figure 2, #10). Histophysiological examination of the bovine and monkey pineal gland provoked Lukaszyk and Reiter[133] to propose a generalized schema for the release of a peptide-neuroepiphysin complex from the epiphysis. Adapting this hypothesis to a possible AVT-neuroepiphysin complex (Figure 2, #1) the peptide hormone (Figure 2, #14) would be exchanged for Ca^{++} in a process of "complex dissociation." (Figure 2, #13). In some species, a potential buildup of the neuroepiphysin-calcium debris would complex with lipoprotein microvesicles to form multilayered calcareous deposits known as corpora arenacea.[133]

The biosynthetic pathway for the production of the indoleamine melatonin has been studied in detail.[134-136] Briefly, norepinephrine release (Figure 2, #1) from sympathetic nerve endings stimulates β-adrenergic receptors on pinealocytes and increases intracellular cyclic AMP (Figure 2, #2) which, in turn, stimulates synthesis of the enzyme *N*-acetyltransferase (NAT). Through a cascade of well-defined steps, the amino acid tryptophan is converted to a family of indole compounds including serotonin (Figure 2, #3) and melatonin (Figure 2, #4). Sartin and co-workers[137] investigated the interaction of AVT and norepinephrine upon pineal indoleamine synthesis in vitro. Serotonin, *N*-acetylserotonin and melatonin (Figure 2, #5 and Figure 4) biosynthesis was decreased by incubation of rat pineal glands for 10 hr with AVT. The neuropeptide also decreased the conversion of (^{14}C)-serotonin to 5-methoxyindoleacetic acid (5MIAA) (Figure 2, #7) and to 5-hydroxytryptophol. In combination with Pavel's theory on the role of melatonin in the release of AVT, a feedback loop could be tentatively proposed. In this hypothetical schema, melatonin stimulates the release of AVT (Figure 2, #6a) which, in turn, feeds back at various points in the biosynthetic pathway of indole biosynthesis to inhibit melatonin production and ultimately AVT release.

C. Effects of AVT in the Female Mammal

The reproductive consequences of administration of synthetic AVT have been extensively investigated in well-defined female reproductive model systems whose endpoints include direct visual assessment of ovulation, sexual receptivity and/or fertility or by other quantitative methods such as gonadal weights or hormonal profiles. One advantage conferred by the use of immature rodents is the initial infantile status of the reproductive organs which can dramatically and rapidly be brought to an adult functioning state by the administration of exogenous gonadotrophin-like substances such as pregnant mare's serum (PMS) and human chorionic gonadotrophin (HCG). The precisely timed sequence of hormonal and gonadal changes induced by PMS has been used repeatedly to demonstrate the suppressive influence of potential antigonadotrophic pineal substances.

The first concrete evidence for an antigonadotrophic effect of synthetic AVT in mammals was published in 1966 by Pavel and Petrescu.[138] Ovarian and uterine weights of immature mice primed with 0.2 IU PMS were significantly inhibited by 3 days of

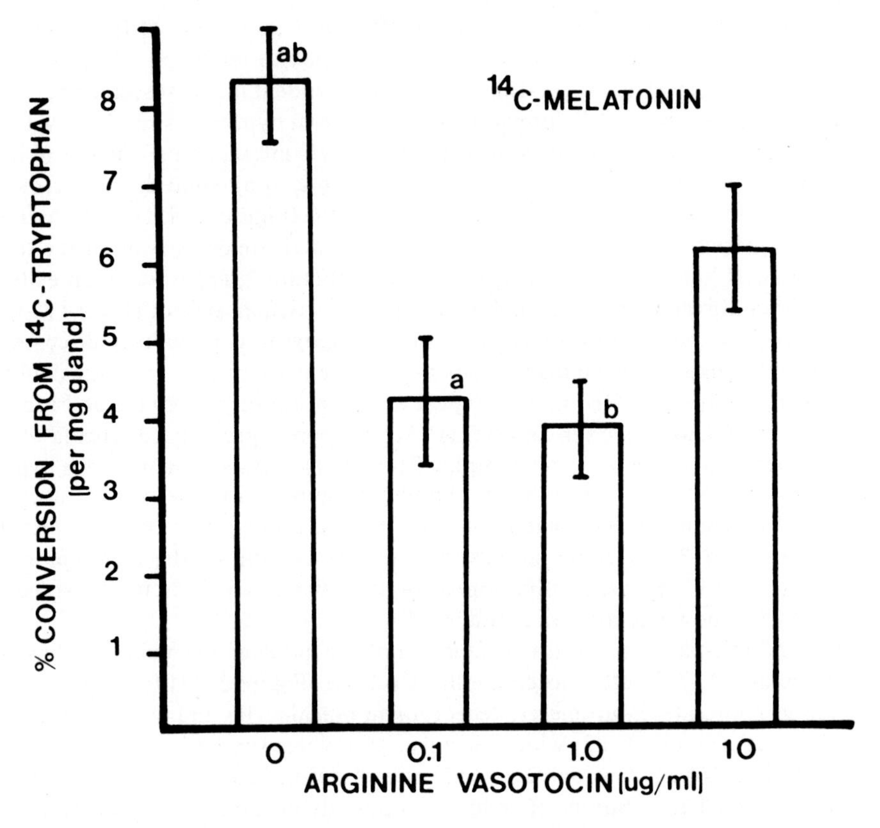

FIGURE 4. The effect of arginine vasotocin upon norepinephrine-stimulated conversion of (^{14}C)-tryptophan to melatonin released into the incubation media. All glands were treated with norepinephrine (10^{-6} *M*) and the indicated dose of AVT. Like letters indicate statistically significant differences: a = $p < 0.001$; n = 6 rat pineal glands per treatment group. (From Sartin, J. L., Bruot, B. C., and Orts, R. J., *Mol. Cell. Endocrinol.*, 11, 7, 1978. With permission.)

subcutaneous injections of 0.1 IU of either a partially purified pineal peptide or synthetic AVT. Corroborating these earlier results, several investigators have noted a similar impairment of gonadal growth in PMS-treated mice[139] and rats[140,141] after AVT treatment (Figure 5). Additionally, injections of the nonapeptide significantly reduced the percentage of rats ovulating (82% — PMS vs 29% — PMS + AVT) and the number of ova shed; the afternoon preovulatory surge of plasma LH, however, was magnified as was the pituitary LH content.[141] The causative factors involved in the diminished growth of the gonads could potentially involve all levels of the neuroendocrine-gonadal axis. Since the preovulatory hormonal surges of LH, FSH, and prolactin were not eliminated, the ovary must be considered a prime site of action of the nonapeptide. Johnson and co-workers[149] speculated that AVT might possibly inhibit follicular growth by interfering with the action of pituitary hormones at the ovarian level or perhaps by disrupting some steroid-controlled mechanism of follicular growth. Subsequently, they noted[142] that injections of AVT were effective in blocking ovulation only when peptide administration began prior to the time of the expected preovulatory surge of LH. This additional evidence suggests a discrete temporal relationship between AVT and ovulatory inhibition which might involve central AVT interaction at the hypothalamic and/or pituitary level. The latter observation may also explain why Cheesman and Forsham[143] were unable to inhibit ovulation in PMS-HCG treated mice with a single injection of AVT given concomitantly with HCG. Alternatively, AVT

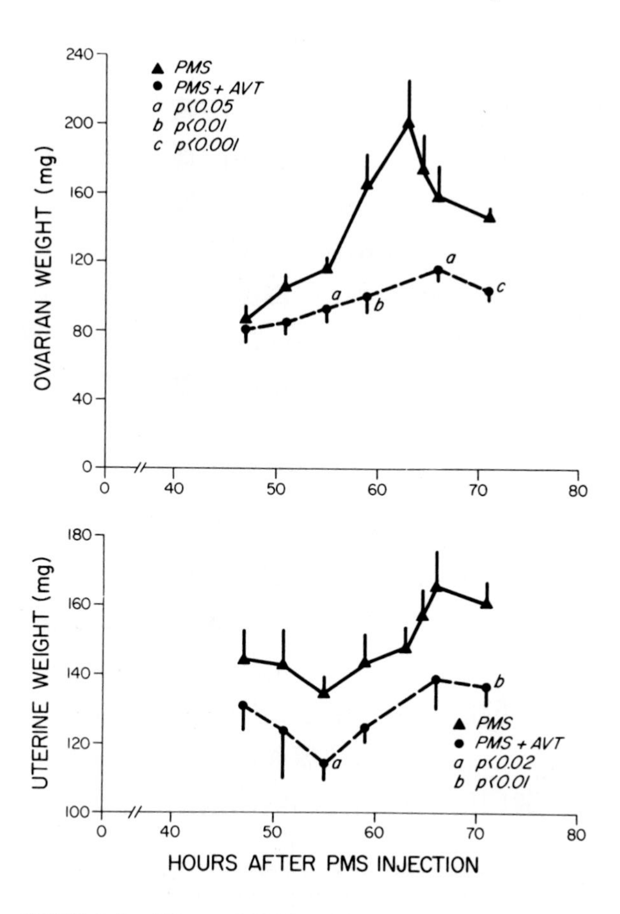

FIGURE 5. Effects of PMS or PMS + AVT treatment on ovarian weight (top) and uterine weight (bottom). Letters indicate significant differences between PMS-treated and PMS + AVT-treated rats at the same intervals of time following PMS injection. (From Johnson, L. Y., Vaughan, M. K., Reiter, R. J, Blask, D. E., and Rudeen, P. K., *Acta Endocrinol.*, 87, 367, 1978. With permission.)

might not be capable of inhibiting a pharmacological dose of HCG in a PMS-primed ovary within a 10-hr period. Indirect substantiation of this latter hypothesis is drawn from the complicated combination in vitro—in vivo bioassay system of Moszkowska and Ebels.[144] In their experiment, AVT was either incubated with hypophyses derived from adult male rats or added after incubation to the media. The resultant mixtures (containing both gonadotrophins and AVT) were injected in 5 doses over a 3-day period to 21-day-old HCG-treated mice. Under this set of circumstances, multiple injections of the hypophyseal media incubated with AVT was able to significantly inhibit ovarian and uterine weight; the ovulatory response was not assessed. In a much simpler reproductive paradigm, other investigators[145-147] have inhibited ovarian and uterine growth in AVT-treated immature female mice primed with just HCG. Under these stimulatory circumstances, injections of AVT spaced at 12-hr intervals failed to block the initial growth spurt of the reproductive organs.[145] However, the ovaries and uteri of AVT-treated mice dramatically ceased growing at 36 and 48 hr respectively and remained repressed until at least 72 hr.[145,146] Similarly, Cheesman and Fariss[147] noted that both AVT and its natural neurohypophyseal counterpart, oxytocin, significantly inhibited uretine hypertrophy following HCG injection. Endogenous gonadotrophin

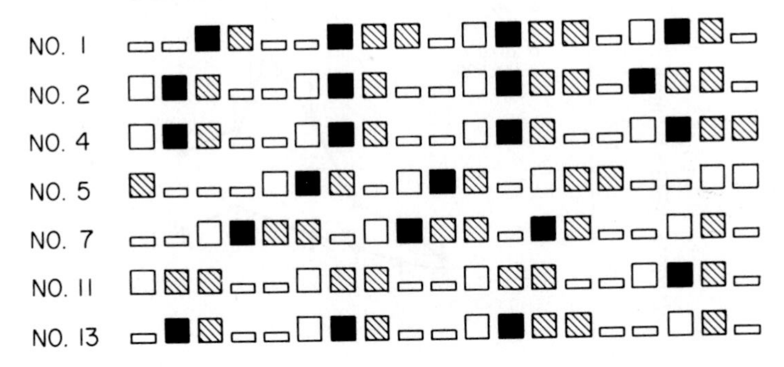

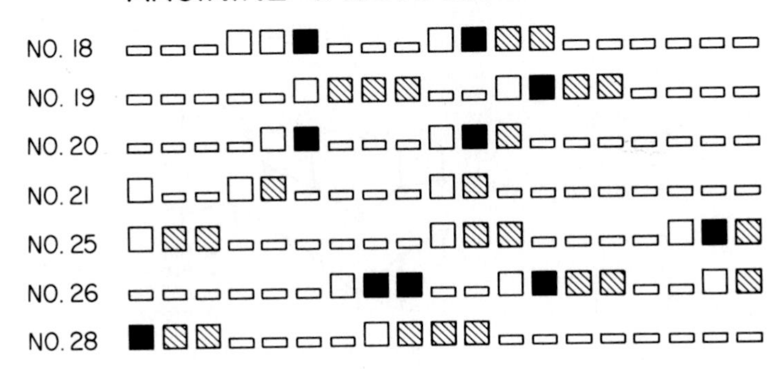

FIGURE 6. Estrous cycles from representative mice of control or arginine vasotocin-treated groups. Mice received either diluent or 2 μg of the peptide per day. The stages are depicted as follows: open square, proestrus; solid square, estrus; cross-hatched square, metestrus; and open rectangle, diestrus. (From Vaughan, M. K., Reiter, R. J., and Vaughan, G. M., *Int. J. Fertil.*, 21, 65, 1976. With permission.)

secretion is thought to contribute significantly during the latter stages of HCG stimulation. Thus, the inhibition of reproductive organs seen in these latter experiments may not have been due to the direct inhibition of the initial effect of HCG but rather to secondary sequelae probably involving a modulatory action of AVT at the hypothalamic and/or pituitary level.

The annual cycle in day length cues the periodic or seasonal breeding of photosensitive vertebrate species such that parturition occurs under optimal environmental conditions for growth and development of the offspring. In their natural habitat, mature adults of polyestrous species undergo short ovulatory cycles until mating and fertilization occur. In the laboratory, these rapid cyclic gonadal changes are easily predicted by daily lavage of the vaginal epithelium; the cornification index of this tissue vacillates in murine rodents in concert with cyclic ovarian events. Under a 12:12 LD cycle, the estrous cycle length of the laboratory mouse is approximately 4.5 days. Daily administration of AVT lengthens the cycle by increasing the proportion of days spent in diestrus (Figure 6).[148] Once AVT-treated mice come into estrus, however, they are receptive to fertile males and carry a normal complement of conceptuses to term provided the peptide is not given during the latter stages of gestation. Administration of AVT after implantation occurs is not conducive to the delivery of viable litters;[148] the causative factors that may contribute to this abortifacient effect have been previously discussed in a recent review.[149] As mentioned earlier, AVT is the uterotonic principle

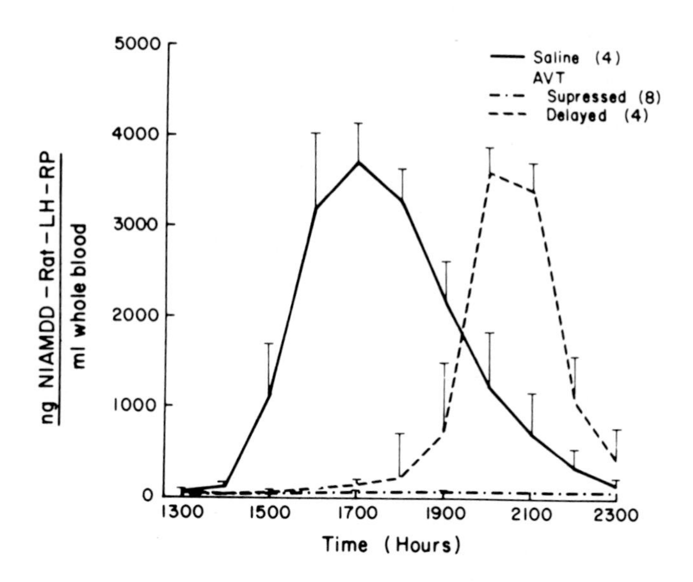

FIGURE 7. The suppression or delay by AVT of the preovulatory surge of LH in proestrous rats. AVT was administered hourly in the third ventricle at a dose of 0.1 μg from 1300 to 1600 hr. (From Osland, R. B., Cheesman, D. W., and Forsham, P. H., *Endocrinology*, 101, 1203, 1977. With permission.)

responsible for oviposition in domestic fowl and stimulation of parturition in viviparous lizards. Exogenous administration in the fowl induces premature oviposition; a prostaglandin-mediated mechanism has been proposed since indomethacin, a prostaglandin inhibitor, blocks the response.[61] If, indeed, the mechanisms are similar between vertebrate classes, then an interesting followup experiment might include inhibition of the abortifacient effects of AVT in the mouse by indomethacin.

Cheesman and co-workers[150] pursued the hormonal mechanisms involved in the anovulatory effects of AVT in adult cycling rodents. Multiple intraventricular injections of AVT inhibited or delayed the endogenous preovulatory surge of LH at doses ranging from 0.2 to 0.001 μg hourly from 1300 to 1600 hr in cannulated proestrous female rats (Figure 7);[151] intravenous low doses of AVT were less effective.[150] A correlative study subsequently published by this same group demonstrated that AVT also had a marked inhibitory effect on the afternoon surge of prolactin in proestrus rats (Figure 8); plasma FSH values were unaffected by administration of AVT, AVP, or oxytocin in this experimental paradigm.[152]

The inhibition of the preovulatory LH surge by AVT[150] prompted additional studies to characterize the neuronal, glandular and/or gonadal site of action of the neuropeptide. Since AVT failed to block LHRH-induced rises in LH secretion, a direct localized action at the level of the pituitary was consequently ruled out. In subsequent experiments, ovariectomized steroid-primed female rats were given either intravenous (0.5, 1, and 2 μg) or intraventricular (0.1 μg) injections of AVT every 30 min for 90 min prior to prostaglandin E_2 (PGE_2) injection; PGE_2 has been shown to induce a surge of blood LH ostensibly by action on the central nervous system (presumably, directly on the LHRH neuron) to release LHRH. AVT failed to inhibit the rise in LH induced by PGE_2. A second method of attacking the problematic site of action utilized electrochemical stimulation of the medial preoptic area (mPOA) to effect an LH surge. Recall the previous discussion that the preoptic area in the killifish (*Fundulus heteroclitus*) has been implicated in the effects of AVT on spawning behavior. However, several doses of AVT failed to block the rise in LH induced by electrochemical stimulation of

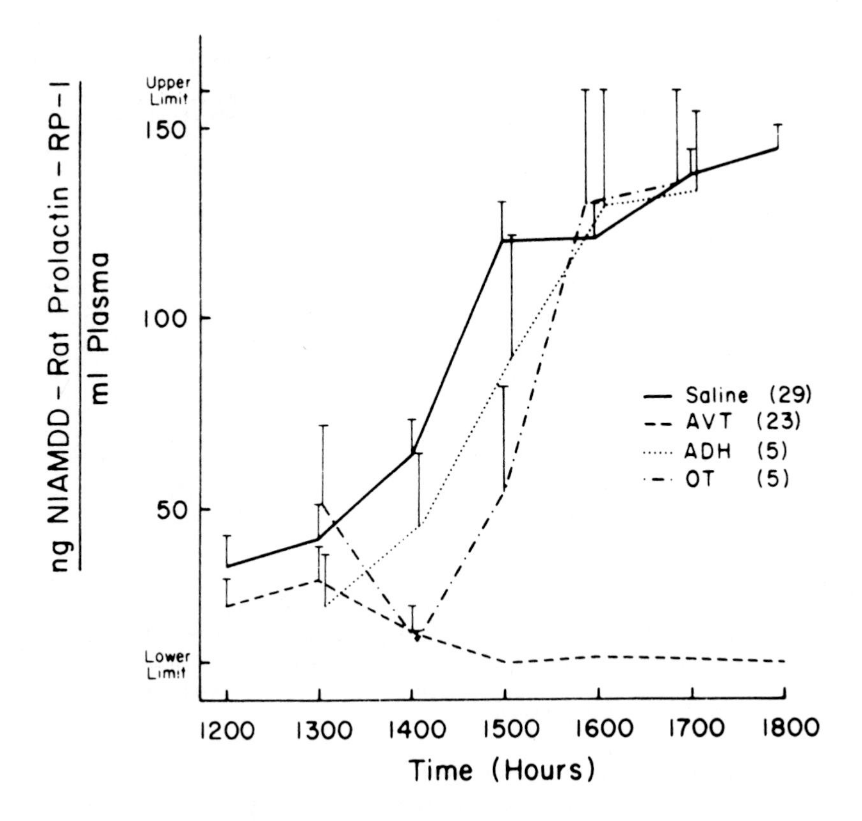

FIGURE 8. Effects of AVT on the surge of prolactin in proestrous rats. AVT and the control compounds arginine vasopressin (ADH) and oxytocin (OT) were administered at 1 μg every 30 min from 1300 to 1600 hr. The number of animals in each treatment group is in parentheses; vertical bars represent means ± S.E. The prolactin values of AVT-treated animals were suppressed ($p < 0.001$) when compared with the ADH-, OT-, and the saline-treated controls at 1500, 1600, 1700, and 1800 hr. (From Cheesman, D. W., Osland, R. B., and Forsham, P. H., *Proc. Soc. Exp. Biol. Med.*, 156, 369, 1977. With permission.)

the mPOA in proestrous female rats. The authors concluded that the inhibition of the proestrous LH surge by AVT occurs at a higher neuronal level than the LHRH release into the hypophyseal portal system; they could not dogmatically exclude an action of AVT confined within the mPOA. A central site of action had previously been tendered by Pavel and colleagues[153] using the compensatory ovarian hypertrophy model. The results of those experiments have been considered at length elsewhere and will not be reiterated here.[139,149]

The positive feedback of estrogen at the level of the POA is largely responsible for the preovulatory surge of gonadotrophic hormones and subsequent ovulation in mammals. An injection of estradiol benzoate in ovariectomized rats induces an afternoon surge of plasma LH and FSH 2 days after steroid treatment.[154] The injection of 1 μg AVT every few hours prior to and into the time frame of the anticipated surge of gonadotrophin completely prevented the peak in plasma LH; the FSH peak was unaffected. The fact that the POA has a high concentration of estrogen receptors and is a site for the positive feedback of estrogen in the rat prompted Blask and co-workers to speculate that AVT might exert its inhibitory action in this hypothalamic area. A precedent for such an AVT-estrogen receptor interaction has been forthcoming. In an in vitro model, all concentrations of AVT ranging from 5×10^{-7} to 5×10^{-9} *M* significantly inhibited binding of ^{3}H-estradiol to estrogen receptors in the cytosol fraction of uteri from immature female rats either during the light or dark phase of the photoperiod.[155] Albeit the estrogen receptors for the above-mentioned experiment were derived from the uterus rather than the hypothalamus, it does provide tentative evidence that an AVT-steroid receptor interaction is hypothetically feasible.

In reflex ovulators such as the rabbit, the neural stimulus elicited by coitus triggers the preovulatory surge of LH and growth of follicles held in abeyance for just such an opportunistic encounter of a reproductively competent male and a fertile receptive female. Besides mating, intravenous injections of LHRH or copper salts induces a surge of LH within 1 hr and ovulation within 10 hr of injection. A single injection of AVT given immediately after copulation or copper salt administration completely inhibited ovulation; AVT given 2 hr after coitus or cupric acetate injection was too late to block the normal ovarian progression to ovulation.[156] From these results, Ying surmized that AVT in a reflex ovulator, as others have claimed for spontaneous ovulators, acts at a neural site higher than the pituitary.[156]

D. Effects of AVT in the Male Mammal

Notwithstanding, the extensive dossier now accumulating on the antigonadotrophic effects of synthetic AVT on gonadotrophic hormones, ovulation, and reproductive organs in the female mammal, only meager evidence is available on organ weight studies in male rodents. At present only a few contradictory reports have attempted to correlate short term injection studies of AVT in immature male rodents with perturbations in gonadal and accessory organ weights; gross,[157] moderate,[158,159] or no[160] inhibition of testicular and accessory organ weights have been elicited after protocols involving 3 to 14 days of nonapeptide injection. One obvious shortcoming now realized by hindsight to the above four papers concerns the unknown time of day at which the injections were administered. Over the last 5 years, one of the major breakthroughs in pineal physiology concerns the daily critical period for melatonin action in the hamster. Administration of the indole to male hamsters in the late afternoon just prior to lights out duplicated the effects of a nonstimulatory photoperiod by inducing total reproductive collapse of the testes and accessory organs, and reducing gonadotrophin levels; morning injections were completely ineffective.[161] The obvious question which must be tendered but has yet to be addressed in a comprehensive manner is whether a daily critical period for AVT also exists. Hypothetically speaking, if such a critical period for AVT also is present, then the seemingly disparate above-mentioned experiments may not necessarily be in conflict.

The hormonal consequences of exogenously administered synthetic AVT are divided among three general types of experimental paradigms which measure its effects both in vivo and in vitro on (1) basal levels of gonadotrophins, (2) castration-induced rises in LH and FSH and, (3) LHRH-induced rise in gonadotrophin secretion. Considering the first model mentioned, several routes of administration have now been evaluated in the normal male rat. By far the most effective, intraventricular injections of AVT induce a dose-dependent fall in plasma LH in urethane-anesthetized male rats which cannot be duplicated at these dose levels by intrapituitary injections[162] or at much higher intravenous doses.[163]

A very elaborate but unproven hypothesis by the Roumanian group details the probable cascading series of events by which AVT interferes with serotonin neurotransmission in the brain to effect lower circulating gonadotrophin levels. To verify a portion of this hypothesis, p-chlorophenylalanine (pCPA), a tryptophan-hydroxylase inhibitor which depletes brain serotonin was injected into the third ventricle of rats 48 hr prior to AVT. After treatment with pCPA, no effects on plasma LH titers could be detected in rats concomitantly treated with the nonapeptide; however, administration of 5-hydroxytryptophan to pCPA-treated rats 1 hr before AVT partially restored the ability of AVT to decrease plasma LH levels. Pavel and colleagues[162] contend that the failure of AVT to decrease plasma gonadotrophin levels in pCPA-treated rats suggests that AVT bypasses the LHRH-containing neuron to act at a higher neuronal level.

Further evidence for this complicated hypothesis was derived from experiments involving immature male rats.[164] In this species, pinealectomy is purported to decrease hypothalamic serotonin content and to increase circulating gonadotrophin titers.[164] Intraventricular injection of 10^{-4} pg of AVT prevented the above-mentioned sequelae of pineal ablation presumably by increasing serotonin levels at polysynaptic receptor sites.[164] The hypothesis further envisions that CSF flows caudally from the third ventricle, through the aqueduct, and into the fourth ventricle; thus intraventricularly injected or endogenously produced AVT potentially could act on the serotoninergic neurons in the raphe nuclei (having been borne to that location by the CSF). By interfering with serotonin release at serotoninergic nerve terminals, AVT presumably blocks neuronal transmission in LHRH-containing neurons, resulting in decreased LHRH release and, thereby, ultimately causing an inhibition of LH release. One seemingly inocuous variable used in the experiments which provided substantial groundwork for this theory may pose a major threat to the viability of the hypothesis. Throughout these experiments and, indeed, those previously considered on the release of AVT by various humoral or intraventricularly administered neuroactive substances, the anesthetic agent used was ethylcarbamate (urethane). It is becoming increasingly apparent and will be considered at length later that this anesthetic modifies the actions of AVT as well as other hormones. Thus, data and hypotheses arising from experimental paradigms in which urethane has been administered should be reevaluated in awake, unrestrained animals to ascertain that the anesthetic is not unduly biasing or masking important hormonal or behavioral events.

Corroborating the above-mentioned hypothesis, Demoulin and co-workers[165] observed that various concentrations of AVT ranging from 10^{-7} to 10^{-18} mol/ℓ of AVT had no effect on basal or LHRH-induced release of LH on male rat anterior pituitary cells in monolayer culture; no effects would have indeed been predicted if the site of action of the nonapeptide is in the midbrain raphe nuclei. However, higher doses of AVT incubated with hemipituitaries in short-term culture conditions have a marked stimulatory effect on LH release under both basal and stimulated (LHRH) conditions.[139,166] These results militate against the Roumanian hypothesis and suggest that AVT has some direct action at the level of the pituitary.

Coincidentally, the latter results correspond to the synergism between the hypothalamic-releasing hormone and AVT observed in acute in vivo studies in intact or castrated, urethane-anesthetized adult male rats.[163] A single injection of AVT + LHRH significantly augmented the plasma titers of LH compared to levels observed in LHRH-treated control rats (Figure 9);[163] a second injection of both peptides 60 min later elicited the same results.[163] The unique synergistic pattern in hormonal profiles was also noted in similarly treated castrated rats; however, the magnitude of the induced surge was greater and the basal levels of LH at which the surge started were higher. Similarly, a potentiation of the LH surge was observed with AVP but not oxytocin. Thus, the synergistic augmentation of LH has been observed in vitro and in vivo in urethane-anesthetized rats; in unanesthetized animals, preliminary evidence indicates that AVT blocks the LHRH-induced rise in LH secretion.[204] A plausible explanation for the observed effects is not easy. Examining first the results from the in vitro experiment, the isolated gonadotrophic cells must be interpreting the incubation medium to be highly stimulatory since the net effect is an increased release of LH. Potential hypotheses to accommodate this phenomenon include: (1) an action of AVT on LHRH receptor sites, or perchance, on its own hypophyseal receptors or (2) decreased inactivation of either LHRH and/or AVT resulting in a prolonged stimulatory re-

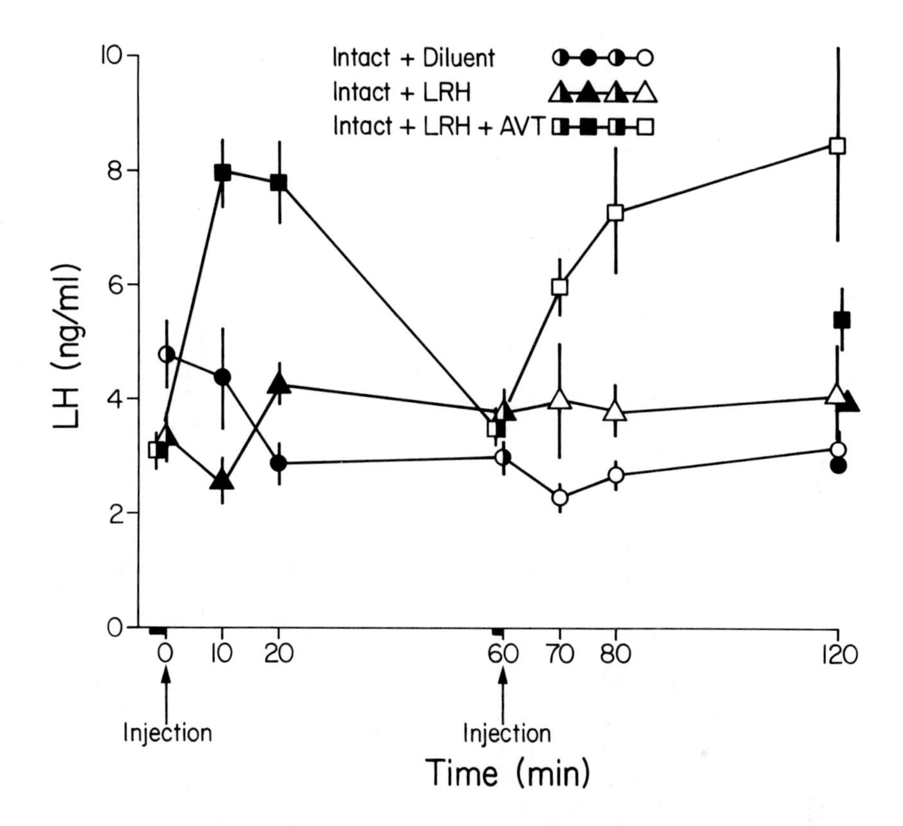

FIGURE 9. Plasma LH (ng/mℓ) values of two groups of intact rats each of which was subdivided into three treatment subgroups which received either diluent, 50 ng LRH, or 50 ng LRH + 5 μg AVT. Rats from group 1 are represented by solid circles, squares or triangles; these rats were bled by venipuncture, injected with diluent or test solution, bled again at 10, 20, and 60 min, and decapitated at 120 min. Following the removal of an initial blood sample (0 min) rats of group 2 (represented by open circles, squares, and triangles) were injected with their respective diluent or test solutions, bled again at 70 and 80 min and sacrificed at 120 min. Corresponding values of similar treatment subgroups from Group 1 and 2 were combined at the 0 and 60 min time points (represented by half-solid and half-open circles, squares, and triangles). Means ± S. E. M. are indicated. (From Vaughan, M. K., Trakulrungski, C., Petterborg, L. J., Johnson, L. Y., Blask, D. E., Trakulrungski, W., and Reiter, R. J., *Mol. Cell. Endocrinol.*, 14, 59, 1979. With permission.)

sponse by effectively higher concentrations. Precedence for this latter hypothesis can be derived from previous experiments by Griffiths and Hooper.[167] They proposed that hypothalamic peptidases inactivating oxytocin, a naturally occurring analogue of AVT, might be involved in LHRH metabolism. From kinetic studies, it was found that there was competitive inhibition between the two polypeptides for the same degrading enzyme. If such a competition exists within the in vitro pituitary, then one might speculate that effectively higher levels of LHRH are maintained due to its slower rate of inactivation. A variant of this hypothesis might also apply in an in vivo situation in which the liver-degrading capacity of the animals has also been severely compromised by the known toxic effects of urethane. A further complicating variable concerns the modification of the LHRH response by hemipituitaries of male rats in vitro by urethane as recently suggested by Carter and Dyer.[168] Of course, one cannot completely exclude the hypothesis that urethane somehow dampens or inhibits hypothalamic influences on the pituitary, thus allowing the *in situ* pituitary to behave similarly to an in vitro preparation .

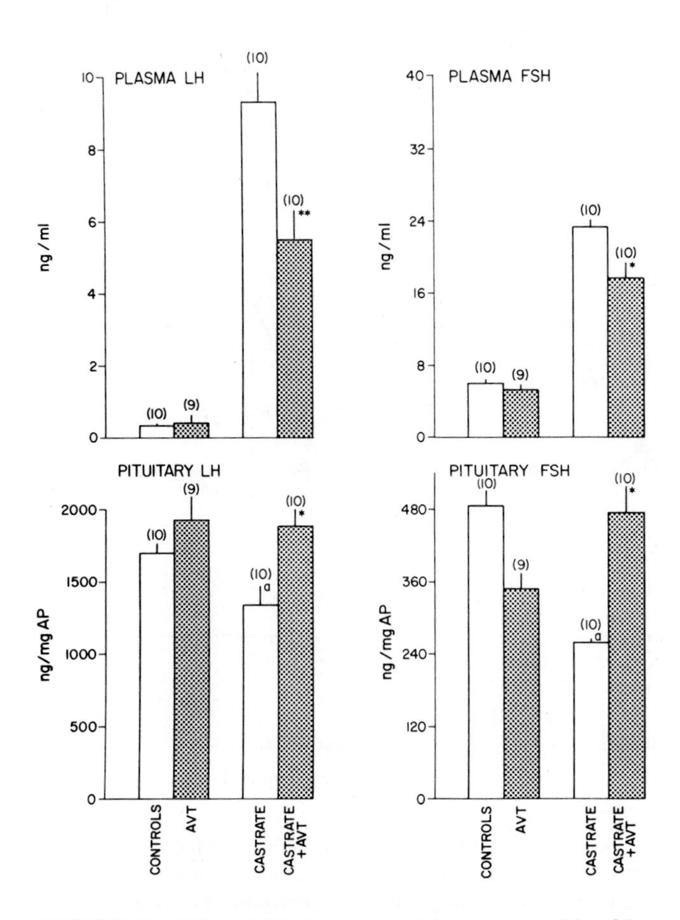

FIGURE 10. Effect of 16 subcutaneous injections every 3 hr of 5 μg AVT on plasma (ng/mℓ) and pituitary (ng/mg anterior pituitary [AP]) levels of LH and FSH. Means ± S. E. M. are indicated. Number of rats is given in parentheses. Castrated group vs *, $p < 0.001$; (a) $p < 0.05$ vs intact controls). (From Vaughan, M. K., Blask, D. E., Johnson, L. Y., and Reiter, R. J., *Endocrinology*, 104, 212, 1979. With permission.)

In the unanesthetized male rat, there is little question that the net effects of AVT injections on circulating gonadotrophin levels is inhibitory (Figure 10). The acute effects of castration depletes the readily releasable pool of stored gonadotrophin in the hypophysis; multiple injections of AVT completely prevent this depletion.[169] Although these experiments do not bespeak of a specific site of action of the neuropeptide, the results are consistent with the Roumanian hypothesis discussed earlier.

Other investigators have observed results which are indicative of a pituitary site of action of AVT. Yamashita and co-workers[170,171] use the concentrations of 17-oxosteroids in the spermatic venous blood of anesthetized dogs as an index of LH release after intracarotid injections of LHRH. In both mature[170] and immature[171] dogs, administration of AVT 3 hr prior to LHRH injection inhibited the testicular secretion of 17-oxosteroids (Figure 11); pretreatment with AVT did not affect the testicular response to HCG. These authors conjectured that AVT inhibited the LHRH-induced release by acting directly on the hypophysis.

E. Effects of AVT on Prolactin Secretion

The galactogenic nature of pineal extracts has been recognized since 1911 when

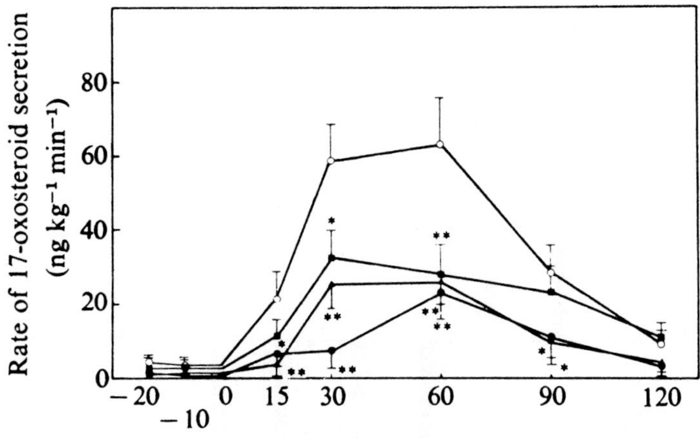

FIGURE 11. Output of 17-oxosteroids (means ± S. E. M.) by one testis in response to the injection of luteinizing hormone releasing hormone (LH-RH, 5 μg/kg body weight) into the carotid artery of dogs pretreated wtih 0.01 (solid squares), 0.1 (solid triangles), or 1.0 (solid circles) μg arginine vasotocin per kg body weight or isotonic saline (open circles) injected into the carotid artery 3 hr previously. $*p < 0.05$; $**p < 0.01$: significantly different from the corresponding value when LHRH was given without pretreatment with arginine vasotocin. (From Yamashita, K., Mieno, M., and Yamashita, E., *J. Endocrinol.,* 81, 103, 1979. With permission.)

Mackenzie[172] reported that intravenous injections of boiled ovine pineal extracts stimulated mammary secretion. More recently, prolactin-releasing and/or inhibiting factor (PRF and PIF, respectively) activity has been described in the bovine, human,[173,174,176] and in vivo[173,174] rat,[173,175] and ovine[176] preparations using both in vitro[173,174,176] and in vivo[173,175] techniques. One pertinent reason for this sudden upsurge in interest about the pineal prolactin factor concerns its possible crucial role in the mediation of the dramatic reproductive collapse of the gonads and accessory organs initiated by short natural photoperiodic conditions, simulated short days in the laboratory, or bilateral orbital enucleation. A consistent reduction in pituitary prolactin levels has been observed in sight-deprived male hamsters with pineal-stimulated gonadal regression.[177] Moreover, Bartke and colleagues[178] observed that injection of prolactin restores plasma testosterone levels and stimulates testicular growth in hamsters in short daily photoperiods. Thus, the identity of pineal factors which modulate prolactin secretion either directly on synthesis and secretion or indirectly on (1) hypothalamic releasing inhibiting factors or (2) on other neural sites may indeed be a key to unlocking the mysteries to gonadal regression and recrudescence in photosensitive species.

The release of both prolactin and one neurohypophyseal hormone, oxytocin, during lactation and machine-milking in the cow is well established. In dairy cows, Forsling and co-workers[179] favor the hypothesis of independent release of oxytocin and the adenohypophyseal hormone. This hypothesis was seemingly borne out by other groups of scientists who failed to note a causal relationship between oxytocin administration and prolactin release in male[180] and female[181] rats. Additionally, trials to modify basal prolactin release and the amount of milk produced by postpartum women with exogenous administration of oxytocin failed.[182]

The prolactin-releasing activity of AVT was first recognized by Vaughan and co-workers[166] in in vitro explants of normal male rat hemipituitaries; this effect has since been verified in bovine pituitary cell cultures[183] (Figure 12) and in two subclones of

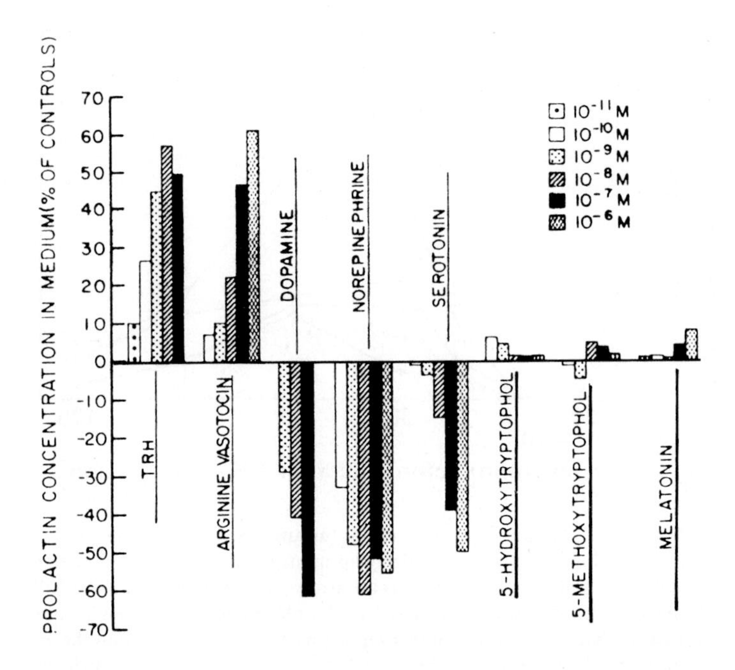

FIGURE 12. Prolactin concentration in medium following incubation of bovine pituitary cells with varying concentrations of pineal compounds. Values are expressed as percent increase or decrease relative to controls. PRL concentrations (ng/mℓ) for control cultures were: TRH, 445; AVT, 444; DA, 576; NE, 214; melatonin, 275; serotonin, 268; and, 5-hydroxytryptophol and 5-methoxytryptophol; 441. Coefficients of variation for the controls ranged from 8.3 to 11.2% and were not different among groups. (From Padmanabhan, V., Convey, E. M., and Tucker, H. A., *Proc. Soc. Exp. Biol. Med.*, 160, 340, 1979. With permission.)

prolactin-producing cells.[184] As seen from Figure 12, both AVT- and TRH-induced prolactin release by a direct effect on the bovine pituitary; release was linearly related to the log of the concentrations used. Interestingly, the two rat pituitary prolactin-producing subclones (T2B8CA4 — TRH-induced and TE2B8CF11 — TRH- and E_2-induced) which respond to AVT stimulation had been induced by TRH. Further experiments by Blask and colleagues,[185] however, demonstrated that in vitro AVT-induced prolactin secretion was significantly suppressed by the β-receptor antagonist, propranolol; conversely, the α and dopamine antagonist, haloperidol, had no effect. They concluded that AVT directly stimulated pituitary prolactin secretion via a receptor mechanism not related to dopamine but rather of the β-adrenergic type.

As the tale of the effect of AVT on prolactin secretion evolved over the last 5 years it has become increasingly apparent that the story is more complicated than originally perceived. Early experiments in urethane-anesthetized estrogen-progesterone treated rats corroborated in vitro data; intravenous injections of AVT stimulated the secretion of prolactin in a dose-dependent manner.[186,187] Neither time of day, age of solution, nor presence of an intact pineal gland were found to be critical factors in this response.[186,187] On the surface, the accumulated data both in vitro and in vivo suggested that AVT was a prolactin-releasing factor.

The initial disturbing thread of evidence which threatened the above hypothesis concerned the inhibition of the prolactin surge in proestrus female rats published by Cheesman and co-workers[152] in 1977 (Figure 8). Further experiments by Johnson and colleagues[188] attacked the urethane-AVT-prolactin question specifically in PMS-treated rats early on the day before the expected prolactin surge. In urethane-anesthetized PMS treated rats, 1 or 5 μg dose caused a significant increase at 15 min followed by a significant depression at 45, 60, and 120 min. Similarly, in unanesthetized estrogen-progesterone-treated male rats, a depressive effect on prolactin could be elicited by 1 μg AVT whereas only the stimulatory effect of AVT was observed in anesthetized rats. This biphasic effect of AVT on prolactin opens new questions concerning the endocrine modulating effects of urethane as well as which effect may be the more physiological role for the neuropeptide. Certainly, the release of the nonapeptide into the vascular system or CSF would be in ranges far below those administered in the above studies; thus, it cannot be unequivocably ruled out that both effects may be pharmacological. However, very low doses (0.001 pg AVT or ≃ 60,000 molecules per μℓ), which probably are within the physiological range, injected into the third ventricle of the mouse on the day of pinealectomy prevented the increase in pituitary prolactin which occurred 7 days after pineal ablation.[190] Unfortunately, a major drawback to this experiment is that the pigeon-crop bioassay was implemented rather than the more sensitive radioimmunoassay; thus, no plasma levels were available to determine, once and for all, whether the net secretory effect of the hypophysis to AVT adminstration is stimulatory, inhibitory, or biphasic.

Albeit recognizing the pitfalls inherent in the following experiments due to the use of urethane anesthesia (UA), some potentially relevant data are indicative of a positive interaction between the steroidal status of the male rat and the ability of AVT to modify prolactin release. Initial experiments on the prolactin-releasing activity of AVT were conducted either in normal male rats or in males sensitized by subcutaneous injection of estrogen and progesterone. Further experiments revealed that UA-castrated male rats are insusceptible to stimulatory effects of AVT administration.[191] Subcutaneous injections of testosterone or androsterone completely restored the susceptibility of castrated rats to the induction of prolactin release by AVT; androstenedione and dihydrotesterone were completely ineffective. Obviously, under the circumstances outlined above, the in vivo response of prolactin required the presence of the gonads, or in the case of castrated rats, the proper steroidal milieu. Castration itself eventually causes a decrease in pituitary and plasma prolactin levels due, presumably, to an increase in hypothalamic PIF;[192] conversely, testosterone is thought to stimulate prolactin by decreasing PIF.[192] One plausible explanation includes the notion that a rise in PIF was an overriding factor in negating the stimulatory effects of AVT in castrated rats; in androgen-treated animals, however, PIF synthesis and/or release may have been curtailed sufficiently to permit the stimulatory action of AVT to go unopposed.

Endorphins and enkephalins, endogenous opiates, possess potent prolactin-releasing effects;[193,194] opiate receptor blockers such as naloxone negate this effect.[194] A tenable connection between the opiates and the neurohypophysis has been advocated by several investigators;[195-198] opiate receptors are even found in the posterior lobe.[197] The aforementioned observations prompted Blask and Vaughan[199] to determine if AVT stimulated prolactin release via an interaction with opiate receptors. Injection of 1 μg AVT to UA male rats provoked a twofold rise in plasma prolactin; naloxone completely prevented the response (Figure 13). Realizing the problems mentioned earlier with the anesthetic, the experiment was repeated in unanesthetized rats. In rats treated only with AVT, a biphasic (first stimulatory and then inhibitory) effect on prolactin release was noted; naloxone completely blocked both effects.[200] One encouraging note is that the effects of naloxone were the same in both the stimulatory phase of awake rats and anesthetized rats. The anaesthetic may block only the depressive effects of AVT on prolactin secretion; this may be due to the fact that basal levels of the adenohypophyseal hormone are already maximally depressed. Some support for this latter hypothesis is derived from the reported inability of l-dopa to further suppress prolactin titers in UA rats.[201]

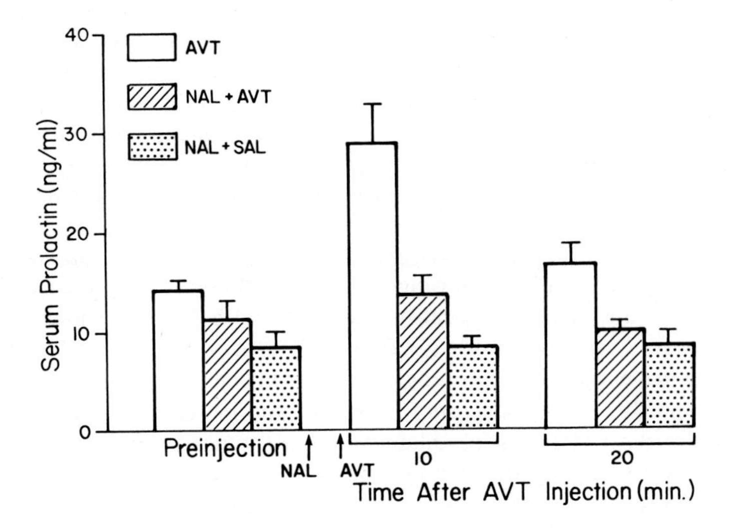

FIGURE 13. Effect of a single intravenous injection of either diluent (SAL) or 1 μg of arginine vasotocin (AVT) on serum prolactin (ng/mℓ) in naloxone (NAL) (0.2 mg/kg BW)-treated, urethane-anesthetized male rats. NAL was administered i.p. 5 min prior to AVT or SAL injection. Vertical lines from top of bars signify SEM. *$p < 0.01$ vs AVT + NAL at 10 min., preinjection, and 20 min. samples; *$p < 0.01$ vs AVT + NAL at 20 min. (From Blask, D. E. and Vaughan, M. K., *Neurosci. Lett.*, in press. With permission.)

IV. CONCLUDING REMARKS

The undeniable existence of AVT in the neurohypophysis of all inframammalian species tested and its questional presence in mammals is noteworthy tribute to the tremendous diversity of roles which the peptide has been pressed to assume throughout vertebrate evolution. In the lowest vertebrates, the neuropeptide may be involved in the spawning migration and perhaps various behavioral aspects of the spawning reflex itself. Neurohypophyseal lobe extracts and hormones induced parturition and oviposition in viviparous, oviparous, and ovoviparous members of the three major groups of living reptiles. Finally, in birds, AVT has been implicated not only in egglaying but also other behavioral activities such as waltzing, comb-grabbing, mounting, and copulation. In reviewing this literature, one is immediately struck by the spottiness and overall lack of knowledge and research devoted to the vast number of species in these classes. Surely, a comprehensive knowledge of the basic mechanisms by which the neuropeptide affects the behavioral, hormonal, and reproductive system in species where it is absolutely known to exist would be valuable information around which to design meaningful mammalian experiments. A case in point concerns behavioral reproductive effects of AVT which are prevalent in fish, amphibians, and birds; no comprehensive studies have yet been designed to examine potential behavioral effects of the peptide on reproduction in mammals. Yet, AVT is known to provoke other behavioral responses in mice (hyperactivity, extensive foraging, increased grooming, stereotyped scratching, squeaking, and occasional barrel rolling)[202] and to affect consolidation and retrieval of memory.[204]

The pineal gland of several lower vertebrates as well as mammalian species contains AVT or an AVT-like substance which has the same biological, radioimmunoassayable and electrophoretic characteristics of the synthetic nonapeptide; the exact nature of the peptide is one of consuming controversy as of this writing. Just this year, in the case of the nonapeptide vasopressin, a prohormone and/or storage form was shown to exist within the bovine neurohypophysis; the newly discovered peptides had natriuretic activity and were indistinguishable from AVP in current radioimmunoassay systems. If such compounds exist for AVT in the pineal, it is conceivable that either false positive or negative results could be obtained using various detection techniques. That such a compound could exist has been tentatively shown in the extraction of a 14-amino acid peptide (E_5) with biological characteristics of AVT.

Accumulating evidence from several laboratories indicates that the release of AVT or the AVT-like peptide from the pineal can be induced by numerous hormonal (LHRH, SRIF, MIF-I), neurotransmitter (AcH, NE), or nonhormonal stimuli (hypertonic saline) or by the pineal indole melatonin. Since all of the above stimulating agents also release AVP from the neurohypophysis, it might be difficult to separate the effects due to pineal peptide release from those of the neurohypophyseal hormone in vivo without doing neural lobectomy.

The weight of the evidence to date indicates that the net effect of synthetic AVT administration is inhibitory to various reproductive paradigms in both immature and mature rodents; interestingly, where tested, AVP has many of the same effcts as AVT. Although the general concensus acclaims a neural site of action in some models, some evidence supports the pituitary and/or gonads as ancillary sites at which definite modulatory effects can be elicited by the synthetic nonapeptide. While early evidence purported a stimulatory action on AVT on prolactin secretion, it is now surmised that the anesthetic urethane may have biased these initial results; in unanesthetized animals, AVT appears to have a biphasic effect.

In conclusion, it is manifestly evident after 2 decades of research that the nonapeptide AVT directly or indirectly affects the reproductive system of all classes of vertebrates. Although most investigators ostensibly agree that an antigonadotrophic peptide compound is found in the pineal of mammals, considerable controversy still resounds over whether it is AVT, AVP, oxytocin, E_5, and/or as yet another unidentified peptide. Clarification of these pressing issues hopefully will be resolved in the near future so that a forthright attack on the effect of the endogenous peptide(s) on the reproductive economy of mammals can proceed.

REFERENCES

1. **Heller, H.,** Differentiation of an (amphibian) water balance principle from the antidiuretic principle of the posterior pituitary gland, *J. Physiol., (London),* 100, 124, 1941.
2. **Henderson, N. E.,** Ultrastructure of the neurohypophysial lobe of the hagfish, *Eptatretus stouti* (Cyclostomata), *Acta Zool.,* 53, 243, 1972.
3. **Rurak, D. W. and Perks, A. M.,** The pharmacological characterization of arginine vasotocin in the pituitary of the Pacific hagfish (*Polistrotrema stoutii*), *Gen. Comp. Endocrinol.,* 22, 480, 1974.
4. **Sawyer, W. H.,** Oxytocic, antidiuretic and vasopressor activities in the neurohypophysis of the sea lamprey, *Petromyzon marinus, Fed. Proc. Fed. Am. Soc. Exp. Biol.,* 14, 130, 1955.
5. **Sawyer, W. H.,** Active neurohypophysial principles from a cyclostome (*Petromyzon marinus*) and two cartilaginous fishes (*Squalus acanthias* and *Hydrolagus colliei*), *Gen. Comp. Endocrinol.,* 5, 427, 1965.
6. **Sawyer, W. H., Munsick, R. A., and Van Dyke, H. B.,** Pharmacological evidence for the presence of arginine vasotocin and oxytocin in neurohypophysial extracts from cold-blooded vertebrates, *Nature (London),* 184, 1464, 1959.
7. **Sawyer, W. H., Munsick, R. A., and Van Dyke, H. B.,** Pharmacological characteristics of the active principles in neurohypophysial extracts from several species of fishes, *Endocrinology,* 68, 215, 1961.

8. **Lanzing, W. J. R.,** The occurrence of a water-balance, a melanophore expanding and an oxytocic principle in the pituitary gland of the river lamprey (*Lampetra fluviatilis* L.), *Acta Endocrinol.*, 16, 277, 1954.
9. **Follett, B. K. and Heller, H.,** The neurohypophysial hormones of bony fishes and cyclostomes, *J. Physiol. (London)*, 172, 74, 1964.
10. **Goossens, N., Dierickx, K., and Vandesande, F.,** Immunocytochemical demonstration of the hypothalamo-hypophysial vasotocinergic system of *Lampetra fluviatilis, Cell. Tissue Res.*, 177, 317, 1977.
11. **Bentley, P. J. and Follett, B. K.,** Fat and carbohydrate reserves in the river lamprey during spawning migration, *Life Sci.*, 4, 2003, 1965.
12. **Bentley, P. J. and Follett, B. K.,** The effects of hormones on the carbohydrate metabolism of the lamprey *Lampetra fluviatilis, J. Endocrinol.*, 31, 127, 1965.
13. **John, T. M., Thomas, E., George, J. C., and Beamish, F. W. H.,** Effect of vasotocin on plasma free fatty acid level in the migrating anadromous sea lamprey, *Arch. Int. Physiol. Biochim.*, 85, 865, 1977.
14. **McKeown, B. A., John, T. M., and George, J. C.,** Effect of vasotocin on plasma GH, free fatty acids and glucose in Coho salmon, *Endocr. Exp.*, 10, 45, 1976.
15. **John, T. M., McKeown, B. A., and George, J. C.,** Effect of vasotocin and glucagon on plasma growth hormone levels in the pigeon, *Comp. Biochem. Physiol.*, 48A, 521, 1974.
16. **Vaughan, M. K., Johnson, L. Y., Little, J. C., Vaughan, G. M., and Reiter, R. J.,** Stimulation of rat growth hormone secretion by arginine vasotocin *in vivo* and *in vitro, Neuroendocrinology Lett.*, 2, 19, 1980.
17. **Sawyer, W. H.,** Phylogenetic aspects of the neurohypophysial hormones, in, *Handbook of Experimental Pharmacology*, Vol. 23, Berde, B., Ed., Springer-Verlag, New York, 1968, 717.
18. **LaPointe, J.,** Comparative physiology of neurohypophysial hormone action on the vertebrate oviduct-uterus, *Am. Zool.*, 17, 763, 1977.
19. **Heller, H.,** The effect of neurohypophysial hormones on the female reproductive tract of lower vertebrates, *Gen. Comp. Endocrinol. Suppl.*, 3, 703, 1972.
20. **Heller, J., Leathers, D. H. G., and Lane, G. J.,** The effect of neurohypophysial hormones on the oviduct of an elasmobranch fish, *Scyliorhinus caniculus, J. Endocrinol.*, 50, 357, 1971.
21. **Heller, H. and Pickering, B. T.,** Identification of a new neurohypophysial hormone, *J. Physiol. (London)*, 152, 56, 1960.
22. **Heller, H. and Pickering, B. T.,** Neurohypophysial hormones of nonmammalian vertebrates, *J. Physiol. (London)*, 155, 98, 1961.
23. **Follett, B. K. and Heller, H.,** The neurohypophysial hormones of lungfishes and amphibians, *J. Physiol. (London)*, 172, 92, 1964.
24. **Goosens, N., Dierickyx, K., and Vandesande, F.,** Immunocytochemical study of the neurohypophysial hormone producing system of the lungfish, *Protopterus aethiopicus, Cell. Tissue Res.*, 190, 69, 1978.
25. **Goosens, N., Dierickyx, K., and Vandesande, F.,** Immunocytochemical localization of vasotocin and isotocin in the preopticohypophysial neurosecretory systems of Teleosts, *Gen. Comp. Endocrinol.*, 32, 371, 1977.
26. **Gill, V. E., Burford, G. D., and Lederis, K.,** An immunocytochemical investigation for arginine vasotocin and neurophysin in the pituitary gland and the caudal neurosecretory system of *Catostomus commersoni, Gen. Comp. Endocrinol.*, 32, 506, 1977.
27. **Schreibman, M. P. and Halpern, L. R.,** The demonstration of neurophysin and arginine vasotocin by immunocytochemical methods in the brain and pituitary gland of the platyfish, *Xiphophorus maculatus, Gen. Comp. Endocrinol.*, 40, 1, 1980.
28. **Berlind, A., Lacanilao, F., and Bern, H. A.,** Teleost caudal neurosecretory system: II. Effect of osmotic stress on urophysial proteins and active factors, *Comp. Biochem. Physiol.*, 42A, 345, 1972.
29. **Lacanilao, F.,** The urophysial hydrosmotic factor of fishes. I. Characteristics and similarity of neurohypophysial hormones, *Gen. Comp. Endocrinol.*, 19, 405, 1972.
30. **Moore, G., Buford, G., and Lederis, K.,** Properties of urophysial proteins (urophysins) from the white sucker *Catostomus commersoni, Mol. Cell. Endocrinol.*, 3, 297, 1975.
31. **Lacanilao, F. and Bern, H.,** The urophysial hydrosmotic factor of fishes. III. Survey of fish caudal spinal cord regions for hydrosmotic activity, *Proc. Soc. Exp. Biol. Med.*, 140, 1252, 1972.
32. **Lacanilao, F.,** The urophysial hydrosmotic factor of fishes. II. Chromatographic and pharmacologic indications of similarity to arginine vasotocin, *Gen. Comp. Endocrinol.*, 19, 413, 1972.
33. **Holder, F. C., Schroeder, M. D., Guerne, J. M., and Vivien-Roels, B.,** A preliminary comparative immunohistochemical, radioimmunological, and biological study of arginine vasotocin (AVT) in the pineal gland and urophysis of some teleostei, *Gen. Comp. Endocrinol.*, 37, 15, 1979.
34. **Goosens, N.,** Immunohistochemical evidence against the presence of vasotocin in the trout urophysis, *Gen. Comp. Endocrinol.*, 30, 231, 1976.

35. **Pickford, G. E.**, Induction of a spawning reflex in hypophysectomized killifish, *Nature (London)*, 170, 807, 1952.
36. **Wilhelmi, A. E., Pickford, G. E., and Sawyer, W. H.**, Initiation of the spawning reflex response in Fundulus by the administration of fish and mammalian neurohypophysial preparations and synthetic oxytocin, *Endocrinology*, 57, 243, 1955.
37. **Pickford, G. E. and Strecker, E. L.**, The spawning reflex response of the killifish, *Fundulus heteroclitus:* isotocin is relatively inactive in comparison with arginine vasotocin, *Gen. Comp. Endocrinol.*, 32, 132, 1977.
38. **Morel, F. and Jard, S.**, Actions and functions of the neurohypophysial hormones and related peptides in lower vertebrates, in, *Handbook of Experimental Pharmacology*, Vol. 23, Berde, B., Ed., Springer-Verlag, New York, 1968, 655.
39. **Macey, M. J., Pickford, G. E., and Peter, R. E.**, Forebrain localization of the spawning reflex response to exogenous neurohypophysial hormones in the killifish, *Fundulus heteroclitus, J. Exp. Zool.*, 190, 269, 1974.
40. **Hayward, J. N.**, Physiological and morphological identification of hypothalamic magnocellular neuroendocrine cells in goldfish preoptic nucleus, *J. Physiol.*, 239, 103, 1974.
41. **Heller, H., Ferreri, E., and Leathers, D. H. G.**, The effect of neurohypophysial hormones on the amphibian oviduct *in vitro* with some remarks on the histology of this organ, *J. Endocrinol.*, 47, 495, 1970.
42. **Munsick, R. A.**, The effect of neurohypophysial hormones and similar polypeptides on the uterus and other extravascular smooth muscle tissue, in, *Handbook of Experimental Pharmacology*, Vol. 23, Berde, B., Ed., Springer-Verlag, New York, 1968, 443.
43. **Diakow, C.**, Hormonal basis for breeding behavior in female frogs: vasotocin inhibits the release call of *Rana pipiens, Science*, 199, 1456, 1978.
44. **Vivien-Roels, B., Guerne, J. M., Holder, F. C., and Schroeder, M. D.**, Comparative immunohistochemical, radioimmunological and biological attempt to identify arginine vasotocin (AVT) in the pineal gland of reptiles and fishes, *Prog. Brain Res.*, 52, 459, 1979.
45. **Clausen, H. J.**, Studies on the effect of ovariotomy and hypophysectomy on gestation in snakes, *Endocrinology*, 27, 700, 1940.
46. **Panigel, M.**, Contribution a l'etude de la ovoviparite chez les reptiles; gestation et parturition chez le lezard vivipare, *Zootoca vivpara, Ann. Soc. Nat. (Zool.)*, 18, 569, 1965.
47. **LaPointe, J. L.**, Induction of oviposition in lizards with the hormone oxytocin, *Copeia*, 1964, 451, 1964.
48. **Ewert, M. A. and Legler, J. M.**, Hormonal induction of oviposition in turtles, *Herpetologica*, 34, 314, 1978.
49. **Munsick, R. A., Sawyer, W. H., and Van Dyke, H. B.**, Avian neurohypophysial hormones: pharmacological properties and tentative identification, *Endocrinology*, 66, 860, 1960.
50. **La Point, J. L.**, Effect of ovarian steroids and neurohypophyseal hormones on the oviduct of the viviparous lizard, *Klauberina riversiana, J. Endocrinol.*, 43, 197, 1964.
51. **Guillette, L. J., Jr.**, Stimulation of parturition in a viviparous lizard (*Sceloporus jarrovi*) by arginine vasotocin, *Gen. Comp. Endocrinol.*, 38, 457, 1979.
52. **Callard, I. P. and Hirsch, M.**, The influence of oestradiol-17b and progesterone on the contractility of the oviduct of the turtle, *Chrysemys picta, in vitro, J. Endocrinol.*, 68, 147, 1976.
53. **Fernstrom, J. D., Fisher, L. A., Cusack, B. M. and Gillis, M. A.**, Radioimmunologic detection and measurement of nonapeptides in the pineal gland, *Endocrinology*, 106, 243, 1980.
54. **Negro-Vilar, A., Sanchez-Franco, F., Kwiatkowski, M., and Samson, W. K.**, Failure to detect radioimmunoassayable arginine vasotocin in mammalian pineals, *Brain Res. Bull.*, 4, 789, 1979.
55. **Rzasa, J. and Ewy, Z.**, Effect of vasotocin and oxytocin on oviposition in the hen, *J. Reprod. Fertil.*, 21, 549, 1970.
56. **Brezezinska-Slebodzinska, E. and Rzasa, J.**, Inactivation of vasotocin by hen's tissues in vitro, *Experientia*, 29, 111, 1973.
57. **Rzasa, J. and Ewy, Z.**, Effect of vasotocin and oxytocin on intrauterine pressure in the hen, *J. Reprod. Fertil.*, 25, 115, 1971.
58. **Tanaka, K. and Nakajo, S.**, Oxytocin in the neurohypophysis of the laying hen, *Nature (London)*, 187, 245, 1960.
59. **Tanaka, K. and Nakajo, S.**, Participation of neurohypophysial hormones in oviposition in the hen, *Endocrinology*, 70, 453, 1962.
60. **Sturkie, P. D. and Lin, Y. C.**, Release of vasotocin and oviposition in the hen. *J. Endocrinol.*, 35, 325, 1966.
61. **Rzasa, J.**, Effects of arginine vasotocin and prostaglandin E_1 on the hen uterus, *Prostaglandins*, 16, 357, 1978.
62. **Opel, H.**, Release of oviposition-inducing factor from the median eminence pituitary stalk region in neural lobectomized hens, *Anat. Rec.*, 154, 396, 1966.

63. **Khilstrom, J. E. and Danninge, I.,** Neurohypophysial hormones and sexual behavior in males of the domestic fowl (*Gallus domesticus* L.) and the pigeon (*Columba livia* Gmel.), *Gen. Comp. Endocrinol.,* 18, 115, 1972.
64. **Niezgoda, J.,** The effect of noradrenaline and acetylcholine on blood arginine vasotocin level in the domestic fowl, *Acta Physiol. Pol.,* 27, 269, 1976.
65. **Niezgoda, J.,** The effect of adrenergic blocking and stimulating agents on arginine vasotocin release in birds, *Acta Physiol. Pol.,* 27, 275, 1976.
66. **Katsoyannis, P. G. and Du Vigneaud, V.,** Arginine vasotocin, a synthetic analog of the posterior pituitary hormones containing the ring of oxytocin and the side chain of vasopressin, *J. Biol. Chem.,* 233, 1352, 1958.
67. **Katsoyannis, P. G. and Du Vigneaud, V.,** Arginine vasotocin, *Nature (London),* 184, 1465, 1959.
68. **Kitay, J. I. and Altschule, M. D.,** *The Pineal Gland,* Harvard University Press, Cambridge, 1954.
69. **Milcu, S. M., Pavel, S., and Neacsu, C.,** Biological and chromatographic characteristics of a polypeptide with pressor and oxytocic activities isolated from bovine pineal gland, *Endocrinology,* 72, 563, 1963.
70. **Pavel, S.,** Evidence for the presence of lysine vasotocin in the pig pineal gland, *Endocrinology,* 77, 812, 1965.
71. **Boissonnas, R. A. and Huguenin, R. L.,** Synthese de la Lys[8]-oxytocine (lysine-vasotocine) et nouvelle synthese de la lysine-vasopressine, *Helv. Chim. Acta,* 43, 182, 1960.
72. **Kimbrough, R. D., Jr. and Du Vigneaud, V.,** Lysine-vasotocin, a synthetic analogue of the posterior pituitary hormones containing the ring of oxytocin and the side chain of lysine vasopressin, *J. Biol. Chem.,* 236, 778, 1961.
73. **Cheesman, D. W. and Fariss, B. I.,** Isolation and characterization of a gonadotrophin-inhibiting substance from the bovine pineal gland, *Proc. Soc. Exp. Biol. Med.,* 133, 1254, 1970.
74. **Cheesman, D. W.,** Structural elucidation of a gonadotropin-inhibiting substance from the bovine pineal gland, *Biochim. Biophys. Acta,* 207, 247, 1970.
75. **Vizsolyi, E. and Perks, A. M.,** New neurohypophyseal principle in foetal mammals, *Nature (London),* 223, 1169, 1969.
76. **Pavel, S.,** Tentative identification of arginine vasotocin in human cerebrospinal fluid, *J. Clin. Endocrinol.,* 31, 369, 1970.
77. **Coculescu, M. and Pavel, S.,** Arginine vasotocin-like activity of cerebrospinal fluid in diabetes insipidus, *J. Clin. Endocr. Metab.,* 36, 1031, 1973.
78. **Popoviciu, L., Corfariu, O., Foldes, A., Farkas, E., Goldstein, R., and Pavel, S.,** REM sleep dependent release of vasotocin into cerebrospinal fluid of narcoleptics, *Waking and Sleeping,* 3, 341, 1979.
79. **Pavel, S., Goldstein, R., Popoviciu, L., Corfariu, O., Foldes, A., and Farkas, E.,** Pineal vasotocin: REM sleep dependent release into cerebrospinal fluid of man, *Waking and Sleeping,* 3, 347, 1979.
80. **Pavel, S.,** Presence of relatively high concentrations of arginine vasotocin in the cerebrospinal fluid of newborns and infants, *J. Clin. Endocr. Metab.,* 50, 271, 1980.
81. **Pavel, S.,** Evidence for the ependymal origin of arginine vasotocin in the bovine pineal gland, *Endocrinology,* 89, 613, 1971.
82. **Skowsky, W. R. and Fisher, D. A.,** Fetal neurohypophyseal arginine vasopressin and arginine vasotocin in man and sheep, *Pediatr. Res.,* 11, 627, 1977.
83. **Legros, J. J., Louis, F., Demoulin, A., and Franchimont, P.,** Immunoreactive neurophysins and vasotocin in human foetal pineal glands, *J. Endocrinol.,* 69, 289, 1976.
84. **Rosenbloom, A. A. and Fisher, D. A.,** Arginine vasotocin in the rabbit subcommissural organ, *Endocrinology,* 96, 1038, 1970.
85. **Pavel, S.,** Vasotocin biosynthesis by neurohypophysial cells from human fetuses. Evidence for its ependymal origin, *Neuroendocrinology,* 19, 150, 1975.
86. **Pavel, S., Goldstein, R., and Calb., M.,** Vasotocin content in the pineal gland of foetal, newborn and adult male rats, *J. Endocrinol.,* 66, 283, 1975.
87. **Reinharz, A. G., Czernichow, P., and Vallotton, M. B.,** Neurophysin-like protein in bovine pineal gland, *J. Endocrinol.,* 62, 35, 1974.
88. **Reinharz, A. C., Czernichow, P., and Vallotton, M. B.,** Neurophysins I and II from the bovine posterior pituitary lobe and neurophysin-like proteins from bovine pineal gland, *Ann. N. Y. Acad. Sci.,* 248, 172, 1975.
89. **Reinharz, A. C. and Vallotton, M. B.,** Presence of two neurophysins in the human pineal gland, *Endocrinology,* 100, 994, 1977.
90. **Legros, J. J., Louis, F., Grotschel-Stewart, U., and Franchimont, P.,** Presence of immunoreactive neurophysin-like material in human target organs and pineal gland: physiological meaning, *Ann. N.Y. Acad. Sci.,* 248, 157, 1975.
91. **Reinharz, A. C., Pavel, S., and Vallotton, M. B.,** Evidence for in vitro release of neurophysin by the rat pineal gland, *Experientia,* 34, 1232, 1978.

92. **Legros, J. J., Louis, F., Demoulin, A., and Franchimont, P.,** Immunoreactive neurophysins and vasotocin in human foetal pineal glands, *J. Endocrinol.,* 69, 289, 1976.
93. **Bargmann, W.,** Neurosekretion und hypothalamisch-hypophysäres System, *Verh. Anat. Ges.,* 51, 30, 1954.
94. **Suomalainen, P.,** Stress and neurosecretion in the hibernating hedgehog, *Dall. Mus. Comp. Zool. Harvard Coll.,* 124, 271, 1960.
95. **Oksche, A.,** Survey of the development and function of the epiphysis cerebri, *Prog. Brain Res.,* 10, 627, 1965.
96. **Barry, J.,** Les voies extra-hypophysaires de la neurosecretion diencephalique, *Assoc. Anatomistes,* 89, 264, 1956.
97. **Bowie, E. P. and Herbert, D. C.,** Immunocytochemical evidence for the presence of arginine vasotocin in the rat pineal gland, *Nature (London),* 261, 5555, 1976.
98. **Pévet, P., Dogterom, J., Buijs, R. M., Ebels, I., Swabb, D. F., and Arimura, A.,** Presence of α-MSH-, AVT-, and LHRH-like compounds in the mammalian pineal and subcommissural organ and their relationship with the UMO5R pineal fraction, X Conf. Europ. Comp. Endocrinol., Sorrento, Italy, May 21, 1979.
99. **Pévet, P. and Swaab, D. F.,** Immunocytochemical evidence for the presence of an α-MSH-like compound in the rat pineal gland, *J. Physiol. (Paris),* 75, 101, 1979.
100. **Dogterom, J., Snijdewint, F. G. M., Pévet, P., and Swaab, D. F.,** Studies on the presence of vasopressin, oxytocin and vasotocin in the pineal gland, subcommissural organ and fetal pituitary gland: failure to demonstrate vasotocin in mammals, *J. Endocrinol.,* 84, 115, 1980.
101. **Pévet, P., Dogterom, J., Buijs, R. M., and Reinharz, A.,** Is it the vasotocin or a vasotocin-like peptide which is present in the mammalian pineal and subcommissural organ?, *J. Endocrinol.,* 80, 49P, 1978.
102. **Buijs, R. M. and Pévet, P.,** Vasopressin- and oxytocin-containing fibers in the pineal gland and subcommissural organ of the rat, *Cell. Tissue Res.,* 205, 11, 1980.
103. **Pévet, P., Reinharz, A. C., and Dogterom, J.,** Neurophysins, vasopressin and oxytocin in the bovine pineal gland, *Neuroscience Lett.,* 16, 301, 1980.
104. **Coculescu, M., Zaoral, M., and Matulevicius, V.,** Tentative identification of arginine vasotocin in the bovine pineal extract prepared by the Milcu-nanu method, *Rev. Roum. Med. Endocrinol.,* 15, 27, 1977.
105. **Gitelman, H. J., Klapper, D. G., Alderman, F. R., and Blythe, W. B.,** Ala-Gly and Val-Asp- [Arg[8]]-vasopressin: bovine storage forms of arginine vasopressin with natriuretic activity, *Science,* 207, 893, 1980.
106. **Neacsu, C.,** The mechanism of antigonadotrophic action of a polypeptide extracted from a bovine pineal gland, *Rev. Roum. Physiol.,* 9, 161, 1972.
107. **Pavel, S., Goldstein, R., Ghinea, E., and Calb, M.,** Chromatographic evidence for vasotocin biosynthesis by cultured pineal ependymal cells from rat fetuses, *Endocrinology,* 100, 205, 1977.
108. **Pavel, S., Ghinea, E., Goldstein, R., and Matulevicius, V.,** Vasotocin biosynthesis by cultured pineal glands from adult male rats, *J. Endocrinol.,* 77, 147, 1978.
109. **Coculescu, M., Matulevicius, V., Goldstein, R., and Pavel, S.,** Presence and synthesis of vasotocin in the pineal gland of Brattleboro rats, *J. Endocrinol.,* 77, 145, 1978.
110. **Pavel, S., Dorcescu, M. Petrescu-Holban, R., and Ghinea, E.,** Biosynthesis of a vasotocin-like peptide in cell cultures from pineal glands of human fetuses, *Science,* 181, 1252, 1973.
111. **Rosenbloom, A. A. and Fisher, D. A.,** Radioimmunoassayable AVT and AVP in adult mammalian brain tissue: comparison of normal and Brattleboro rats, *Neuroendocrinology,* 17, 354, 1975.
112. **Sartin, J. L., Bruot, B. C., and Orts, R. J.,** Neurotransmitter regulation of arginine vasotocin release from rat pineal glands *in vitro, Acta Endocrinol.,* 91, 571, 1979.
113. **Romijn, H. J.,** Structure and innervation of the pineal gland of the rabbit, *Oryctolagus cuniculus* (L.). III. An electron microscopic investigation of the innervation, *Cell. Tissue Res.,* 157, 24, 1975.
114. **Wartman, S. A., Branch, B. J., George, R., and Taylor, A. N.,** Evidence for a cholinergic influence on pineal hydroxyindole *O*-methyltransferase activity with changes in environmental lighting, *Life Sci.,* 8(7), 1263, 1969.
115. **Eränko, O., Rechard, L., Eränko, L., and Cunningham, A.,** Light and electron microscopic histochemical observations on cholinesterase-containing sympathetic nerve fibers in the pineal body of the rat, *Histochem. J.,* 2, 479, 1970.
116. **Machado, A. B. M. and Lemos, V. P. J.,** Histochemical evidence for a cholinergic sympathetic innervation in the rat pineal body, *J. Neuro-Visc. Relat.,* 32, 104, 1971.
117. **Cusack, B. M., Fisher, L. A., and Fernstrom, J. D.,** Increase in plasma vasotocin levels following norepinephrine injection, *Fed. Proc. Fed. Am. Soc. Exp. Biol.,* 37, 437, 1978.
118. **Kühn, E. R.,** Cholinergic and adrenergic release mechanism for vasopressin in the male rat: a study with injections of neurotransmitters and blocking agents into the third ventricle, *Neuroendocrinology,* 16, 255, 1974.

119. **Calb, M., Goldstein, R., and Pavel, S.,** Diurnal rhythm of vasotocin in the pineal of the male rat, *Acta Endocrinol.,* 84, 523, 1977.
120. **Sartin, J., Bruot, B., and Orts, R.,** Blinding induced alterations in pineal gland arginine vasotocin and pituitary and plasma LH, FSH and prolactin, *Fed. Proc. Fed. Am. Soc. Exp. Biol.,* 37, 437, 1978.
121. **Coculescu, M., Serbanescu, A., and Temeli, E.,** Influence of arginine vasotocin administration on nocturnal sleep of human subjects, *Rev. Roum. Morphol. Embryol. Physiol.,* 16, 173, 1979.
122. **Pavel, S., Psatta, D., and Goldstein, R.,** Slow-wave sleep induced in cats by extremely small amounts of synthetic and pineal vasotocin injected into the third ventricle of the brain, *Brain Res. Bull.,* 2, 251, 1977.
123. **Mendelson, W. B., Gillin, J. C., Pisner, G., and Wyatt, R. J.,** Arginine vasotocin and sleep in the rat, *Brain Res.,* 182, 246, 1980.
124. **Pang, P. K. T.,** Osmoregulatory functions of neurohypophysial hormones in fishes and amphibians, *Am. Zool.,* 17, 739, 1977.
125. **Moses, A. and Miller, M.,** Osmotic influences on the release of vasopressin, in, *Handbook of Physiology,* Vol. 4 (Part I), American Physiological Society, Washington, D. C., 1974, 225.
126. **Pavel, S. and Coculescu, M.,** Arginine vasotocin-like activity of cerebrospinal fluid induced by injection of hypertonic saline into the third cerebral ventricle of cats, *Endocrinology,* 91, 925, 1972.
127. **Goldstein, R. and Pavel, S.,** Vasotocin release into the cerebrospinal fluid of cats induced by luteinizing hormone releasing hormone, thyrotrophin releasing hormone and growth hormone release-inhibiting hormone, *J. Endocrinol.,* 75, 175, 1977.
128. **Pavel, S., Goldstein, R., Gheorghiu, C., and Calb, M.,** Pineal vasotocin: release into cat cerebrospinal fluid by melanocyte-stimulating hormone release-inhibiting factor, *Science,* 197, 179, 1977.
129. **Skowsky, E. and Swan, L.,** Effects of hypothalamic releasing hormones on neurohypophyseal arginine vasopressin (AVP) secretion, Abstr. Vth Int. Cong. Endocrinology, Hamburg, 197, 1976.
130. **Pavel, S.,** Arginine vasotocin release into cerebrospinal fluid of cats induced by melatonin, *Nature (London),* 246, 183, 1973.
131. **Pavel, S. and Goldstein, R.,** Further evidence that melatonin represents the releasing hormone for pineal vasotocin, *J. Endocrinol.,* 82, 1, 1979.
132. **Lemay, A., Brouillette, A., Denizeau, F., and Lavoie, M.,** Melatonin- and serotonin-stimulated release of vasopressin from rat neurohypophysis perifused in vitro, Abstr. 61st Ann. Mtg. Endocrine Soc., Endocrine Society, Bethesda, Md., 161, 1979.
133. **Lukasazyk, A. and Reiter, R. J.,** Histophysiological evidence for the secretion of polypeptides by the pineal gland, *Am. J. Anat.* 143, 451, 1975.
134. **Zatz, M.,** Sensitivity and cyclic nucleotides in the rat pineal gland, *J. Neural Transm.,* Suppl. 13, 97, 1978.
135. **Oleshansky, M. A. and Neff, N. H.,** Studies on the control of pineal indole synthesis: cyclic nucleotides, adenylate cyclase and phosphodiesterase, *J. Neural Transm.,* Suppl. 13, 81, 1978.
136. **Wurtman, R. J. and Ozaki, Y.,** Physiological control of melatonin synthesis and secretion: mechanisms generating rhythms in melatonin, methoxytryptophol, and arginine vasotocin levels and effects on the pineal of endogenous catecholamines, the estrous cycle and environmental lighting, *J. Neural Transm.,* Suppl. 13, 59, 1978.
137. **Sartin, J. L., Bruot, B. C., and Orts, R. J.,** Interaction of arginine vasotocin and norepinephrine upon pineal indoleamine synthesis in vitro, *Mol. Cell. Endocrinol.,* 11, 7, 1978.
138. **Pavel, S. and Petrescu, S.,** Inhibition of gonadotrophin by a highly purified pineal peptide and by synthetic arginine vasotocin, *Nature (London),* 212, 1954, 1966.
139. **Vaughan, M. K., Vaughan, G. M., Blask, D. E., Barnett, M. P., and Reiter, R. J.,** Arginine vasotocin: structure-activity relationships and influence on gonadal growth and function, *Am. Zool.,* 16, 23, 1976.
140. **Smith, M. L.,** Pineal modulation of the rodent ovulatory cycle, *Diss. Abstr.,* 33, 4476, 1973.
141. **Johnson, L. Y., Vaughan, M. K., Reiter, R. J., Blask, D. E., and Rudeen, P. K.,** The effects of arginine vasotocin on pregnant mare's serum-induced ovulation in the immature female rat, *Acta Endocrinol.,* 87, 367, 1978.
142. **Johnson, L. Y. and Waring, P. J.,** Effect of arginine vasotocin (AVT) on pregnant mare's serum (PMS)-induced ovulation in the immature female rat, *Anat. Rec.,* 184, 437, 1976.
143. **Cheesman, D. W. and Forsham, P. H.,** Inhibition of induced ovulation by a highly purified extract of the bovine pineal gland, *Proc. Soc. Exp. Biol. Med.,* 146, 722, 1974.
144. **Moszkowska, A. and Ebels, I.,** A study of antigonadotrophic action of synthetic arginine vasotocin, *Experientia,* 24, 610, 1968.
145. **Vaughan, M. K., Vaughan, G. M., and Reiter, R. J.,** Inhibition of human chorionic gonadotrophin-induced ovarian and uterine growth in the mouse by synthetic arginine vasotocin, *Experientia,* 31, 862, 1975.

146. **Vaughan, M. K., Vaughan, G. M., and Reiter, R. J.,** Inhibition of human chorionic gonadotrophin-induced hypertrophy of the ovaries and uterus in immature mice by some pineal indoles, 6-hydroxy-melatonin and arginine vasotocin, *J. Endocrinol.*, 68, 397, 1976.
147. **Cheesman, D. W. and Fariss, B. L.,** Isolation and characterization of a gonadotrophin-inhibiting substance from the bovine pineal gland, *Proc. Soc. Exp. Biol. Med.*, 133, 1254, 1970.
148. **Vaughan, M. K., Reiter, R. J., and Vaughan, G. M.,** Fertility patterns in female mice following treatment with arginine vasotocin or melatonin, *Int. J. Fertil.*, 21, 65, 1976.
149. **Vaughan, M. K. and Blask, D. E.,** Arginine vasotocin — a search for its function in mammals, *Prog. Reprod. Biol.*, 4, 90, 1978.
150. **Cheesman, D. W., Osland, R. B., and Forsham, P. H.,** Suppression of the preovulatory surge of luteinizing hormone and subsequent ovulation in the rat by arginine vasotocin, *Endocrinology,* 101, 1194, 1977.
151. **Osland, R. B., Cheesman, D. W., and Forsham, P. H.,** Studies on the mechanism of the suppression of the preovulatory surge of luteinizing hormone in the rat by arginine vasotocin, *Endocrinology,* 101, 1203, 1977.
152. **Cheesman, D. W., Osland, R. B., and Forsham, P. H.,** Effects of 8-arginine vasotocin on plasma prolactin and follicle-stimulating hormone surges in the proestrous rat, *Proc. Soc. Exp. Biol. Med.*, 156, 369, 1977.
153. **Pavel, S., Petrescu, M., and Vicoleanu, N.,** Evidence of a central gonadotropin inhibiting activity of arginine vasotocin in the female mouse, *Neuroendocrinology,* 11, 370, 1973.
154. **Blask, D. E., Vaughan, M. K., Reiter, R. J., and Johnson, L. Y.,** Influence of arginine vasotocin on the estrogen-induced surge of LH and FSH in adult ovariectomized rats, *Life Sci.*, 23, 1035, 1978.
155. **Vaughan, M. K., Buchanan, J., Blask, D. E., Reiter, R. J., and Sheridan, P. J.,** Diurnal variation in uterine estrogen receptors in immature female rats — inhibition by arginine vasotocin, *Endocr. Res. Commun.*, 6, 191, 1979.
156. **Ying, S. Y.,** Inhibition of ovulation with a single injection of arginine vasotocin in the rabbit, Abstr. 61st Ann. Mtg. Endocrine Soc., Endocrine Society, Bethesda, Md., 124, 1979.
157. **Pavel, S.,** Effects of pineal and synthetic arginine vasotocin on the gonads of prepuberal male mice, 5th Conf. Eur. Comp. Endocrinol., (Abstr.), Utrecht, 129, 1969.
158. **Vaughan, M. K., Reiter, R. J., McKinney, T., and Vaughan, G. M.,** Inhibition of growth of gonadal dependent structures by arginine vasotocin and purified bovine pineal fractions in immature mice and hamsters, *Int. J. Fertil.*, 19, 103, 1974.
159. **Vaughan, M. K., Vaughan, G. M., and Klein, D. C.,** Arginine vasotocin: effects on development of reproductive organs, *Science,* 186, 938, 1974.
160. **Young, L. L., Hadley, M. E., and Barrow, W. O.,** Failure to demonstrate antigonadotrophic activities of arginine vasotocin in the mouse, *Hormone Res.*, 10, 88, 1979.
161. **Reiter, R. J., Blask, D. E., Johnson, L. Y., Rudeen, P. K., Vaughan, M. K., and Waring, P. J.,** Melatonin inhibition of reproduction in the male hamster: its dependency on time of day of administration and on an intact and sympathetically innervated pineal gland, *Neuroendocrinology,* 22, 107, 1976.
162. **Pavel, S., Luca, N., Calb, M., and Goldstein, R.,** Inhibition of release of luteinizing hormone in the male rat by extremely small amounts of arginine vasotocin: further evidence for the involvement of 5-hydroxytryptamine-containing neurons in the mechanism of action of arginine vasotocin, *Endocrinology,* 104, 517, 1979.
163. **Vaughan, M. K., Trakulrungsi, C., Petterborg, L. J., Johnson, L. Y., Blask, D. E., Trakulrungsi, W., and Reiter, R. J.,** Interaction of luteinizing hormone-releasing hormone, cyproterone acetate and arginine vasotocin on plasma levels of luteinizing hormone in intact and castrated adult male rats, *Mol. Cell. Endocrinol.*, 14, 59, 1979.
164. **Pavel, S., Luca, N., Calb, M., and Goldstein, R.,** Reversal by arginine vasotocin of the effects of pinealectomy on the amount of 5-hydroxytryptamine in the hypothalamus and the concentrations of luteinizing hormone and follicle-stimulating hormone in the plasma of immature male rats, *J. Endocrinol.*, 84, 159, 1980.
165. **Demoulin, A., Hudson, B., Franchimont, P., and Legros, J. J.,** Arginine vasotocin does not affect gonadotrophin secretion *in vitro, J. Endocrinol.*, 72, 105, 1977.
166. **Vaughan, M. K., Blask, D. E., Johnson, L. Y., and Reiter, R. J.,** Prolactin-releasing activity of arginine vasotocin in vitro, *Hormone Res.*, 6, 342, 1975.
167. **Griffiths, E. C. and Hooper, K. C.,** Competitive inhibition between oxytocin and luteinizing hormone-releasing factor (LRF) for the same enzyme system in the rat hypothalamus, *Acta Endocrinol.*, 75, 435, 1974.
168. **Carter, D. A. and Dyer, R. G.,** Inhibition by pentobarbitone and urethane on the *in vitro* response of the adenohypophysis to luteinizing hormone-releasing hormone in male rats, *Br. J. Pharmacol.*, 67, 277, 1979.

169. **Vaughan, M. K., Blask, D. E., Johnson, L. Y., and Reiter, R. J.**, The effect of subcutaneous injections of melatonin, arginine vasotocin, and related peptides on pituitary and plasma levels of luteinizing hormone, follicle-stimulating hormone, and prolactin in castrated adult male rats, *Endocrinology*, 104, 212, 1979.
170. **Yamashita, K., Mieno, M., and Yamashita, E.**, Suppression of the luteinizing hormone releasing effect of luteinizing hormone releasing hormone by arginine-vasotocin, *J. Endocrinol.*, 81, 103, 1979.
171. **Yamashita, K., Mieno, M., and Yamashita, E.**, Suppression of the luteinizing hormone releasing effect of luteinizing hormone releasing hormone by arginine-vasotocin in immature male dogs, *J. Endocrinol.*, 84, 449, 1980.
172. **Mackenzie, K.**, An experimental investigation of the mechanism of milk secretion with special reference to the action of animal extracts, *Q. J. Exp. Physiol.*, 4, 305, 1911.
173. **Blask, D. E., Vaughan, M. K., Reiter, R. J., Johnson, L. Y., and Vaughan, G. M.**, Prolactin-releasing and release-inhibiting factor activities in the bovine rat and human pineal gland: *in vitro* and *in vivo* studies, *Endocrinology*, 99, 152, 1976.
174. **Chang, N., Ebels, I., and Benson, B.**, Preliminary characterization of bovine pineal prolactin releasing (PPRF) and release-inhibiting factor (PPIF) activity, *J. Neural Transm.*, 46, 139, 1979.
175. **Zrjakov, O. N. and Kornjushenko, N. P.**, Effect of epiphysis factors on the lactotrophic function of the adenohypophysis, *Proc. Ukrainian Acad. Sci. (PCP), Ser. B*, 263, 1976.
176. **Demoulin, A., Hudson, B., Legros, J. J., and Franchimont, P.**, Influence d'un extrait de glandes pinéales ovines sur al libération de prolactien *in vitro*, *C. R. Seances Soc. Biol. Paris*, 171, 1134, 1977.
177. **Reiter, R. J.**, Photoperiods, pineal and reproduction, in, *The Pineal Vol. 3*, Eden Press, Montreal, 1978.
178. **Bartke, A., Croft, B. T., and Dalterio, S.**, Prolactin restores plasma testosterone levels and stimulates testicular growth in hamsters exposed to short daylength, *Endocrinology*, 97, 1601, 1975.
179. **Forsling, M. L., Reindardt, V., and Himmler, V.**, Neurohypophysial hormones and prolactin release, *J. Endocrinol.*, 63, 579, 1974.
180. **Vaughan, M. K., Little, J. C., Johnson, L. Y., Blask, D. E., Vaughan, G. M., and Reiter, R. J.**, Effects of melatonin and natural and synthetic analogues of arginine vasotocin on plasma prolactin levels in adult male rats, *Hormone Res.*, 9, 236, 1978.
181. **Shani, J., Urbach, L., Terkel, J., and Goldhaber, G.**, Serum prolactin and LH in chronically-cannulated cycling rats after intra-atrial administration of oxytocin, *Arch Int. Pharmacodyn. Ther.*, 221, 323, 1976.
182. **Del Pozo, E., Kleinstein, J., Del Re, R. B., Derrer, F., and Martin-Perez, J.**, Failure of oxytocin and lysine-vasopressin to stimulate prolactin release in humans, *Horm. Metab. Res.*, 12, 26, 1980.
183. **Padmanabhan, V., Convey, E. M., and Tucker, H. A.**, Pineal compounds alter prolactin release from bovine pituitary cells, *Proc. Soc. Exp. Biol. Med.*, 160, 340, 1979.
184. **Shiino, M., Ishikawa, H., and Rennels, E. G.**, Four types of prolactin producing clonal strains derived from Rathke's pouch epithelium, *Abstr. 59th Ann. Mtg. Endocr. Soc.*, Endocrine Society, Bethesda, Md., 123, 1977.
185. **Blask, D. E., Vaughan, M. K., and Reiter, R. J.**, Modification by adrenergic blocking agents of arginine vasotocin-induced prolactin secretion *in vitro*, *Neurosci. Lett.*, 6, 91, 1977.
186. **Vaughan, M. K., Blask, D. E., Vaughan, G. M., and Reiter, R. J.**, Dose-dependent prolactin releasing activity of arginine vasotocin in intact and pinealectomized estrogen-progesterone treated adult male rats, *Endocrinology*, 99, 1319, 1976.
187. **Vaughan, M. K., Little, J. C., Johnson, L. Y., Vaughan, G. M., and Reiter, R. J.**, Stimulation of rat prolactin secretion *in vivo* by arginine vasotocin: influence of age of solution, nighttime administration, and dose, *Neuroendocrinology*, 24, 35, 19.
188. **Johnson, L. Y., Vaughan, M. K., and Reiter, R. J.**, The effects of arginine vasotocin, a pineal peptide, on prolactin secretion and the influence of urethane anesthesia, *Anat. Rec.*, 193, 577, 1979.
189. **Vaughan, M. K., Johnson, L. Y., Petterborg, L. J., and Ferguson, B. N.**, Effect of arginine vasotocin on prolactin secretion in unanesthetized and urethane-anesthetized estrogen-progesterone treated adult male rats, *Fed. Proc. Fed. Am. Soc. Exp. Biol.*, 38(2), 983, 1979.
190. **Pavel, S., Calb, M., and Georgescu, M.**, Reversal of the effects of pinealectomy on the pituitary prolactin content in mice by very low concentrations of vasotocin injected into the third cerebral ventricle, *J. Endocrinol.*, 66, 289, 1975.
191. **Vaughan, M. K., Blask, D. E., Johnson, L. Y., Trakulrungsi, C., and Reiter, R. J.**, Effect of several androgens, cyproterone acetate or estrogen-progesterone on the prolactin-releasing activity of arginine vasotocin in castrated male rats, *Mol. Cell. Endocrinol.*, 12, 309, 1978.
192. **Meites, J. and Clemens, J. A.**, Hypothalamic control of prolactin secretion, in, *Vitamins and Hormones: Advances in Research and Application*, Harris, R. S., Munson, P., Diczfalusky, E., and Glover, J. Eds., Churchill Livingstone, London, 1972, 165.

193. **Lamberts, S. W. J. and MacLeod, R. M.**, *Physiological and Pathological Aspects of Prolactin Secretion,* Vol. 1, Eden Press, Montreal, 1977, 149.
194. **Meites, J., Bruni, J. F., and Van Vugt, D. A.**, Effects of endogenous opiate peptides on release of anterior pituitary hormones, in *Central Nervous System Effects of Hypothalamic Hormones and Other Peptides,* Collu, R., Ed., Raven Press, New York, 1979, 261.
195. **de Bodo, R. C.**, The antidiuretic action of morphine and its mechanism, *J. Pharmacol. Exp. Ther.,* 82, 74, 1944.
196. **Duke, H. N., Pickford, M., and Watt, J. A.**, The antidiuretic action of morphine: its site and mode of action in the hypothalamus of the dog, *Q. J. Exp. Physiol.,* 36, 149, 1951.
197. **Simantov, R. and Synder, S.**, Opiate receptor binding in the pituitary gland, *Brain Res.,* 124, 177, 1977.
198. **Firemark, H. M. and Weitzman, R. E.**, Effects of β-endorphin, morphine and nalaxone on arginine vasopressin secretion and the electroencephalogram, *Neuroscience,* 4, 1895, 1979.
199. **Blask, D. E. and Vaughan, M. K.**, Naloxone inhibits arginine vasotocin (AVT)-induced prolactin release in urethane-anesthetized male rats *in vivo, Neurosci. Lett.,* 18, 181, 1980.
200. **Vaughan, M. K., Johnson, L. Y., Dinh, D. T., Blask, D. E., Guerra, J. C., De los Santos, R., and Reiter, R. J.**, Effect of arginine vasotocin and/or naloxone treatment on plasma prolactin, LH and FSH secretion in unanesthetized adult male rats, *Anat. Rec.,* 196, 195A, 1980.
201. **Smythe, G. A. and Lazarus, L.**, Blockade of the dopamine-inhibitory control of prolactin secretion in rats by 3, 4-dimethoxyphenylethylamine (3, 4-di-*O*-methyldopamine), *Endocrinology,* 93, 147, 1973.
202. **Delanoy, R. L., Dunn, A. J., and Tintner, R.**, Behavioral responses to intracerebroventricularly administered neurohypophyseal peptides in mice, *Horm. Behav.,* 11, 348, 1978.
203. **de Wied, D. and Bohus, B.**, The modulation of memory processes by vasotocin, the evolutionarily oldest neurosecretory principle, *Prog. Brain Res.,* 48, 327, 1978.
204. **Vaughan, M. K.**, unpublished results.

Chapter 7

OTHER PINEAL PEPTIDES AND RELATED SUBSTANCES — PHYSIOLOGICAL IMPLICATIONS FOR REPRODUCTIVE BIOLOGY

Bryant Benson and Ielskina Ebels

TABLE OF CONTENTS

I. MORPHOLOGICAL EVIDENCE FOR PINEAL SECRETION

It is most likely that the mammalian pinealocyte develops from secretory rudimentary photoreceptor cells and are not to be considered as neurons. Pinealocyte morphology reveals cells which appear to be specialized for endocrine secretion. Abundant, rough and smooth endoplasmic reticulum, as well as free ribosomes and dense-cored vesicles manufactured by active Golgi complexes, predict the production of proteinaceous and other compounds. Ueck and Wake[1] consider the pinealocyte as a paraneuron, a cellular concept based on the ability of certain cells to possess mechanisms for amine precursor uptake and decarboxylation (APUD), after the definition of the APUD cell as described by Pearse.[2] According to Fujita[3] such cells would produce neurotransmitters or related substances and proteins or peptidic hormonal substances in response to nervous sensory input.

A. Pineal Innervation

The innervation of the pineal of vertebrates including mammals has been reviewed recently by Kappers.[4] It is known that in mammals there is a rich pinealopetal orthosympathetic innervation by nerves originating in the superior cervical ganglion and reaching the gland via the nervii conarii. Other parasympathetic nerves probably reach the gland from the superficial petrosal nerve via the conarii nerves and synapse for the most part with intramural parasympathetic nerve cell soma. While the function of the latter is not known, that of the sympathetic innervation has been demonstrated to be of primary importance for pineal secretory function. In neither case are there true synaptic contacts with the pinealocytes since the nerve fibers end freely in either the parenchyma or pericapillary spaces.

New evidence indicates also that fibers may reach the gland from other sources, including the habenula nuclei.[5] There are suspected connections with the limbic system via these nuclei.[6] Rarely, have neurosecretory fibers been identified in the pineal, most likely in the pineal stalk.[7-9]

B. Pinealocyte Secretion

In certain animals a type of pinealocyte secretory process has been described by Pévet[10] as ependymal-like. Accumulation of proteinaceous material in the granulated endoplasmic reticulum and the formation of vacuoles by cisternae of the granular endoplasmic reticulum is observed ultrastructurally. It is problematic whether the vacuoles containing floculent material are involved in the transport of newly synthesized substances to the Golgi where it may be packaged for secretion, or whether the material is released directly into the extracellular space.

In certain mammals, including the rabbit, hamster, and mouse, dense-cored or granulated vesicles migrate to the end of long pinealocyte processes where secretory products appear to be delivered to the pericapillary spaces and intercellular spaces.[11] As pointed out by Pévet,[10] dense-cored or granulated vesicle formation by Golgi saccules has been observed in the pinealocytes of most mammals, and the rate of their formation is dependent on different physiological conditions. An increase is observed in vitro after the addition of norepinephrine[12] and cyclic AMP.[13] In vivo, granulated vesicle numbers are seen to change in relationship to photoperiodic conditions and light deprivation by bilateral optic enucleation. In general, prolonged periods of light produce inhibition of pineal functional activity.[14-16] Ultrastructural changes characteristic of decreased metabolic and sympathetic activity, and reduction in the number of granulated vesicles are seen.[17] The results are explained as resulting from sympatholysis by photoinhibition.[18-20]

Animals maintained under diurnal lighting conditions demonstrate, on the other hand, a circadian rhythm in the number of pinealocyte granulated vesicles. Romijn et al.[21] observed in the rabbit highest amounts of granulated vesicles adjacent to Golgi saccules at midphotoperiod, and in pinealocyte terminals at late photoperiod. Lowest numbers were obtained during the dark phase. In our studies in the mouse,[22] a similar rhythm of higher amplitude was observed in pericapillary pinealocyte cytoplasm. The rhythm was abolished by bilateral superior cervical ganglionectomy. Release of the granulated vesicles during the dark phase was inhibited by reserpine, and melatonin treatment resulted in partial reversal of this effect. On the other hand, parachlorophenylalanine, an inhibitor of serotonin synthesis, had no significant effects on the circadian rhythm.[23] Recently, we[24] have observed a similar, yet low amplitude, circadian rhythm in the pinealocytes of hamsters maintained under various conditions of alternating light. The late photoperiod peak in granulated vesicle numbers was reduced as the photoperiod was reduced from 14 to 12 and then to 10 hr, and dark phase minima were lowest in hamsters in the shorter photoperiods. Taken together, these results suggest that the synthesis and packaging of pineal secretory product in granulated vesicles occurs during the photoperiod at a rate increased over the rate of secretion, and accumulation occurs at this time as reflected by increased numbers. During the scotoperiod, when norepinephrine is released at the pinealotrophic nerve terminals,[25] the granulated vesicles are apparently released or the secretory product changed into another form for secretion as evidenced by decrease in numbers.

The use of proteases has established that the compounds present in certain dense-cored vesicles are proteinaceous in nature.[26] Dense-cored vesicles could contain a specific complex peptidergic neurohormone to which a portion of the indole aminergic pool may be bound. In this hypothetical process amino acid uptake and biosynthesis into protein would occur by ribosomal mechanisms and segregation in the cisternal spaces of the rough endoplasmic reticulum. This would be followed by intracisternal transport of protein with concentration in the Golgi apparatus where the dense-cored vesicles are formed. Upon demand, and depending upon the status of the lighting conditions and the mediation of those stimuli via pineal innervation, the dense-cored granulated vesicles would migrate from the cell body to the perivascular processes and there be released by exocytosis or diffusion after disolution into the pericapillary space, with the secretory products eventually reaching the capillary lumen. Simultaneously, the pinealocyte would be actively involved in the uptake of tryptophan and its hydroxylation by monooxygenases located in the mitochondria. The 5-hydroxytryptophan would be converted to serotonin by L-amino acid decarboxylase, and once formed, serotonin could be faculatively taken up by dense-cored vesicles or stored in the cytosol. In addition to the pinealocytes, it is possible that the interstitial cells also may store serotonin. In either case, only a small portion of the serotonin would be converted to melatonin which could act in the formation and storage processes of pineal proteinaceous secretory product.

The two types of secretory processes described may or may not represent the synthesis and release of separate hormones. Pévet[10] has drawn an analogy between pinealocyte secretion and the mechanisms proposed for insulin secretion. In this scheme different forms of a pineal hormone may be secreted by different mechanisms, i.e. the ependymal type or via granular vesicles, and the ratio of the different processes at any one time may depend upon the demand for secretion placed on the pinealocyte, presumably regulated by the periodic synthesis of pinealotropic neurotransmitter(s), which in general is inhibited by light. It is more certain that mammalian pinealocytes contain the organelles requisite for the synthesis and release of proteinaceous hormone by either of the processes mentioned above, and the special morphology of pinealocytes, with processes extending out to perivascular spaces and intercellular cannaliculi, provides ready access for pineal secretions to the perivascular space and hence the blood vascular system.

How the intracellular proteinaceous processes are regulated is of course not understood. Only a small portion of pineal serotonin is directly involved in the synthesis of melatonin, suggesting that serotonin must then play another role in the gland. Autoradiographic and other studies in nonmammalian species[26] show that serotonin coexists with proteinaceous compounds in the granulated vesicles. A recent cytochemical study by Lu and Lin[27] has shown that granulated vesicles in pinealocytes of the hamster contain serotonin. Using the modified chromaffin reaction we also have found reaction product over granulated vesicles in the pinealocytes of mice, implicating the presence of serotonin.[151] Pinealocytes, if considered "paraneurons",[1] could release peptidergic hormones by a mechanism in which serotonin would change membrane permeability.

In recent years certain authors have suggested the possibility that one of the target organs of melatonin or other methoxyindoles could be the pineal itself.[28-30] Ultrastructural studies by Freire and Cardinali[31] and Benson and Krasovich[22] support this view since the characteristics of pinealocytes of animals treated with melatonin are similar to those observed in light-deprived animals. In these animals, granulated vesicle synthesis and release are stimulated. It is possible that similar to other amines[32] the methoxyindole amines, including melatonin, may be functionally related to the formation, storage, and/or release of pineal hormone(s).

C. Pineal Secretion In Vitro

Only a relatively few experiments have been carried out in which pineal secretion of nonindolic substances have been studied in vitro. In 1972 Benson et al.[33] reported on the presence of an antigonadotrophic substance in rat pineal incubation. When injected into adult mice, media from short-term incubations were observed to block compensatory ovarian hypertrophy. The substance in the incubation media was inactivated by treatment with proteolytic enzymes. When injected i.v. into urethane-anesthetized male rats, certain ultrafiltrates of rat pineal incubation media were seen to reduce serum prolactin levels,[34] suggesting that the pineal gland may produce prolactin regulating substances.

An LRF-like substance was identified in media from long-term rat pineal organ culture.[35] Apparently, the pineal substance is capable of inhibiting LRF-stimulated release of LH in vitro, and reacts with antibodies to LRH, but these investigations are silent with respect to methods of purification or chemical characteristics of the putative pineal secretory product.

D. Nature of Pineal Secretory Products

Prevailing views of pineal function implicate a regulatory role in the control of most adenohypophysial hormones and a number of other parameters.[4,36] Through experiments in which a variety of animal models were employed, it is now apparent that the pineal gland secretes substances which influence the development and function of the reproductive system.[37,38]

In a recent review Pévet[10] states that the previous view of melatonin as a substance unique to the pineal must be discarded since it is now known that melatonin is detected in the blood of pinealectomized animals[39-41] and is synthesized in the Harderian gland,[42-44] intestine,[45,46] intestinal enterochromaffin cells,[47] and retina.[44,48-53] In 1978 Pévet et al.[54] showed that the atrophied eyes of the mole produce more methoxyindoles than the pineal, and concluded therefore, that many of the effects of synthetic melatonin and related 5-methoxy derivatives observed in numerous experiments may not be relevant to unique pineal physiology, but more likely may represent in his opinion pharmacological effects on the central nervous system and hypothalamus. If correct, this line of reasoning gives more importance to the physiological implications of the biologically active, nonindolic substances discussed below. What, then, is the chemical nature of pineal secretory products?

II. PINEAL EXTRACTS AND UNIDENTIFIED PINEAL FRACTIONS

A. Effects on Gonadotrophin Secretion In Vitro

Pioneer experiments by Moszkowska (see Ebels[55]) demonstrated the ability of fresh pineal bodies of rats and sheep, or acetone-dried sheep pineal powder, to diminish or inhibit gonadotrophin secretion by rat pituitaries in vitro. In 1965, Ebels et al.[56] obtained two fractions after gel filtration of an aqueous extract of sheep pineals on Sephadex G-25. One fraction (F_2) stimulated, and a second low molecular fraction (F_3) inhibited, follicle-stimulating hormone activity in vitro. These biologically active fractions were further studied using Amberlite® IRC-5, XE64 and Sephadex G-10.[57] Control cortical and cerebellar extracts subjected to gel filtration on Sephadex G-25 showed no antigonadotrophic activity. In other experiments, it was shown that the antigonadotrophic substance in sheep pineals was not melatonin, 5-methoxytryptophol, serotonin, or arginine vasotocin.[58,59] It was concluded that the antigonadotrophic activity in the Sephadex G-25 fraction F_3 was specific for the pineal body, but that it was labile under the conditions used.[59,60]

Subsequent studies utilizing methods for the isolation of indoleamines and antigonadotrophic substances[61-64] revealed the presence of a substance showing special fluorescence characteristics in the low molecular weight, Sephadex G-25 fractions.[65] This substance, with an excitation maximum at 360 nm and a fluorescence maximum at 440 nm, was further purified by ultrafiltration and subsequent separation on Sephadex G-10 and localized by paper electrophoresis[66] and preparative paper chromatography.[67] It was not found in sheep cerebral cortex. One compound has been identified in the antigonadotropic preparations as a pteridine, i.e., 6-L-erythrobiopterin.[68] These substances are discussed in subsequent sections of this chapter.

B. Effects on the Hypothalamus In Vitro

The ability of pineal factors to alter hypothalamic gonadotrophin-releasing factor activity in vitro was first demonstrated by Moszkowska and her colleagues.[57] In more recent studies hypothalami from mice were incubated with an ultrafiltrate (UM2-R) of a sheep pineal fraction derived from Sephadex G-25 gel filtration yielding a highly significant decrease in hypophysiotrophic activity. This UM2-R was without direct effect on pituitary secretion of gonadotrophin. Certain subfractions of the UM2-R, prepared by Sephadex G-10 chromatography, decreased the gonadotrophin-releasing activity of the hypothalami.[66,69,70] Hypothalami incubated with the residue of the Amicon ultrafiltration UMO5 membrane (UMO5-R), derived from low molecular Sephadex G-25 fractions, demonstrated an increased hypothalamic hypophysiotrophic activity. No significant, direct effects on anterior pituitary secretion of gonadotrophin were observed with the UMO5-R. Similarly prepared pineal UMO5-R fractions are known to possess significant antigonadotrophic effects in vivo[71-75] and are discussed in a following section of this chapter. It may be important to note at this point that a pteridine-like fluorescence has sometimes been noted in the UMO5-R fractions. The possible relationship of pteridines to pineal metabolism was discussed recently by Ebels et al.[76]

Using an isobutanol extraction method[77] the remaining water layer was ultrafiltered through diaflo membranes PM10, UM2, and UMO5. The PM10 and UM2 residues demonstrated a stimulating effect on the hypophysiotrophic activity on the median eminence of the hypothalamus. PM10 residues (PM10-R) contain substances with MW $> 10,000$. Further ultrafiltration of this PM10-R on the ultrafilters XM100 and PM30, and subsequent testing of these residues have shown that substances with MW $>$ 100,000 and those with a MW between 30,000 and 100,000 contain the stimulating activity. After these purification steps nearly no activity remained in the PM10-R fraction of the PM30 filtrate. These experiments were carried out in red light (wavelength

> 585 nm) and as much as possible in darkness. This was the first time that we have found biologically active, high-molecular weight fractions (for details see Slama-Scemama et al.[78]). However, Reiss et al. reported in 1963[79] that a high molecular weight bovine pineal fraction showed stimulating activity on gonads. Therefore, it may be that under the experimental lighting conditions used in the last years by Ebels et al. to protect the light-sensitive pteridines, it was possible to isolate and preserve high molecular weight substances which stimulated the gonadotrophic releasing activity of male rats, and that this was not possible when normal fluorescent light was used.

In some preliminary experiments attempts were made to precipitate the high molecular weight substances present in XM100-R and PM30-R fractions by changing the pH of the solution. It was observed that while the precipitate did not show biological activity the supernatant did. Therefore, it may be that XM100-R and PM30-R contain active "complexes" which can be destroyed easily by chemical manipulations and perhaps by normal light. This supposition is supported by the results of Frowein and Lapin,[80] who have compared the effects of sham pinealectomy performed under white and red light.

C. Effects on Gonadotrophins In Vivo

The early work of certain European investigators revealed stimulatory and inhibitory effects of crude and partially purified pineal extracts on the reproductive systems of experimental animals.[81] Among these initial studies were those of Moszkowska[82] who showed that aqueous pineal extracts inhibited compensatory ovarian hypertrophy (COH) and the formation of corpora lutea in guinea pigs after unilateral ovariectomy. Similar results were obtained in the rat by Reiss et al. in 1963.[79]

Thiéblot and his collaborators[83-88] have studied pineal fractions which reduced gonadotrophin levels in the pituitaries of castrated rats. Melatonin was without effects in their test in amounts ranging between 10 and 500 μg/day over a period of 12 days, and after examination it was learned that active pineal fractions were devoid of indoles which were apparently removed by the organic solvents used during the extraction procedures.

When it was learned that an intact pineal gland is required for bilateral optic enucleation to inhibit COH in female rats,[89,90] it was reasoned that inhibition of COH could be used to detect and perhaps estimate quantities of pineal antigonadotrophin in pineal extracts. Using this test in female mice, a nonmelatonin substance was partially separated from bovine pineal gland extracts using Sephadex G-25 gel filtration and ultrafiltration through Amicon membranes.[91,92] A similar activity was identified in aqueous extracts of human pineal glands[93] and in rat pineal incubation media after short-term incubation in oxygenated Krebs-Ringer buffer containing glucose.[33] Inactivation of the anti-COH substance in rat pineal incubation media and bovine gland extracts was produced by proteolytic enzymes.[94]

These results and the previous studies of others prompted us to postulate at that time that the nonmelatonin pineal antigonadotrophin (PAG) was a small polypeptide in the approximate molecular weight range of 500 to 1000. These results were summarized by Benson and Orts[72] in a review in which the hypothesis was advanced that pineal antigonadotrophic substances act by altering the sensitivity of the hypothalamo-hypophysial axis to the feedback effects of gonadal steroids. This hypothesis was supported by experiments carried out by Orts and Benson in 1973,[74] in which it was shown that fractions of aqueous extracts of bovine pineal glands, partially purified to exclude melatonin, inhibited the postcastration rise in LH in orchidectomized rats. In studies reported the following year[95] similar LH-inhibitory properties were observed in aqueous extracts of rat pineals.

In 1972, a breeding colony of Charles River CD-1 mice was transported by the authors from Texas to Utrecht in order to search for COH-inhibiting activity in sheep pineal extracts. Dr. I. Ebels and her colleagues had extensive, previous experience with sheep pineal fractionation. In this study partial purification of an antigonadotrophic substance(s) was accomplished by gel filtration of aqueous extracts on Sephadex G-25. Active fractions were ultrafiltered through Amicon Diaflo ultrafiltration membranes UM2 and UMO5. The UMO5 residue (UMO5-R, MW $\simeq$ 500 to 1000) was found to contain a substance which inhibited COH and reduced ventral prostate weights in mice, and retarded vaginal opening in young female mice. When the UMO-5 was gel filtered on a Sephadex G-10 column equilibrated and eluted with distilled water, an active fraction was eluted long before syntheitc melatonin. Paper chromatographic and electrophoretic studies yielded R_f values for the anti-COH activity in a variety of solvent systems. We published the results in 1973.[64] These studies demonstrated clearly that an antigonadotrophic substance(s) could be separated from melatonin since this indoleamine, as well as 5-methoxytryptophol and 5-hydroxytryptophol are eluted from Sephadex G-25 columns after the fraction in which COH-inhibiting activity is localized.[96]

Similarly prepared, melatonin-free fractions obtained from extracts of bovine pineal glands were shown also to inhibit COH, delay vaginal opening time, and reduce the incidence of estrous vaginal smears in rats and reduce serum LH levels in long-term orchidectomized rats.[97]

Benson et al.[97] studied the effects of antigonadotrophic fractions derived from bovine pineal extracts on fertility and ovulation. When given over a period of 10 days to female mice, COH-inhibitory fractions were observed to reduce fertility. Single intravenous injections of this material given at various times during the estrous cycle were observed to inhibit ovulation in female rats. The pineal antigonadotrophic substance(s) was more effective in the blockage of ovulation when given throughout the cycle, or during diestrus, rather than on the morning of proestrus. In these studies the bovine pineal antigonadotrophic fraction was ineffective in reducing the action of exogenous gonadotrophins in immature female and male mice. Therefore, a gonadal site of action was ruled out and the effects on ovulation and fertility were postulated to be the result of inhibition of LH secretion. In other studies the antigonadotrophin was localized to the pineal parenchyma and vasotocic-like activity to the pineal stalk.[99]

Using bovine pineal extracts similar to those prepared by Benson et al.[91,92] and acetic acid extracts prepared similar to those used in the extraction of hypothalamic peptides, Orts et al.[100] studied their effects on fertility in female rats. Both methods yielded contraceptive fractions which were thought by the authors to be small polypeptides which were inhibitory to LH secretion. Orts[101] extended these studies to male rats in 1977. Acetic acid extracts of defatted bovine pineals were gel filtered on Sephadex G-25 and certain fractions obtained were effective in reducing LH or testosterone levels when injected over a 5-day period.

In recent years we have attempted final purification of bovine PAG using preparations obtained from Sephadex G-25 gel filtration and Amicon ultrafiltration. From these preliminary separatory steps the UMO5 residues were gel filtered on Sephadex G-15 and active fractions further subjected to ion-exchange chromatography on DEAE-Sephadex A-25. Two biological tests, *viz.*, inhibition of COH and reduction of ventral prostate weights in mice, were used to guide the purification.[102] In this extensive study various preparations partially purified in this manner were characterized using UV and fluorescence spectrometry, thin layer and paper chromatography, paper electrophoresis and amino acid analysis. The absorption and fluorescence spectra and amino acid composition did not correspond to those for AVT (for details see Rosenblum et al.[103]). Throughout these studies it was learned that the sulfur-containing amino acid taurine was closely associated with the COH-inhibiting factor. This promoted other studies on pineal taurine discussed in another section of this chapter. Following DEAE-Sephadex ion exchange chromatography, as a final step prior to amino acid analysis, only minute amounts of amino acids were recovered after hydrolysis of active fractions weighing < 1.0 mg. These were obtained from extracts of 100 g, wet-weight quantities of bovine glands and a major component was determined to be oxidized glutathione. When tested in mice, both taurine and glutathione were found not to possess antigonadotrophic activity. It was surprising to learn also that the absorption and fluorescence maxima of antigonadotrophic fractions obtained from both Sephadex G-15 and DEAE-Sephadex were similar, and that absorption maxima (between 255 to 260 nm) did not correspond to amino acid composition since phenylalanine was not detected.

Further studies employing preparative paper chromatography[104] showed that the COH-inhibiting substance could be separated from taurine, hypotaurine, and reduced and oxidized glutathione using the solvent butanol-acetic acid-water (4:1:1). In addition to DEAE-Sephadex, Dowex® AG 50W-X4 was used as a final step prior to amino acid analysis. Both ion-exchange steps gave irregular results and the amount of active material recovered from 100 to 200 g quantities of bovine glands could not be visualized in chromatograms using ninhydrin. Consequently, paper strips were sprayed with *O*-phthalaldehyde reagent and visualized under UV light. In this study also all antigonadotrophic fractions showed an absorption maximum at approximately 260 nm and amino acid analyses were negative for phenylalanine. The R_f values obtained in this study for bovine PAG compare well with those in our previous studies which localized COH-inhibiting activity in ovine pineals.[64]

In 1973 Bensinger et al. described in abstract[77] a method for aqueous-acetone extraction of bovine pineal and the further removal of fats and indoleamines by chloroform/methanol extraction. An antigonadotrophic substance(s) was subsequently extracted into isobutanol and further purified by gel filtration on Sephadex G-15 and thin layer chromatography. Although the details of the methods and all results were not published in detail, the authors state that the active principle was probably a small, non-tryptophan-containing peptide. In a recent study[105] we have extracted bovine glands according to the method of Bensinger[77] and confirmed the presence of COH-inhibiting activity. Certain fractions obtained from Sephadex G-15 gel filtration, and further separated by paper chromatography, showed different R_f values than the COH-inhibiting materials obtained from acetic extracts of ovine and bovine pineals.[64,104]

Recent studies by Ebels and her colleagues[46,76,78] have identified a pteridine-fluorescence in active fractions prepared similar to those in the study by Ebels et al.[104] Mass spectral studies identified the material as neopterin. The pteridines identified in pineal extracts are discussed in another section of this chapter.

The antigonadotrophic substance(s) obtained by both acid[104] and isobutanol extractions[105] could be localized using high pressure liquid chromatography. Refer also to the review by Benson and Ebels.[106] Since the amount of residues derived are extremely small this method is currently being used with the hope of obtaining pure material for structural determination by mass spectrometry.

D. Effects on Prolactin

In 1975 Bartke et al.[107] induced pineal-mediated gonadal regression in male hamsters by exposure to short days and discovered that early recrudescence resulted after treatment with exogenous prolactin. Other results by Matthews et al.,[108] Bex et al.,[109] Reiter and Ferguson,[110] and Benson and Matthews,[111] in which the transplantation of pituitaries beneath the kidney capsules either partially or totally prevented gonadal atrophy in light-deprived hamsters, have suggested a possible role for prolactin and pineal prolactin regulating factors in pineal-mediated gonadal atrophy.

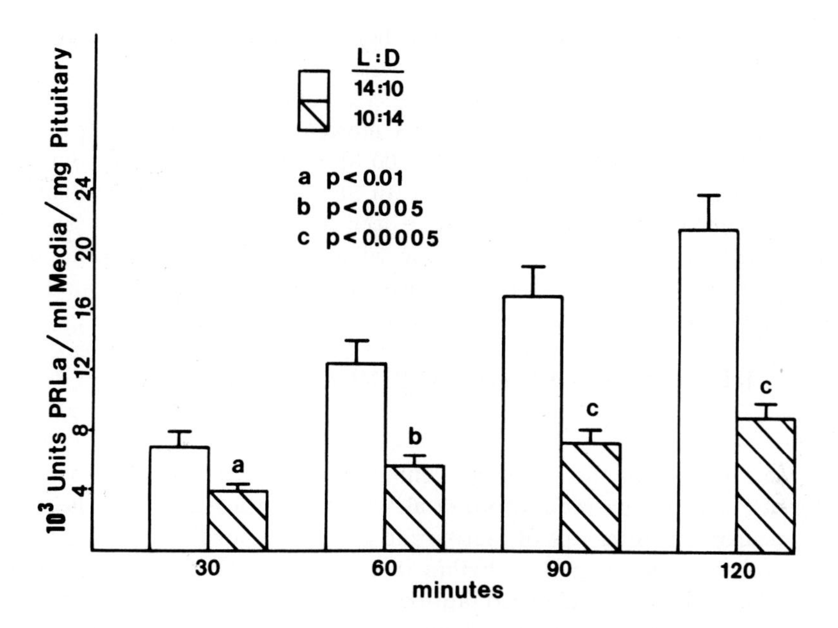

FIGURE 1. Hemipituitaries from adult male hamsters housed for 13 weeks on either long (L:D = 14:10 hr) or short (L:D = 10:14 hr) photoperiods were incubated in medium 199 for 2 hr. At the intervals indicated, samples were attained for prolactin radioimmunological assay using antisera against rat prolactin and a standard consisting of pooled hamster sera (1.0 U = 1.0 μl) of standard hamster serum). After 13 weeks in short photoperiod the hamster pituitary donors demonstrated pineal-induced gonadal and accessory organ atrophy and low levels of LH and prolactin in blood. In addition to reduced prolactin response to urethane (not shown) the hemipituitaries demonstrated a significant diminished ability to secrete prolactin in vitro, as shown in the illustration. Bars represent means of eight hemipituitaries and S.E.s are indicated.[112]

More recently, Orstead and Benson[112] have shown that when hamsters are placed in a short photoperiod, the pituitary demonstrates the gradual inability to synthesize and secrete prolactin in response to urethane injection. Also, when hemipituitaries from hamsters kept on short photoperiods for 12 weeks were incubated in oxygenated medium 199 for 2 hours, and samples collected for prolactin determination at 30-min intervals, the ability to secrete prolactin was markedly reduced (refer to Figure 1).

These combined studies prompted Benson and Matthews[111] to postulate that when stimulated, the pineal may secrete a prolactin-regulating substance. Such a substance could act by either regulating the synthesis and/or release of hypothalamic prolactin-releasing factors or prolactin release-inhibiting factors such as dopamine, or act directly on pituitary synthesis and/or release of prolactin.

This possibility is reinforced by reports on the presence of substances in pineal extracts which are capable of affecting the release of prolactin either in vitro or in vivo. In 1976 Blask et al.[113] observed that two partially purified bovine pineal fractions, apparently free of indoleamines, inhibited prolactin release from anterior pituitaries in vitro. These results were confirmed and extended in a recent study by Chang, Ebels, and Benson[114] in which dilute acetic acid extracts of bovine pineal glands were subjected to gel filtration on Sephadex G-25 and subsequently to ultrafiltration using the Diaflo membranes PM10, UM2, and UMO5. The lyophilized fractions derived were added to short-term rat hemipituitary incubations and effects on prolactin secretion determined. Low molecular weight (MW < 500) fractions were observed to inhibit prolactin release in this system in vitro and the active fractions were inactivated by incubation with trypsin. Dopamine could not be detected in these small molecular weight fractions using a relatively sensitive paper chromatographic method employing *O*-phthalaldehyde reagent. Conversely, higher molecular weight Sephadex G-25 fractions, and ultrafiltrates approximately > 10,000 MW, were observed to stimulate the release of prolactin from pituitaries in vitro. In this study, the low molecular weight UMO5-R fractions (MW ≃ 500 to 100) were seen to reduce prolactin in vivo in urethane-anesthetized rats, but had no effects in vitro. These results confirmed those of Larsen and Benson,[115] who have partially purified the bovine prolactin-inhibiting factor using high pressure liquid chromatography as a final step.

III. NEWLY IDENTIFIED PINEAL SUBSTANCES

A. Taurine

High concentrations of the sulfur-containing amino acid taurine are found in the pineal gland.[116] On this basis Bradford et al.[117] suggested a relationship between the postulated neurotransmitter role of taurine and pineal hormonal regulation. Grosso et al.[119] reported a distinct circadian rhythm in the content of this amino acid in rat pineals and the ability of radiolabeled taurine to bind with low affinity to the gland in short-term organ culture. Maximum amounts were observed between the end of the dark period and the middle of the light phase. A sodium-dependent transport system was identified.

In bovine pineal glands large amounts of taurine were discovered in partially purified fractions which showed antigonadotrophic activity in mice.[102] Synthetic taurine demonstrated antigonadotrophic activity only in large amounts. Similar aqueous methods were used to purify antigonadotrophic fractions which, when subjected to high performance liquid chromatography, revealed the presence of taurine.[104]

These related studies raised the questions of whether the pineal gland contains the enzymes required for the biosynthesis of taurine and what role taurine might play. To answer partially these questions, Ebels, Benson, and Larsen[119] incubated rat pineal glands for 3 hr in oxygenated Krebs-Ringer buffer containing glucose and ^{14}C-cystine. Separately, the pineal incubation media and pineal homogenates were subjected to Dowex® AG W50,X4 chromatography at low pH in order to separate ^{14}C-taurine, cysteic acid and other acidic compounds including cystine. The ^{14}C-containing Dowex® eluates were further subjected to one- and two-dimensional paper chromatography in a variety of solvents with taurine, hypotaurine, cystine, cysteic acid, oxidized and reduced glutathione, and cystine sulfinic acid as reference standards. After spraying with orthophthalaldehyde reagent, the paper strips were cut into pieces and radioactivity eluted and counted by scintillation counting methods. ^{14}C-labeling was found in the region with R_f values identical to that for taurine. Additionally, a method utilizing high performance liquid chromatography was developed[120] for the isolation of taurine and was used to confirm the results obtained with ion exchange and paper chromatography. With this method the ^{14}C-labeled taurine was specifically isolated from the ^{14}C-cystine precursor and its metabolites, providing sound evidence that the rat pineal has the synthetic machinery required for taurine synthesis.

The second question, namely that of the physiological significance of pineal taurine synthesis, was not answered by these experiments. In 1973 Baskin and Dagirmanjian,[121] showed that taurine could alter the synthesis and/or release of pineal melatonin. In recent studies using rat pineals in organ culture, Wheler et al.[122] have shown that taurine stimulates *N*-acetyltransferase activity.

B. The Pineal Tripeptide Threonylseryllysine

In 1978 Orts et al.[123] reported on the presence of a bovine pineal tripeptide which was extracted by glacial acetic acid and purified by Sephadex gel filtration, ultrafiltration, and paper electrophoresis. The sequence was determined by subtractive Edman degradation methods to be threonylseryllysine (Thr-Ser-Lys). The tripeptide was synthesized in 1979 by Larsen, Hruby, and Benson[124] and its biological effects are under study.[125]

Similar to the UMO5-R fractions prepared from bovine and ovine glands discussed above in II.C., synthetic Thr-Ser-Lys is capable of blocking compensatory hypertrophy. Also, effects on prolactin secretion have been noted in vivo[125] but questions about the validity of the results have arisen, since the male rats employed in these studies were anesthetized with urethane which apparently in itself produces maximal prolactin inhibiting activity, possibly through dopaminergic mechanisms.[126] Although the effects on prolactin observed with Thr-Ser-Lys are consistent with those of the pineal prolactin regulating substance partially purified by Larsen and Benson,[115] their characteristics of elution using high pressure liquid chromatographic techniques are different, implying different chemical structures.

C. Pteridines

When aqueous extracts of sheep pineal glands and sheep cerebral cortex were separated on Sephadex G-25 and G-10, several distinct peaks could be distinguished showing excitation and fluorescence maxima resembling those of indoles.[62,63] However, in sheep pineal extracts, one high peak was detected with different excitation and fluorescence maxima which was not observed in extracts of sheep cerebral cortex, i.e. an excitation maximum at about 360 nm and a fluorescence maximum at 440 nm.[65] This low molecular weight, sheep pineal fraction, showed antigonadotrophic activity in vitro. It was found that after ultrafiltration and subsequent separation of the UMO5-F fraction (MW $\simeq <$ 500) on Sephadex G-10 columns, the inhibitory activity could be located by paper electrophoresis in the same region where the special fluorescence was found.[127]

In subsequent experiments employing preparative paper chromatography with two different solvent systems, substances were detected in two regions which demonstrated inhibitory action on the gonadotrophic activity of anterior hypophyses of male rats in vitro.[67] One of these inhibitory substances was found to be located in the same region (I) in which substance(s) with special fluorescence characteristics were consistently detected. In a second region (II), the special fluorescence was not always found to be of high intensity. From the main fluorescent band (I) a compound could be isolated which, when subjected to gas-liquid chromatography (GLC) after trimethyl-silylation, showed a retention time identical with that of synthetic biopterin. Combined GLC-mass spectrometry gave a mass spectrum of the isolated compound, which was in good agreement with the spectrum of synthetic biopterin published by Lloyd et al.[128] From thin-layer chromatographic studies and a comparison of the isolated compound and synthetic biopterin in the *Crithidia fasciculata* test, the structure of the isolated pineal compound was concluded to be 6-L-erythrobiopterin (Figure 2). For details see Van der Have-Kirchberg et al.[68] When the isolated fraction was compared with a synthetic preparation of 6-L-erythrobiopterin (with about 18% of 7-L-erythrobiopterin), ten times more material was needed of the synthetic compound to obtain a similar significant inhibiting activity in our bioassays, based on quantitation using the relative fluorescence intensity.[55,68] With another synthetic preparation of more pure 6-L-erythrobiopterin, no inhibiting activity was found. Therefore, it is possible that the isolated paper chromatographic fraction contains besides biopterin (an) other compound(s) which caused the inhibitory activity observed.

The biopterin isolated may exist in vivo in a reduced form (Figures 3 and 4) and perhaps even in a state partly bound to proteins. In the reduced form, tetrahydrobiop-

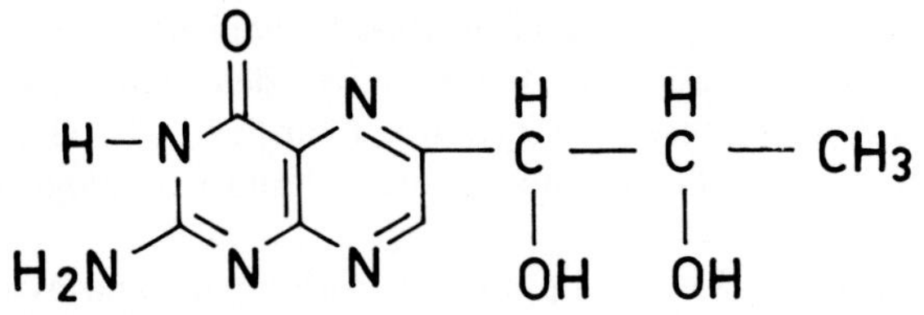

FIGURE 2. 6-L-Erythrobiopterin

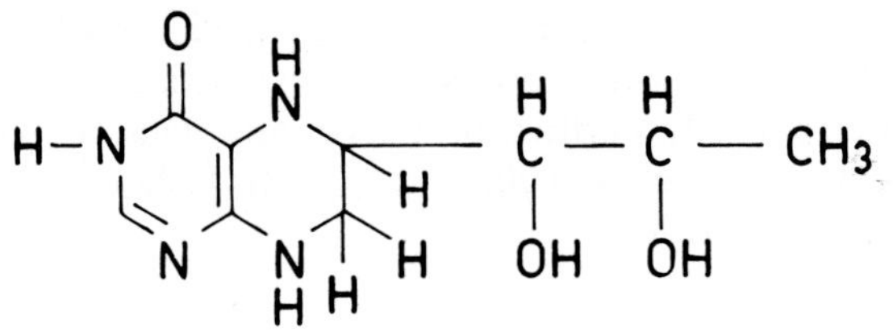

FIGURE 3. 5,6,7,8-Tetrahydrobiopterin

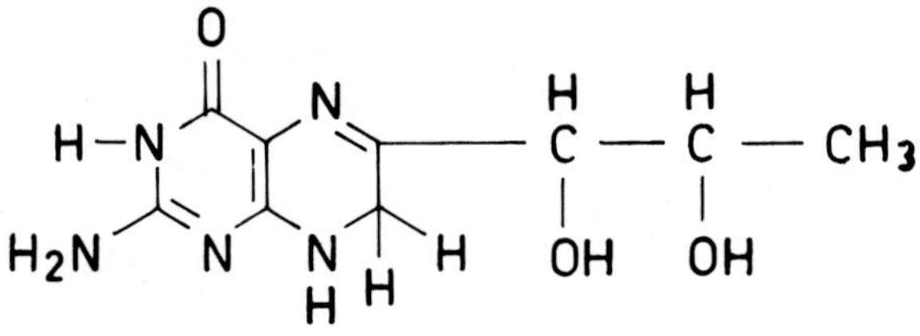

FIGURE 4. 7,8-Dihydrobiopterin

terin (BH_4) can function as a cofactor in the hydroxylation of phenylalanine to tyrosine[129,130] and in that of tyrosine to dihydroxyphenylalanine.[131] A defect in tetrahydrobiopterin metabolism could impair the synthesis of the neurotransmitters dopamine, noradrenalin, adrenalin, and serotonin.[132] On the other hand, C6-subsituted pterins can easily be oxidized to pterin-6-aldehyde and pterin-6-carboxylic acid under normal laboratory lighting conditions.[133] Pterin-6-aldehyde is an inhibitor of the enzyme *Xanthopterin* and *Xanthine* oxidase.[134] Therefore, a number of synthetic pteridines were tested and reduced neopterin and reduced biopterin showed reproducible antigonadotrophic activity in vitro and in vivo.

Recently, two interesting reports appeared describing the presence of biopterin and hydroxylase cofactor activity in rat pineals. Total biopterin distribution in rat tissues was analyzed by Fukushima and Nixon[135] by reverse-phase high performance liquid chromatography. These authors found that the pineal organ contains ten times more biopterin than liver and bone marrow and 100 to 150 times more than in brain and pancreatic tissue. Levine et al.[136] found a hydroxylase cofactor activity (expressed as tetrahydrobiopterin) for the pineal which far exceeds that of all other brain areas measured.

In previous experiments it was shown that the UMO5-R of low molecular weight Sephadex fractions of bovine pineal extracts showed antigonadotrophic activity using hemicastrated mice as test animals.[64] From a similar UMO5-R of sheep pineals a pteridine could be isolated and identified after trimethylsilylation by GLC-MS as neopterin[55] (Figure 5). In paper chromatographic studies on Whatman 3MM, using

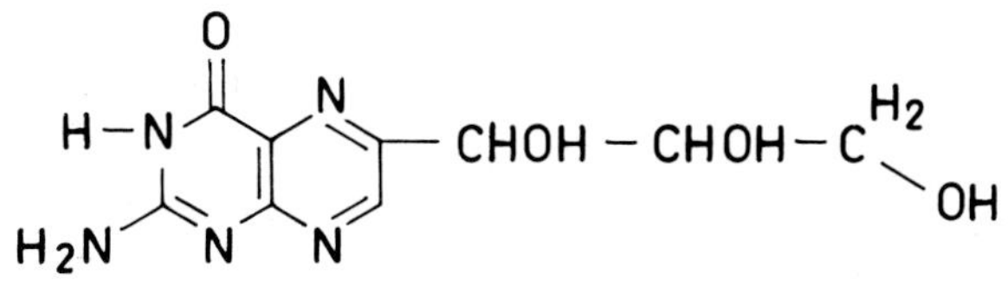

FIGURE 5. Neopterin

ethanol:H_2O = 1:3, as solvent, synthetic neopterin yielded an R_f value of about 0.50, while in a region with an R_f value of about 0.23, the pineal-neopterin was detected. The mass spectrum obtained of the isolated compound was identical with that of synthetic neopterin, and in agreement with that published by Lloyd et al.[128] Therefore, we have concluded that neopterin (we do not know which isomer yet) can be present in this pineal extract as a complex which is rather labile. In the isobutanol fraction from an aqueous sheep pineal extract neopterin could also be detected.[152]

Recently, the influence of three pterins on the diurnal fluctuations of hydroxyindole-*O*-methyltransferase (HIOMT) activity was studied in isolated pineal glands of 42-day-old male Wistar rats. Isoxanthopterin, which can be considered as a metabolic end product, did not exert any influence on the diurnal variation of all the HIOMT activities studied. Reduced neopterin stimulated the HIOMT activities for 5-hydroxytryptophan (5-HTP), 5-hydroxytryptamine (5-HT), and 5-hydroxyindole acetic acid (5-HIAA) during the night. The combined HIOMT activities for H-acetyl-t-HT/5-hydroxytryptophol (5-HTL) was shifted to a later moment in the dark period and pterin-6-aldehyde, the first photolytic oxidation product of 6-substituted pteridines as folic acid, stimulated the HIOMT activity for 5-HT during day time. The HIOMT activities for the substrates 5-HTP, 5-HIAA, and H-ac-5-HT/5-HTL were shifted towards an earlier period (Figure 6). The authors present a physiological explanation for certain of these alterations.[137]

Pineal pterins may play a role in another interesting field of pineal research, i.e., the relation of tumor growth and pineal metabolism. From the work of Lapin[138] it is known that the pineal gland, the pineal indole melatonin, and perhaps other pineal compounds can influence the growth, spread of metastasis, and manifestation of neoplastic disease. Recently, different authors have described a disorder of pterin metabolism in malignant growth in vitro and in vivo. Stea et al.[139] have studied folate and pterin metabolism by cancer cells in culture. Malignant cells grown in culture excrete into their growth media a folate catabolite, which has been identified as 6-hydroxymethylpterin. Moreover, when pterin-6-carboxaldehyde was added to the growth medium of logarithmically growing malignant cells, it is primarily reduced to 6-hydroxymethylpterin, in contrast to normal cell lines in culture, where pterin-6-carboxylate was the principal product formed from added pterin-6-carboxaldehyde. Ziegler and Kokolis[140] found high levels of tetrahydrobiopterin in the blood of human tumor bearing organisms, especially in melanoblastoma of the eye. Wachter et al.[141] detected neopterin in four to six times higher amounts in the urine of patients with malignant tumors than in that of normal persons and concluded that this observation may be important for the diagnosis and control of malignancy.

Recently Lapin[142] observed that after treatment with the tranquilizer reserpine the incidence of dimethylbenzanthracene (DMBA)-induced tumors was found to be significantly higher in pinealectomized rats after reserpine administration. The author suggested that the neuroendocrine disturbance resulting from removal of the pineal gland in the newborn animal were latent, and become evident after reserpine administration as reflected in an increased tumor incidence. Kokolis and Issidoris[143] have seen a relation between reserpine and pteridine metabolism. Reserpine can alter the pteridine pat-

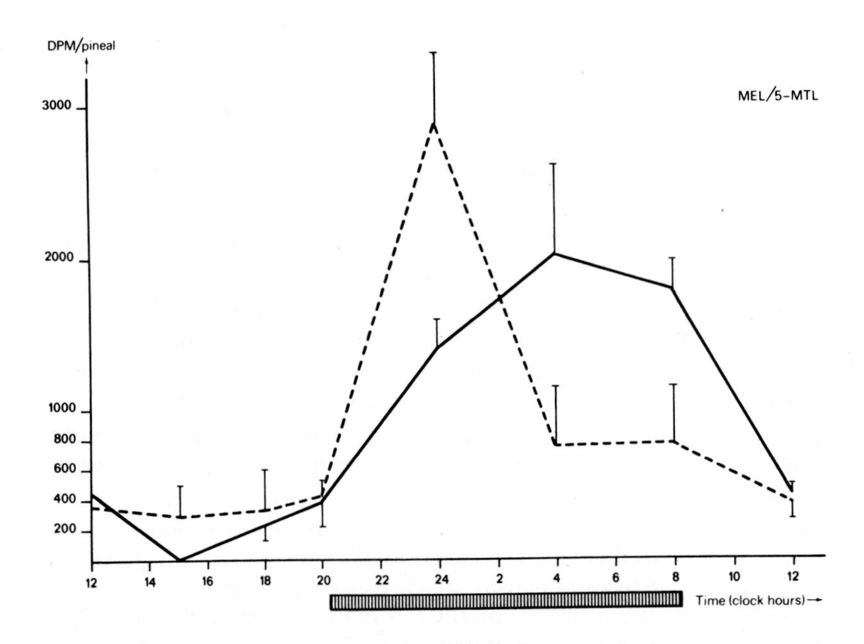

FIGURE 6A. The influence of reduced neopterin (———) on the circadian rhythmicity of the methylation of *N*-acetylserotonin/5-hydroxytryptophol in the pineal of 42-day-old male Wistar rats in September is shown. The circadian methylating activity in the synthesis of melatonin/5-MTL in the pineals which are not incubated with a pterin but with solvent only is indicated as a broken line (------). Vertical lines indicate S.E.s and DPM = desintegration per minute. (From Balemans, M. G. M., van Brenthem, J., Legerstee, W. C., deMorée, A., Noteborn, H. J. P. M., and Ebels, I., J. Neural Transm., in press. With permission.)

tern of the liver of *Triturus cristatus* such that the ratio of tetrahydrobiopterin to isoxanthopterin is increased. It is suggested that the observed changes in nuclear morphology are due to the interaction of the drug with the pteridine metabolism. It may be that the results of Lapin[138,142] can be explained by an interaction of reserpine with pteridine metabolism and that the pineal can influence in one or another way that interaction. Therefore, the study of pteridine metabolism of the pineal may be an interesting field of research in the future in relation to the growth of gonads and to tumor growth.

IV. CONCLUSIONS — PHYSIOLOGICAL IMPLICATIONS

Pinealogists consider the pineal an endocrine gland capable of regulating a wide variety of functions under control of the central nervous system. It is difficult to define a uniform concept in which a single pineal hormone would mediate the numerous functions suggested by the observations of various investigators. Most agree that effects on the reproductive system are the easiest to demonstrate because of the sensitivity of this system to the pineal-mediated effects of light. The model used by Reiter,[144] in which the antigonadotrophic effects of light deprivation in the hamster are reversed by pinealectomy, clearly demonstrate a role for this structure in the regulation of reproduction in this light-sensitive, seasonal breeder. The question remains, however, of the chemical nature of the pineal hormone(s) capable of producing the profound antigonadotrophic effects observed.

It is clear that certain substances isolated from extracts of sheep and bovine pineals can be injected into experimental animals, yielding significant antigonadotrophic ef-

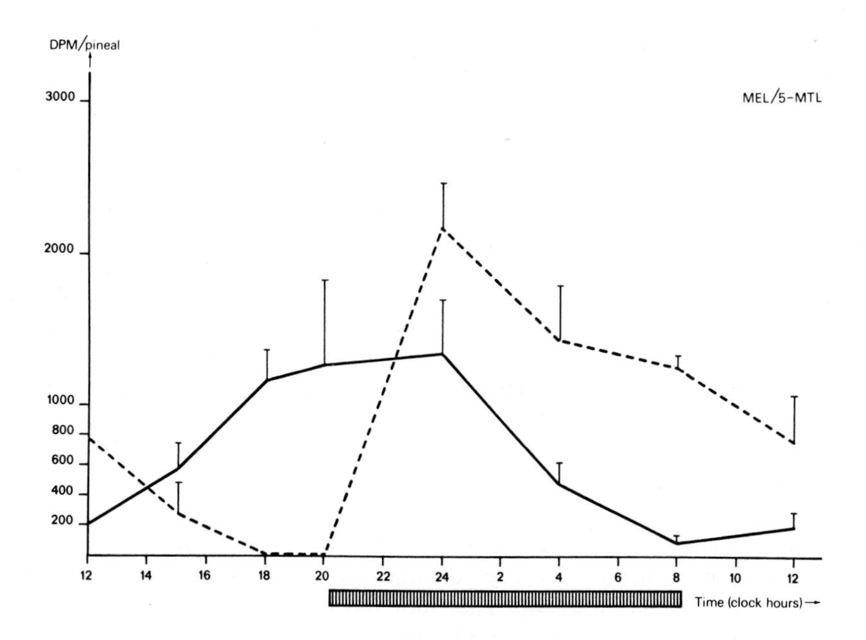

FIGURE 6B. The effect of pterin-6-aldehyde (———) on the circadian rhythmicity of the methylation of *N*-acetylserotonin/5-hydroxytryptophol in the pineals of 42-day-old male Wistar rats in October. In both experiments the animals were housed under natural photoperiod. From Balemans, M. G. M., Brenthem, J., Legerstee, W. C., de Morée, A., Noteborn, H. J. P. M., and Ebels, I., *J. Neural Transm.*, in press. With permission.)

fects. Even though the hypophysis or the gonads have not been ruled out as a site of action, most studies to date point to a hypothalamic site of action of pineal antigonadotrophic substances, at least in the final common pathway to pituitary regulation of gonadotrophins. This probable site of action is reinforced by the fact that isolated pineal substances can increase or reduce directly the apparent gonadotrophin-releasing factor activity of the hypothalamus in vitro. In certain cases it is quite certain that these partially purified materials do not contain melatonin or related indoleamines or neurohypophysial peptides such as oxytocin, vasopressin, or arginine vasotocin.

The purification of biologically active substances have presented a formidable challenge, since extraction methods for indole and peptide purification have failed to yield large quantities sufficient for structural elucidation. The lability of the active compound(s) is legend, and most bioassays carried out in various laboratories around the world have revealed mixtures of stimulatory and inhibitory substances in closely related fractions, and the easy loss of these activities when standard purification methodologies are applied. The problem then of structural identification of a single pineal antigonadotrophin, once thought by these reviewers to be peptidic or to contain a peptide moiety, remains unsolved.

It is clear from recent work[111,112] that pineal regulation of prolactin secretion may be more important than was thought in previous years and evidence is accumulating to suggest that the pineal gland may contain prolactin regulating factors. Certain other unexpected findings have come out of work in the last decade. Small molecular weight antigonadotrophic fractions were found to contain large amounts of the sulfur-containing amino acid taurine, leading to the confirmation of the synthesis by the pineal of this substance[119] and the discovery of an apparent circadian rhythm in its amount in the pineal glands of rats.[118] Years of work by Orts and his colleagues[113] have yielded the structure of a small tripeptide which, in synthetic form, was seen to block COH in mice and which may prove to have other effects on the reproductive system. This compound is of a smaller molecular weight, however, than most pineal antigonadotrophic factors prepared by ourselves in the last years, suggesting that the use of glacial acetic acid may have effectively cleaved this active product from a parent molecule.

Perhaps the most significant, unexpected finding to date is that of the presence of pteridines in the small molecular weight antigonadotrophic fractions.[55,76] Certain of these compounds are photolabile and difficult indeed to work with. Whether or not they, like the pineal tripeptide Thr-Ser-Lys, represent a portion of a parent molecule is problematic at this point, but synthetic compounds have been shown to demonstrate antigonadotropic effects. It is not unlikely that these substances relate to the regulation of catecholamine and hydroxyindole synthesis by the pineal gland itself, but the probability that they are secreted as part of an antigonadotrophic molecule is not yet excluded since there is evidence that they may exist in the pineal as part of a higher molecular weight complex.

In the search for pineal hormones, the now classic experiments of Reiter[144] in the hamster have served as a model in which the antigonadotrophic effects of the pineal could be consistently demonstrated. The extreme sensitivity of the reproductive system of this animal to photoperiodic control, and the altered hormonal events, point to regulation of hypothalamo-hypophysial function by a pineal antigonadotrophic hormone(s). It has now been demonstrated that, in addition to a reduction in gonadotrophin secretion, the stimulated pineal gland of light-deprived hamsters is capable of reducing the synthesis and secretion of prolactin. Recent work in our laboratory suggests that this pineal-induced reduction in prolactin levels precedes the reduction of gonadotrophins and gonadal atrophy.[112] More importantly perhaps is the recent observations that the placement of prolactin secreting pituitary grafts beneath the renal capsules of light-deprived hamsters can reverse the antigonadotrophic effects of the pineal gland.[108-111]

In the light-deprived hamster with atrophied gonads, gonadotrophins in blood do not increase even though blood levels of gonadal hormones are reduced,[145-150] indicating altered sensitivity of the hypothalamo-hypophysial system to the feedback effects of the gonadal steroids. This hypothesis was made by Benson and Orts in 1972,[72] and was based primarily on the ability of the pineal gland to inhibit compensatory ovarian hypertrophy in blinded rats[89,90] and of melatonin-free pineal extracts to inhibit compensatory ovarian hypertrophy in mice. It was thought possible that gonadal function in seasonal-breeding animals could be regulated in part by the ability of the pineal gland to secrete differing amounts of substances at certain times of the year capable of regulating the gonadotrophin response of the hypothalamo-hypophysial unit to negative feedback control by gonadal steroids. The question emerges of whether a single pineal antigonadotrophic hormone is capable of lowering prolactin and gonadotrophin levels and, at the same time, altering the sensitivity of hypothalamo-hypophysial axis to gonadal steroid feedback. Numerous pineal fractions tested in hemicastrated or castrated animals and discussed above would appear to be able to effect these changes.

We will attempt to summarize utilizing the hypothetical scheme in Figure 7. The possible physiological significance of certain of the pineal factors discussed above will be addressed at various possible sites of action in the scheme. To be sure, we have oversimplified by interfacing several systems which in themselves are individually complex, yet such a hypothetical scheme is currently being tested at several levels in the laboratories of the authors of this chapter.

It is postulated that certain environmental stimuli, i.e. olfaction, photoperiod, etc., selectively affect the physiological activity of pinealotrophic nerves. Although other neurotransmitters may be involved, it is rather certain that norepinephrine (NE) stimulates release of pineal secretory product and induces c-AMP mediated indoleamine

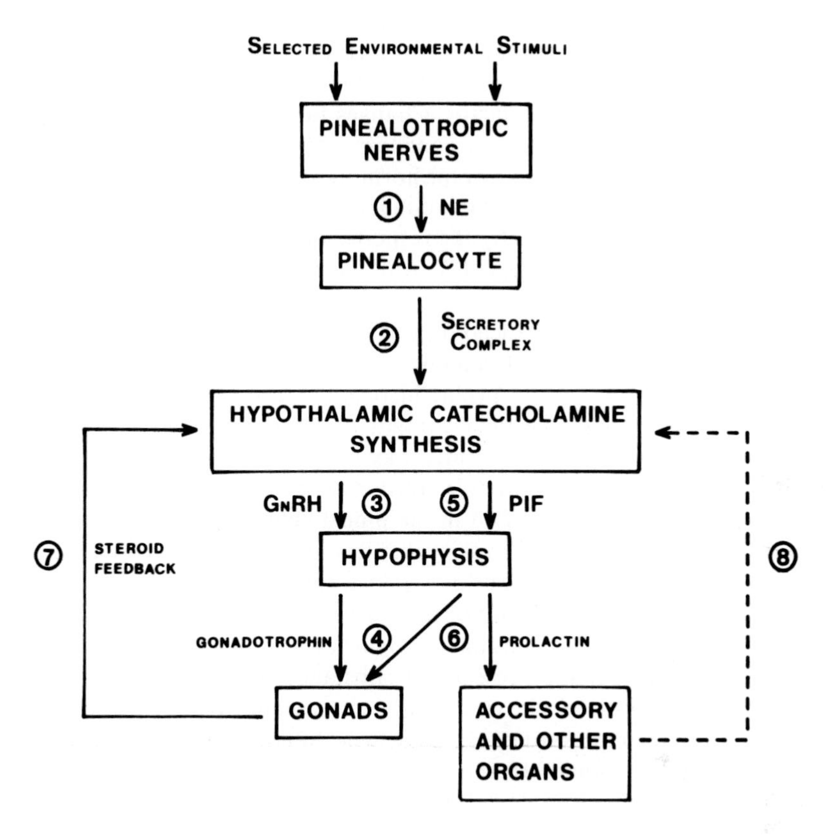

FIGURE 7. Hypothetical mechanism for pineal regulation of hypothalamic control of reproduction (see text for explanation).

synthesis. At this level (1) the pteridines recently identified in pineal extracts could function as cofactors in certain hydroxylation steps in catecholamine biosynthesis. Under the influence of norepinephrine stimulation, newly synthesized (2) secretory complex, observed ultrastructurally as proteinaceous material either in the lamallae of the endoplasmic reticulum or in granulated vesicles, could be secreted into the blood vascular system in a form containing indoles. There is certain histochemical data suggesting the presence of serotonin in the secretory granules. The role that melatonin might play in the secretory process is unknown, but such a role has been suggested since melatonin is known to stimulate both the synthesis and release of pineal secretory product. Also, the amino acid taurine, now known to be synthesized by the pineal, may play a role in the secretory process relative to the permeability of pinealocyte plasma membranes and the mechanisms involved in Ca^{++} exchange. A role for taurine in the regulation of the synthesis and release of norepinephrine by pinealotrophic nerves could also be suggested and effects on *N*-acetyltransferase have been described.[122]

Even though other sites have not been ruled out, the hypothalamus is a likely target for the site of action of pineal secretory products.[4] An alternate hypothesis could of course incorporate other portions of the brain which are known to have strong dopaminergic, noradrenergic, or serotonergic input into the hypothalamus. The secretory product could affect in either case, i.e. directly or indirectly, hypothalamic catecholamine turnover rates. A reduction for example in norepinephrine could account for reduced synthesis and release of gonadotrophin-releasing-factor (3) leading ultimately to reduced gonadotrophin secretion (4) and gonadal atrophy. Similarly, increased rates of synthesis and release of hypothalamic prolactin inhibiting factor (5), or dopamine, would result in alterations in prolactin secretion (6) by the hypophysis and consequently effect blood levels of prolactin. Changes in prolactin could influence significantly the response of the gonads to gonadotrophins and the response of accessory and other organs (e.g., adrenal) to trophic hormones.

Alteration in hypothalamic catecholamine synthesis produced by pineal secretory product could change concomitantly the response of the hypothalamus to the negative feedback effects of gonadal steroids (7). Altered prolactin levels might play a role in this regard also (8).

New data have been presented leading to the hypothesis that the pineal gland contains substances which are in fact capable of affecting GnRH production by the hypothalamus and of changing the prolactin regulating activity of the hypothalamus. Levine et al.[136] believe that the high content of hydroxylase cofactor in the pineal gland, expressed as tetrahydrobiopterine, may indicate in addition to its function as a cofactor of hydroxylase a role for this substance in neuroendocrine function. If the pteridines which have been identified in the pineal gland are transported to the hypothalamus as part of a secretory complex, they could act at that site as cofactors in the regulation of catecholamine synthesis, possibly at specific hydroxylation steps.

When stimulated, the pineal gland of light-deprived hamsters is capable of affecting a reduction in the ability of the pituitary to synthesize and release prolactin.[111,112] This fact, plus the discovery of the presence of substances in the pineal gland which can alter the secretion of prolactin in vivo and in vitro, point to a significant role for the pineal gland in the regulation of prolactin which, in addition to significant effects on the reproductive system, could affect a variety of other tissues in the body.

REFERENCES

1. **Ueck, M. and Wake K.**, The pinealocyte — a paraneuron? A review, *Arch. Hist. Jpn.*, 40, Suppl. 261, 1977.
2. **Pearse, A. G. E.**, The cytochemistry and ultrastructure of polypeptide hormone producing cells of the APUD series and the embryology, physiology and pathologic implications of the concept, *J. Histochem. Cytochem.*, 17, 303, 1969.
3. **Fujita, T.**, The gastro-enteric endocrine cell and its paraneuronic nature, in *Chromaffin, Enterochromaffin and Related Cells*, Coupland, R. E. and Fujita, T., Eds., Elsevier, Amsterdam, 1976, 191.
4. **Kappers, J. A. and Pévet, P.**, Short history of pineal discovery and research, *Prog. Brain Res.*, 52, 3, 1979.
5. **David, G. F. X. and Herbert, J.**, Experimental evidence for a synaptic connection between habenula and pineal ganglion in the ferret, *Brain Res.*, 64, 327, 1973.
6. **Ueck, M.**, Innervation of the vertebrate pineal, *Prog. Brain Res.*, 52, 45, 1979.
7. **Lukaszyk, A. and Reiter, R. J.**, Histophysiological evidence for the secretion of polypeptides by the pineal gland, *Am. J. Anat.*, 143, 451, 1975.
8. **Buijs, R. M. and Pévet, P.**, Vasopressin- and oxytocin-containing fibers in the pineal gland and subcommissural organ of the rat, *Cell. Tissue Res.*, in press.
9. **Pévet, P., Reinharz, A. C., and Dogterom, J.**, Neurophysins, vasopressin and oxytocin in the bovine pineal gland, *Neurosci. Lett.*, in press.
10. **Pévet, P.**, Secretory processes in the mammalian pinealocyte under natural and experimental conditions, *Prog. Brain Res.*, 52, 149, 1979.
11. **Leonhardt, H.**, Über axonähnliche Fortsätze, Sekretbildung und Extrusion der hellen Pinealozyten des Kaninchens, *Z. Zellforsch. Mikrosk. Anat.*, 82, 307, 1967.
12. **Romijn, H. J. and Gelsema, A. J.**, Electron microscopy of the rabbit pineal organ *in vitro*. Evidence of norepinephrine-stimulated secretory activity of the Golgi apparatus, *Cell Tissue Res.*, 172, 365, 1976.

13. **Karasek, M.**, Ultrastructure of rat pineal gland in organ culture; influence of norepinephrine, dibutyril cyclic adenosine 3, 5-monophosphate and adenohypophysis, *Endokrinologie*, 64, 106, 1974.
14. **Ariëns-Kappers, J.**, The mammalian pineal organ, *J. Neuro-Visc. Relat.*, Suppl. 9, 140, 1969.
15. **Ariëns-Kappers, J.**, The mammalian pineal gland, a survey, *Acta Neurochir. Suppl.*, 34, 109, 1976.
16. **Kappers, J. A., Smith, A. R., and DeVries, R. A. C.**, The mammalian pineal gland and its control of hypothalamic activity, *Prog. Brain Res.*, 41, 149, 1974.
17. **Upson, R. H., Benson, B., and Satterfield, V.**, Quantitation of ultrastructural changes in the mouse pineal in response to continuous illumination, *Anat. Rec.*, 184, 311, 1976.
18. **Romijn, H. J.**, The influence of some sympatholytic, parasympatholytic and serotonin-synthesis-inhibiting agents on the ultrastructure of the rabbit pineal organ, *Cell Tissue Res.*, 167, 167, 1976.
19. **Matsushima, S., Kachi, T., Mukai, S., and Morisawa, Y.**, Functional relationship between sympathetic nerves and pinealocytes in the mouse pineal: quantitative electron microscopic observations, *Arch. Histol. Jpn.*, 40, 279, 1977.
20. **Romijn, H. J.**, The ultrastructure of the rabbit pineal gland after sympathectomy, parasympathectomy, continuous illumination and continuous darkness, *J. Neural Transm.*, 36, 183, 1975.
21. **Romijn, H. J. and Wolters, P. S.**, Diurnal variations in numbers of Golgi dense-core vesicles in light pinealocytes of the rabbit, *J. Neural Transm.*, 38, 231, 1976.
22. **Benson, B. and Krasovich, M.**, Circadian rhythm in the number of granulated vesicles in the pinealocytes of mice. Effects of sympathectomy and melatonin treatment, *Cell Tissue Res.*, 184, 499, 1977.
23. **Krasovich, M. and Benson, B.**, Effects of reserpine and *p*-chlorophenylalanine on the circadian rhythm of granulated vesicles in the pinealocytes of mice, *Cell Tissue Res.*, 203, 459, 1979.
24. **Krasovich, M. and Benson, B.**, Studies on perivascular granulated vesicles in pinealocytes of hamsters, *Anat. Rec.*, 196, 104, 1980.
25. **Axelrod, J.**, The pineal gland: a neurochemical transducer, *Science*, 184, 1341, 1974.
26. **Collin, J. P.**, Recent advances in pineal cytochemistry. Evidence of the production of indoleamines and proteinaceous substances by rudimentary photoreceptor cells and pinealocytes of amniota, *Prog. Brain Res.*, 52, 271, 1979.
27. **Lu, K.-S. and Lin, H.-S.**, Cytochemical studies on cytoplasmic granular elements in the hamster pineal gland, *Histochemistry*, 61, 177, 1979.
28. **Quay, W. B.**, *Pineal Chemistry in Cellular and Physiological Mechanisms*, Charles C Thomas, Springfield, Ill., 1974.
29. **Bridges, R., Tamarkin, L., and Goldman, B.**, Effects of photperiod and melatonin on reproduction in the Syrian hamster, *Ann. Biol. Anim. Biochem. Biophys.*, 16, 399, 1976.
30. **Reiter, R. J. and Vaughan, M. K.**, A study of indoles which inhibit pineal antigonadotropic activity in male hamsters, *Endocr. Res. Commun.*, 2, 299, 1975.
31. **Freire, F. and Cardinali, D. P.**, Effects of melatonin treatment and environmental lighting on the ultrastructural appearance, melatonin synthesis, norepinephrine turnover and microtubule protein content of the rat pineal gland, *J. Neural Transm.*, 37, 237, 1975.
32. **Owman, C., Hakanson, R., and Sundler, F.**, Occurrence and function of amines in endocrine cells producing polypeptide hormones, *Fed. Proc. Fed. Am. Soc. Exp. Biol.*, 32, 1785, 1973.
33. **Benson, B., Matthews, M. J., and Orts, R. J.**, Presence of an antigonadotropic substance in rat pineal incubation media, *Life Sci.*, 11, 669, 1972.
34. **Findell, P. R. and Larsen, B. R.**, Presence of prolactin release inhibiting factor in rat pineal incubation media, *Anat. Rec.*, 193, 537, 1979.
35. **Scott, P. M.**, No effect of arginine vasotocin on LRF-stimulated LH release *in vitro*, *Anat. Rec.*, 193, 679, 1979.
36. **Reiter, R. J.**, Pineal-anterior pituitary gland relationships, in *Endocrine Physiology*, McCann, S. M., Ed., Butterworth-University Park Press, London, 1974, 277.
37. **Reiter, R. J.**, Endocrine rhythms associated with pineal gland function, in *Biological Rhythms and Endocrine Functions*, Hedlund, L. W., Franz, M. M., and Kenny, A. A., Eds., Plenum Press, New York, 1975, 43.
38. **Reiter, R. J.**, *The Pineal and Reproduction*, S. Karger, Basel, 1978.
39. **Lynch, H. J., Ozaki, Y., Shakal, D., and Wurtman, R. J.**, Melatonin excretion of man and rats: effect of time of day, sleep, pinealectomy and food consumption. *Int. J. Biometeorol.*, 19, 267, 1975.
40. **Ozaki, Y. and Lynch, H. J.**, Presence of melatonin in plasma and urine of pinealectomized rats, *Endocrinology*, 99, 641, 1976.
41. **Gern, W. A., Owens, O. W., and Ralph, C. L.**, Persistence of the nychtlemeral rhythm of melatonin secretion in pinealectomized or optic tract-sectioned trout *(Salmo gairdneri)*, *J. Exp. Zool.*, 205, 371, 1978.
42. **Wetterberg, L., Geller, E., and Yuwiller, A.**, Harderian gland: an extraretinal photoreceptor influencing the pineal gland in neonatal rats, *Science*, 167, 884, 1970.
43. **Bubenik, G. A., Brown, G. M., Grota, L. J.**, Differential localization of *N*-acetylated indolealkylamines in CNS and the Harderian gland using immunohistology, *Brain Res.*, 118, 417, 1976.

44. **Pang, S. F., Brown, G. M., Grota, L. J., and Rodman, R. L.**, Radioimmunoassay of melatonin in pineal glands, Harderian glands, retinas and sera of rats and chickens, *Fed. Proc. Fed. Am. Soc. Exp. Biol.*, 35, 691, 1976.
45. **Quay, W. B. and Ma, Y. H.**, Demonstration of gastrointestinal hydroxyindole-*O*-methyltransferase, *I.R.C.S. Med. Sci.*, 4, 563, 1976.
46. **Bubenik, G. A., Brown, G. M., and Grota, L. J.**, Immunohistological localization of melatonin in the rat digestive system, *Experientia*, 33, 662, 1977.
47. **Raikhlin, N. T., Kventnoy, I. M., and Tolkachev, V. M.**, Melatonin may be synthesized in enterochromaffin cells, *Nature (London)*, 255, 344, 1975.
48. **Quay, W. B.**, Retinal and pineal hydroxyindole-*O*-methyl transferase activity in vertebrates, *Life Sci.*, 4, 983, 1965.
49. **Cardinali, D. P. and Rosner, J. M.**, Retinal localization of the hydroxyindole-*O*-methyl transferase (HIOMT) in the rat, *Endocrinology*, 89, 301, 1971.
50. **Cardinali, D. P. and Rosner, J. M.**, Ocular distribution of hydroxyindole-*O*-methyl transferase (HIOMT) in the duck *(Anas platyrhinchos)*, *Gen. Comp. Endocrinol.*, 18, 407, 1972.
51. **Bubenik, G. A., Brown, G. M., Uhlir, I., and Grota, L. J.**, Immunohistological localization of *N*-acetyl-indolealkylamines in pineal gland, retina and cerebellum, *Brain Res.*, 81, 233, 1974.
52. **Bubenik, G. A., Brown, G. M., and Grota, L. J.**, Immunohistochemical localization of melatonin in the rat Harderian gland, *J. Histochem. Cytochem.*, 24, 1173, 1976.
53. **Bubenik, G. A., Purtill, R. A., Brown, G. M., and Grota, L. J.**, Melatonin in the retina and the Harderian gland; ontogeny, diurnal variations and melatonin treatment, *Exp. Eye Res.*, 27, 323, 1978.
54. **Pévet, P., Balemans, M. G. M., Bary, F. A. M., and Noordegraaf, E. M.**, The pineal gland of the mole (*Talpa europaea*, L.). V. Activity of hydroxyindole-*O*-methyltransferase (HIOMT) in the formation of melatonin 1'5-hydroxytryptophol in the eyes and the pineal gland, *Ann. Biol. Anim. Biochem. Biophys.*, 18, 259, 1978.
55. **Ebels, I.**, A chemical study of some biologically active pineal fractions, in *The Pineal Gland of Vertebrates Including Man*, Kappers, J. A. and Pevet, P., Eds., Elsevier, Amsterdam, 1979, 309.
56. **Ebels, I., Moszkowska, A., and Slama-Scemama, A.**, Etude *in vitro* des extraits épiphysaires fractionnés. Résultats préliminaires, *C. R. Acad. Sci.*, 260, 5126, 1965.
57. **Moszkowska, A., Ebels, I., and Slama-Scemama, A.**, Etude *in vitro* des extraits fráctionnés d'epiphyse d'agneau, *C. R. Soc. Biol.*, 159, 2298, 1965.
58. **Moszkowska, A. and Ebels, I.**, A study of the antigonadotropic action of synthetic arginine vasotocin, *Experientia*, 24, 610, 1968.
59. **Moszkowska, A. and Ebels, I.**, The influence of the pineal body on the gonadotropic function of the hypophysis, *J. Neuro-Visc. Relat.*, Suppl. 10, 160, 1971.
60. **Ebels, I., Moszkowska, A., and Slama-Scemama, A.**, An attempt to separate a sheep pineal extract fraction showing antigonadotropic activity, *J. Neuro-Visc. Relat.*, 32, 1, 1970.
61. **Balemans, M. G. M., Ebels, I., and Vonk-Visser, D. M. A.**, Separation of pineal extracts on Sephadex G-10. I. A spectrofluorimetric study of indoles in a cockerel pineal extract. *J. Neuro-Visc. Relat.*, 32, 65, 1970.
62. **Ebels, I., Balemans, M. G. M., and Verkleij, A. J.**, Separation of pineal extracts on Sephadex G-10. II. A spectrofluorimetric and thin layer chromatographic study of indoles in a sheep pineal extract, *J. Neuro-Visc. Relat.*, 32, 270, 1972.
63. **Ebels, I., Balemans, M. G. M., and Tommel, D. K. J.**, Separation of pineal extracts on Sephadex G-10. III. Isolation and comparison of extracted and synthetic melatonin, *Anal. Biochem.*, 50, 234, 1972.
64. **Ebels, I., Benson, B., and Matthews, M. J.**, Localization of a sheep pineal antigonadotropin, *Anal. Biochem.*, 56, 546, 1973.
65. **Zurburg, W. and Ebels, I.**, Separation of pineal extracts by gel filtration. I. Isolation from sheep pineals of a substance with special fluorescence characteristics, *J. Neural Transm.*, 35, 117, 1974.
66. **Moszkowska, A., Citharel, A., L'Heritier, A., Ebels, I., and LaPlante, E.**, Some new aspects of a sheep pineal gonadotropic inhibiting activity in *in vitro* experiments, *Experientia*, 30, 964, 1974.
67. **Moszkowska, A., Hus-Citharel, A., L'Heritier, A., Zurburg, W., and Ebels, I.**, Separation of pineal extracts by gel filtration. V. Location by paper chromatography of a sheep pineal principle inhibiting hypophyseal gonadotropic activity, *J. Neural Transm.*, 38, 239, 1976.
68. **Van der Have-Kirchberg, M. L. L., deMorée, A., Van Laar, J. F., Gerwig, G. J., Versluis, C., Ebels, I., Hus-Citharel, A., L'Heritier, A., Roseau, S., Zurburg, W., and Moszkowska, A.**, Separation of pineal extracts by gel filtration. VI. Isolation and identification from sheep pineals of biopterin. Comparison of the isolated compound with some synthetic pteridines and the biological activity in *in vitro* and *in vivo* bioassays, *J. Neural Transm.*, 40, 205, 1977.
69. **Citharel, A., Ebels, I., L'Heritier, A., and Moszkowska, A.**, Epiphyseal-hypothalamic interaction: an *in vitro* study with sheep pineal fractions, *Experientia*, 29, 718, 1973.

70. **Ebels, I., Citharel, A., and Moszkowska, A.**, Separation of pineal extracts by gel filtration. III. Sheep pineal factors acting either on the hypothalamus, or on the anterior hypophysis of mice and rats in *in vitr* experiments, *J. Neural Transm.*, 36, 281, 1975.
71. **Benson, B., Matthews, M. J., and Rodin, A. E.**, Studies on a non-melatonin pineal antigonadotrophin, *Acta Endocrinol.*, 69, 257, 1972.
72. **Benson, B. and Orts, R. J.**, Regulation of ovarian growth by the pineal gland, in *Regulation of Organ and Tissue Growth*, Goss, R., Ed., Academic Press, New York, 1972, 315.
73. **Matthews, M. J. and Benson, B.**, Inactivation of pineal antigonadotropin by proteolytic enzymes, *J. Endocrinol.*, 56, 339, 1973.
74. **Orts, R. J. and Benson, B.**, Inhibitory effects on serum and pituitary LH by a melatonin-free extract of bovine pineal glands, *Life Sci.*, 12, 513, 1973.
75. **Ebels, I. and Benson, B.**, A survey of the evidence that unidentified pineal substances affect the reproductive system in mammals, in *Progress in Reproductive Biology*, Vol. 4, Reiter, R. J., Ed., S. Karger, Basel, 1978, 51.
76. **Ebels, I., deMorée, A., Hus-Citharel, A., and Moszkowska, A.**, A survey of some active sheep pineal fractions and a discussion on the possible significance of pteridines in these fractions in *in vitro* and *in vivo* assays, *J. Neural Transm.*, 44, 97, 1979.
77. **Bensinger, R., Vaughan, M., and Klein, D. C.**, Isolation of a non-melatonin lipophilic antigonadotrophic factor from the bovine pineal gland, *Fed. Proc. Fed. Am. Soc. Exp. Biol.*, 32, 225, 1973.
78. **Slama-Scemama, A., L'Héritier, Moszkowska, A., van der Horst, C. J. G., Noteborn, H. P. J. M., deMorée, A. and Ebels, I.** Effects of sheep pineal fractions on the activity of male rat hypothalami *in vitro*, *J. Neural Transm.*, 46, 47, 1979.
79. **Reiss, M., Davis, R. H., Sideman, M. B., Mauer, I., and Plichta, E. S.**, Action of pineal extracts on the gonads and their function, *J. Endocrinol.*, 27, 107, 1963.
80. **Frowein, A. and Lapin, V.**, Effects of sham-pinealectomy, performed under white and red light, on the melatonin content of rat pineal glands, *Experientia*, 35, 1681, 1979.
81. **Kitay, J. I. and Altschule, M. D.**, *The Pineal Gland, A Review of the Physiologic Literature*, Harvard University Press, Cambridge, Mass., 1954.
82. **Moszkowska, A.**, Contribution a l'etude de l'antagonisme epiphyso-hypophysaire, *J. Physiol.*, 43, 827, 1951.
83. **Thiéblot, L., Alassimone, A., and Blaise, A.**, Etude chromatographique et électrophorétique du facteur antigonadotrope de la glande pinéale, *Ann. Endocrinol.*, 27, 861, 1966.
84. **Thiéblot, L., Blaise, S., and Alassimone, A.**, Essai de caractérisation du principle antigonadotrope de la glande pinéale, *C. R. Soc. Biol.*, 160, 1574, 1966.
85. **Thiéblot, L., Berthelay, J., and Blaise, S.**, Action de la mélatonine sur la sécrétion gonadotrope du rat, *C. R. Soc. Biol.*, 160, 2306, 1966.
86. **Thiéblot, L., Blaise, S., and Couquelet, J.**, Recherche de dérivés indoliques dans des extraits de glande pinéale, *C. R. Soc. Biol.*, 161, 295, 1967.
87. **Thiéblot, L. and Berthelay, S.**, Principe antigonadotrope de la glande pineal, *J. Physiol.*, 60 (Suppl. 2), 554, 1968.
88. **Thiéblot, L. and Menigot, M.**, Acquisitions récentes sur le facteur antigonadotrope de la glande pinéale, *J. Neurovisc. Relat.*, 10 (Suppl. 10), 153, 1971.
89. **Sorrentino, S., Jr. and Benson, B.**, Effects of blinding and pinealectomy on the reproductive organs of adult male and female rats, *Gen. Comp. Endocrinol.*, 15, 242, 1970.
90. **Dickson, K., Benson, B., and Tate, G., Jr.**, The effect of blinding and pinealectomy in unilaterally ovariectomized rats, *Acta Endocrinol.*, 66, 177, 1971.
91. **Benson, B., Matthews, M. J., and Rodin, A. E.**, A melatonin-free extract of bovine pineal with antigonadotropic activity, *Life Sci.*, 10, 607, 1971.
92. **Benson, B., Matthews, M. J., and Rodin, A. E.**, Studies on a non-melatonin pineal antigonadotropin, *Acta Endocrinol.*, 69, 257, 1972.
93. **Matthews, M. J., Benson, B., and Rodin, A. E.**, Antigonadotropic activity in a melatonin-free extract of human pineal glands, *Life Sci.*, 10, 1375, 1971.
94. **Matthews, M. J. and Benson, B.**, Inactivation of pineal antigonadotropin by proteolytic enzymes, *J. Endocrinol.*, 56, 339, 1973.
95. **Orts, R. J., Benson, B., and Cook, B. F.**, LH inhibitory properties of aqueous extracts of rat pineal glands, *Life Sci.*, 14, 1501, 1974.
96. **Ebels, I. and Horwitz-Bresser, A. E. M.**, Separation of pineal extracts by gel filtration. IV. Isolation, location and identification from sheep pineals of three indoles identical with 5-hydroxytryptophol, 5-methoxytryptophol and melatonin, *J. Neural Transm.*, 38, 31, 1976.
97. **Orts, R. J., Benson, B., and Cook, B. F.**, Some antigonadotropic effects of melatonin-free bovine pineal extracts, *Acta Endocrinol.*, 76, 438, 1974.
98. **Benson, B., Matthews, M. J., and Hruby, V. J.**, Characterization and effects of a bovine pineal antigonadotropic peptide, *Am. Zool.*, 16, 17, 1976.

99. **Benson, B., Matthews, M. J., Hadley, M. E., Powers, S., and Hruby, V. J.,** Differential localization of antigonadotropic and vasotocic activities in bovine and rat pineal, *Life Sci.*, 19, 747, 1976.
100. **Orts, R. J., Kocan, K. M., and Yamishani, W. P.,** Fertility control in female rats by bovine pineal gland extracts, *Life Sci.*, 17, 531, 1975.
101. **Orts, R. J.,** Reduction of serum LH and testosterone in male rats by a partially purified bovine pineal extract, *Biol. Reprod.*, 16, 249, 1977.
102. **Rosenblum, I. Y., Benson, B., Bria, C. F., McDonnell, D., and Hruby, V. J.,** Localization and chemical characterization of a partially purified bovine pineal antigonadotropin, *J. Neural Transm.*, 44, 197, 1979.
103. **Rosenblum, I. Y., Benson, B., and Hruby, H. J.,** Chemical differences between bovine pineal antigonadotropin and arginine vasotocin, *Life Sci.*, 18, 1367, 1976.
104. **Ebels, I., Benson, B., Bria, C. F., McDonnell, D., Chang, S. Y., and Hruby, V. J.,** Location by paper chromatography of compensatory ovarian hypertrophy (COH) inhibiting activity in acetic acid extracts from bovine pineals, *J. Neural Transm.*, 42, 275, 1978.
105. **Ebels, I., Benson, B., Bria, C. F., Richardson, D., Larsen, B. R., and Hruby, V. J.,** Location by paper chromatography of compensatory ovarian hypertrophy (COH) inhibiting activity in isobutanol extracts of bovine pineals., *J. Neural Transm.*, 45, 43, 1979.
106. **Benson, B. and Ebels, I.,** Pineal peptides, *J. Neural Transm.*, Suppl. 13, 157, 1978.
107. **Bartke, A., Croft, B. T., and Dalterio, S.,** Prolactin restores plasma testosterone levels and stimulates testicular growth in hamsters exposed to short daylength, *Endocrinology*, 97, 1601, 1975.
108. **Matthews, M. J., Benson, B., and Richardson, D. L.,** Partial maintenance of testes and accessory organs in blinded hamsters by homoplastic anterior pituitary grafts or exogenous prolactin, *Life Sci.*, 23, 1131, 1978.
109. **Bex, F., Bartke, A., Goldman, B. D., and Dalterio, S.,** Prolactin, growth hormone, luteinizing hormone receptors, and seasonal changes in testicular activity in the golden hamster, *Endocrinology*, 103, 2069, 1978.
110. **Reiter, R. J. and Ferguson, B. N.,** Delayed reproductive regression in male hamsters bearing intrarenal pituitary homografts and kept under natural winter photoperiods, *J. Exp. Zool.*, 209, 175, 1979.
111. **Benson, B. and Matthews, M. J.,** Possible role of prolactin and pineal prolactin-regulating substances in pineal-mediated gonadal atrophy in hamsters, *Hormone Res.*, 12, 139, 1980.
112. **Orstead, M. and Benson, B.,** Evidence for decreased release of PRL in hamsters placed in short photoperiod, *Anat. Rec.*, 196, 141, 1980.
113. **Blask, D. E., Vaughan, M. K., Reiter, R. J., Johnson, L. Y., and Vaughan, G. M.,** Prolactin-releasing and release-inhibiting factor activities in the bovine, rat and human pineal gland: *in vitro* and *in vivo* studies, *Endocrinology*, 99, 152, 1976.
114. **Chang, N., Ebels, I., and Benson, B.,** Preliminary characterization of bovine pineal prolactin releasing (PPRF) and release-inhibiting factor (PPIF) activity, *J. Neural Transm.*, 46, 139, 1979.
115. **Larsen, B. R. and Benson, B.,** Purification of bovine pineal prolactin inhibiting factor, *Anat. Rec.*, 193, 598, 1979.
116. **Barbeau, A., Tsukada, Y., and Inone, N.,** Neuropharmacologic and behavioral effects of taurine, in *Taurine*, Huxtable, R. and Barbeau, A., Eds., Raven Press, New York, 1976, 253.
117. **Bradford, J. F., Davison, A. N., and Wheler, G. H. T.,** Taurine and synaptic transmission, in *Taurine*, Huxtable, R. and Barbeau, A., Eds., Raven Press, New York, 1976, 303.
118. **Grosso, D. S., Bressler, R., and Benson, B.,** Circadian rhythm and uptake of taurine by the rat pineal gland, *Life Sci.*, 22, 1789, 1978.
119. **Ebels, I., Benson, B., and Larsen, B. R.,** Biosynthesis of taurine by rat pineals *in vitro*, *J. Neural Transm.*, 48, 101, 1980.
120. **Larsen, B. R., Grosso, D., and Chang, S. Y.,** A rapid method for taurine quantitation using high performance liquid chromatography, *J. Chromatogr.*, 18, 233, 1980.
121. **Buskin, S. I. and Dagirmanjian, R.,** The effect of taurine on the pigmentation of the bullfrog tadpole, *Comp. Biochem. Physiol.*, 44A, 297, 1973.
122. **Wheler, G. H. T., Weller, J. L., and Klein, D. C.,** Taurine: stimulation of pineal *N*-acetyltransferase activity and melatonin production via a beta-adrenergic mechanism, *Brain Res.*, 166, -65, 1979.
123. **Orts, R. J., Laio, T.-H., Sartin, J. L., and Bruot, B.,** Purification of a tripeptide with anti-reproductive properties isolated from bovine pineal glands, *Physiologist*, 21, 87, 1978.
124. **Larsen, B. R., Hruby, V. J., and Benson, B.,** Synthesis and effects on prolactin of a bovine pineal tripeptide, in *Peptides, Structure and Biological Function*, Pierce Chemical Co., Rockford, Ill., 1979, 795.
125. **Larsen, B. R., Benson, B., and Hruby, V. J.,** Synthesis and effects on prolactin of threonylseryllysine, *Abstr. 61st Ann. Mtg. Endocrine Soc.*, Endocrine Society, Bethesda, Maryland, 1979, 242.
126. **Findell, P. R., Larsen, B. R., Benson, B., and Blask, D. E.,** Influence of urethane anesthesia on the *in vitro* and *in vivo* secretion of pituitary prolactin in the rat, *Life Sci.*, accepted.

127. **Moszkowska, A., Citharel, A., L'Heritier, A., Ebels, I., and LaPlante, E.,** Some new aspects of a sheep pineal gonadotropic inhibiting activity in *in vitro* experiments, *Experientia,* 30, 964, 1974.
128. **Lloyd, T., Markey, S., and Weiner, N.,** Identification of 2-amino-4-hydroxy substituted Pteridines by gas-liquid chromatography and mass spectrometry, *Anal. Biochem.,* 42, 108, 1971.
129. **Blakley, R. J.,** The biochemistry of folic acid and related pteridines, in *Frontiers of Biology,* Vol. 13, Neuberger, A. and Tatum, E. L., Eds., North-Holland, Amsterdam, 1969, 1.
130. **Abelson, H. T., Spector, R., Gorka, C., and Fosburg, M.,** Kinetics of tetrahydrobiopterin synthesis by rabbit brain dihydrofolate reductase, *Biochem. J.,* 171, 267, 1978.
131. **Lloyd, T. and Weiner, N.,** Isolation and characterization of a tyrosine hydroxylase co-factor from bovine adrenal medulla, *Mol. Pharmacol.,* 7, 569, 1971.
132. **Leeming, R. J., Blair, J. A., Green, A., and Raine, D. N.,** Biopterin derivatives in normal and phenylketonuric patients after oral loads of L-phenylalanine, L-tyrosine and L-tryptophan, *Arch. Dis. Child.,* 51, 771, 1976.
133. **Lowry, O. H., Bessey, O. A., and Crawford, E. J.,** Photolytic and enzymatic transformations of pteroyl-glutamic acid, *J. Biol. Chem.,* 180, 389, 1949.
134. **Kalcnar, H. M. and Klenow, H.,** Milk xanthopterin oxidase and pteroylglutamic acid, *J. Biol. Chem.,* 172, 349, 1948.
135. **Fukushima, T. and Nixon, J. C.,** Reverse-phase high-performance liquid chromatographic separation of unconjugated pterins and pteridines, in *Chemistry and Biology of Pteridines,* Kisliuk, R. L. and Brown, G. M., Eds. North-Holland, Amsterdam, 1979, 35.
136. **Levine, R. A., Kuhn, D. M., and Lovenberg, W.,** The regional distribution of hydroxylase cofactor in rat brain, *J. Neurochem.,* 32, 1575, 1979.
137. **Balemans, M. G. M., van Brenthem, J., Legerstee, W. C., deMorée, A., Noteborn, H. J. P. M., and Ebels, I.,** The influence of some pterins on the circadian rhythmicity of HIOMT in the pineal gland of 42 day old male rats, *J. Neural Transm.,* in press.
138. **Lapin, V.,** Pineal gland and malignancy, *Öst. Z. Onkol.,* 3, 51, 1976.
139. **Stea, B., Backlund, P. S., Jr., Berkey, P. B., Cho, A. K., Halpern, B. C., Halpern, R. M., and Smith, R. A.,** Folate and pterin metabolism by cancer cells in culture, *Cancer Res.,* 38, 2378, 1978.
140. **Ziegler, I. and Kokolis, N.,** *In vivo* metabolism of deutero-L-phenylalanine and deutero-L-tyrosine and levels of tetrahydrobiopterin in the blood of tumor bearing organisms, in *Chemistry and Biology of Pteridines,* Kisliuk, R. L. and Brown, G. M., Eds., North-Holland, Amsterdam, 1979, 165.
141. **Wachter, H., Hausen, A., Grassmay, K.,** Erhöhte Ausseheidung von Neopterin im Harn von Patienten mit malignen Tumoren und mit Viruserkrankungen, *Hoppe-Seyler's Z. Physiol. Chem.,* 360, 1957, 1979.
142. **Lapin, V.,** Effects of reserpine on the incidence of 9,10-dimethyl-1,2-benzanthracene-induced tumors in pinealectomised and thymectomised rats, *Oncology,* 35, 132, 1978.
143. **Kokolis, N. and Issiodorides, M.** Effects of reserpine and amphetamine on pteridine pattern and nuclear morphology of *Triturus Cristatus* liver, *Exp. Cell Res.,* 65, 186, 1971.
144. **Reiter, R. J.,** Interaction of photoperiod, pineal and seasonal reproduction as exemplified by findings in the hamster, *Prog. Reprod. Biol.,* 4, 169, 1978.
145. **Berndtron, W. E. and Desjardins, D.,** Circulating LH and FSH levels and testicular function in hamsters during light deprivation and subsequent photoperiodic stimulation, *Endocrinology,* 95, 195, 1974.
146. **Turek, F. W., Elliott, J. A., Alvis, J. D., and Menaker, M.,** Effect of prolonged exposure to non-stimulatory photoperiods on the activity of the neuroendocrine-testicular axis of golden hamsters, *Biol. Reprod.,* 13, 475, 1975.
147. **Turek, F. W.,** The interaction of the photoperiod and testosterone in regulating serum gonadotropin levels in castrated male hamsters, *Endocrinology,* 101, 1210, 1977.
148. **Turke, F. W.,** Role of the pineal gland in photoperiod-induced changes in hypothalamic-pituitary sensitivity to testosterone feedback in castrated in male hamsters, *Endocrinology,* 104, 636, 1979.
149. **Ellis, G. B. and Turek, F. W.,** Time course of the photoperiod-induced change in sensitivity of the hypothalamic pituitary axis to testosterone feedback in castrated male hamsters, *Endocrinology,* 104, 625, 1979.
150. **Ellis, G. B., Losee, S. H., and Turek, F. W.,** Prolonged exposure of castrated male hamsters to a nonstimulatory photoperiod: spontaneous change in sensitivity of the hypothalamic-pituitary axis to testosterone feedback, *Endocrinology,* 104, 631, 1979.
151. **Benson, B. and Krasovich, M.,** unpublished results.
152. **Ebels, I.,** unpublished results.

Chapter 8

POTENTIAL SITES OF ACTION OF PINEAL HORMONES WITHIN THE NEUROENDOCRINE-REPRODUCTIVE AXIS

David E. Blask

TABLE OF CONTENTS

I. INTRODUCTION

It is somewhat ironic that the pineal gland, probably the first organ of internal secretion to be described, has only recently gained acceptance as a bona fide endocrine gland. Indubitably, the pineal gland plays a pivotal role in the regulation of reproductive physiology, particularly seasonal phenomena, under conditions of altered environmental photoperiod. In fact, many pinealologists regard this particular function of the pineal as its *raison d'être*.[1] An immense amount of effort has been expended in attempts to elucidate possible targets of pineal antigonadotrophic secretory activity. In a heuristic sense, all of the components of the neuroendocrine-reproductive axis collectively represent the target site for pineal autocoids. Hence, each component may be thought of as a subtarget whose endocrine activity is modulated to a different extent by pineal hormones. The differential and integrated responses of these subtargets to pineal substances may ultimately determine the reproductive status of the organism with respect to the photoperiod.

The greatest bulk of both direct and indirect data imputes the hypothalamus as the most likely primary target for epiphyseal compounds.[1-3] On the other hand, a number of studies would argue that specific extrahypothalamic loci within the CNS are important sites of antigonadotrophic activity.[4] Still, others would claim that the anterior pituitary gland and even the gonads themselves are the chief recipients of pineal antigonadotrophic information.[5] This chapter will review the evidence implicating each level in the neuroendocrine-reproductive hierarchy as a potential site of action of pineal constituents.

II. GONADS AND ACCESSORY SEX ORGANS

A severe reduction in both gonadal and accessory organ size and function is perhaps the most dramatic biological response associated with increased pineal activity.[1-5] So impressive is the gonadal response to pineal secretions that a hypothesis has emerged denoting the gonads as the principal site of action of the pineal antigonadotrophin(s). Early support for this hypothesis derived from the observation that pinealectomy either partially or in some cases totally prevented gonadal regression in hypophysectomized rats.[6,7] An obvious implication of such a striking finding is that even in the absence of pituitary gonadotrophins the pineal hormone is capable of suppressing the gonads, presumably by acting directly at this level.[5]

A. Effects of Indolic Compounds

A few studies have shown that the injection of melatonin (Mel) or one of its cogeners, 5-methoxytryptophol (5-Mtol) into hypophysectomized rats or mice, inhibits gonadal and accessory organ stimulation by human chorionic gonadotrophin (HCG).[8,9] Even in hypophysectomized rats with hypotrophic reproductive organs, Mel causes a further reduction in gonadal and accessory organ weights.[9] However, it is not clear whether Mel inhibits the gonadal uptake of and responsiveness to gonadotrophins, or exerts its effects irrespective of the presence or absence of gonadotrophins. With respect to the former possibility, it has been shown that the ovarian uptake of ^{3}H-LH, but not ^{3}H-FSH, is reduced in rats pretreated with Mel, suggesting that LH binding to its receptors may be inhibited. Conversely, serotonin (5-HT) stimulates the accumulation of both gonadotrophins in ovarian and uterine tissues.[10] Similarly, Mel restricts the growth of the accessory organs in response to testosterone administered as either an injection or chronic implant to orchidectomized animals.[11-13] The mechanism by which this is accomplished is presently unknown.

More convincing testimony of a peripheral site of action of Mel comes from in vitro

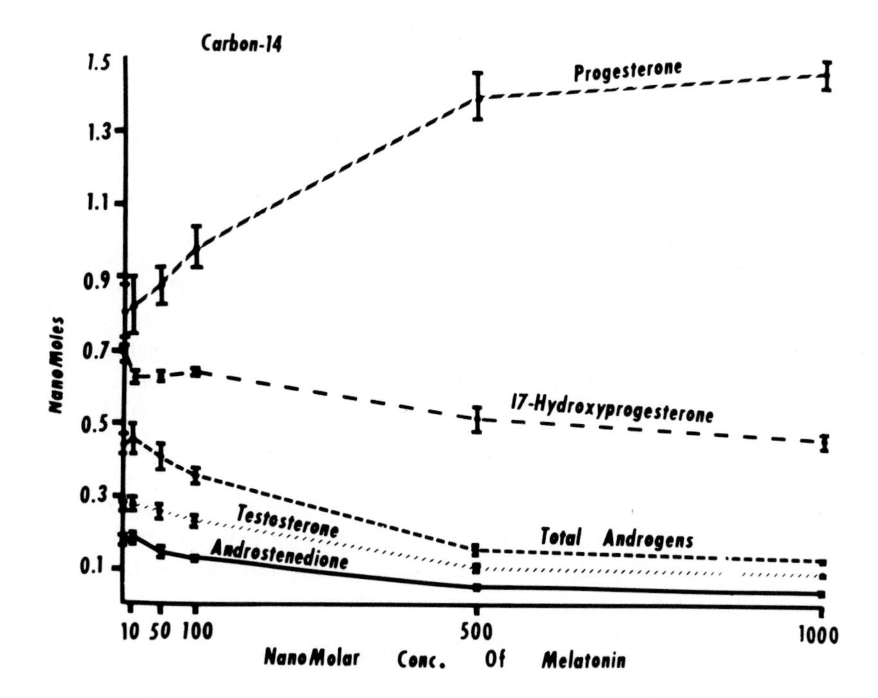

FIGURE 1. Effect of melatonin on the conversion of progesterone-4-^{14}C into 17-hydroxyprogesterone, androstenedione, and testosterone by rat testicular tissue in vitro. (From Ellis, L. C., *Endocrinology*, 100, 17, 1972. With permission.)

experiments demonstrating the efficacy with which this indole impedes the synthesis of testicular androgens in different species (Figure 1).[14-16] Apart from its effects on testicular hormone production, Mel may also hinder certain biomechanical events such as seminiferous tubule contractility, possibly via cyclic GMP-mediated mechanisms.[17-18] Less consistent is the influence of Mel on ovarian steroidogenesis in female mammals. For example, depending upon the species examined, Mel may either promote or prohibit either the basal or HCG-induced biosynthesis of ovarian steroids.[19-21]

Of course, any effect of Mel at the gonadal level infers that the molecule is taken up by the tissue and interacts with a specific receptor. In certain species the uptake and concentration of ^{3}H-Mel by ovarian tissue is significant while in others it is virtually nil.[21-23] The presence of alleged cytoplasmic Mel receptors in the gonadal and accessory tissues of several species, including humans,[24] further strengthens the hypothesis that Mel may exert its antireproductive effects directly upon peripheral reproductive target organs. Widespread acceptance of this hypothesis, however, awaits its further conformation.

B. Effects of Nonindolic Compounds

Unlike their indolic counterparts, there is less substantial evidence pointing to a peripheral reproductive endocrine role for nonindolic pineal compounds. For instance, depending upon the extent of purification, isobutanol extracts of bovine pineal tissue reportedly either stimulate or inhibit the ovarian uptake of ^{32}P in response to exogenous gonadotrophin stimulation.[25] The varying effects of these extracts may relate to their possible contamination with Mel. Recently, another partially purified bovine pineal fraction designated A_1 was shown to inhibit pregnant mare's serum (PMS)-induced ovulation in immature female rats without affecting the preovulatory surge of LH, suggesting an ovarian site of action.[26]

The antigonadotrophic effects of arginine vasotocin (AVT) may also be in part, exerted at the peripheral reproductive level.[27] Various doses of AVT, mixed with anterior pituitary incubation fluid containing secreted gonadotrophins, significantly compromise the subsequent stimulatory effect of the fluid on uterine and follicle size in HCG-treated immature mice.[28] Like bovine pineal fraction A_1, AVT might block ovulation by acting at the level of the ovary, in PMS-treated immature female rats since the preovulatory plasma peaks of LH, FSH, and prolactin (Prl) escape its influence.[29]

Another example of a peripheral influence exerted by AVT is its recently demonstrated ability to decrease ^{3}H-estradiol binding by cytoplasmic estrogen receptors from the uteri of immature rats.[30] Apparently AVT, like Mel, can alter the biomechanical physiology of certain reproductive tissues. This is perhaps best illustrated in the hen's uterus whereby femtomolar quantities of AVT provoke uterine contractility, possibly via prostaglandin-mediated mechanisms.[31,32] This effect of AVT is particularly interesting in view of the abortifacient properties of this peptide in pregnant mice.[27]

Thus it appears that both pineal indoles and peptides may exert part of their influence on the reproductive tract through direct gonadal and accessory organ mechanisms. Hence, these may involve either alterations in the metabolism of the target organ, responsiveness or sensitivity of the target organ to gonadotrophin or steroid stimulation, antagonism of receptors, inhibition of biomechanical contractility or any combination of the above.

III. PARS DISTALIS OF THE PITUITARY GLAND

At least two possible mechanisms exist by which pineal hormones may directly alter the secretion of gonadotrophic hormones from the pars distalis. These include changes in either or both the basal and/or releasing hormone-stimulated output of hypophyseal gonadotrophins.

A. Pituitary Gonadotrophin Release

1. Effects of Indolic Compounds

There is virtually no evidence to support the view that pineal indolic compounds directly affect basal pituitary gonadotrophin synthesis and/or release. Intrapituitary implants or infusions of Mel and related indoles have no effect on either pituitary LH, FSH, or Prl release.[33-35] Similarly, no direct effects of Mel on basal LH or Prl have been demonstrable in vitro.[36-38]

2. Effects of Nonindolic Compounds

In marked contrast to the indoleamines, several lines of evidence indicate that peptidic substances from the pineal gland either stimulate or inhibit gonadotrophic secretory responses directly from the pituitary gland. In a variety of in vitro paradigms AVT directly stimulates Prl secretion from anterior pituitary tissue.[38-40] With regard to LH secretion, AVT exhibits progonadotrophic properties by its ability to markedly stimulate LH release from male rat hemipituitaries in vitro.[27] Other investigators, however, have been unsuccessful in demonstrating an effect of AVT on the basal secretion of either LH or FSH from monolayer pituitary cultures.[41]

Long before the stimulatory effects of AVT were observed on pituitary LH secretion it had been shown that a low molecular weight ovine pineal fraction (F_2), obtained through Sephadex® G-25 chromatography, stimulated the release of FSH from rat pituitaries in vitro. Subsequently, other purification and separation methods yielded pineal fractions that stimulated pituitary gonadotrophin release.[42] Interestingly, White and associates[43] demonstrated that crude extracts of either porcine, bovine, or ovine pineal glands stimulate LH and FSH release from rat hemipituitaries in vitro. Further analysis revealed that the stimulatory nature of these extracts was attributable to the

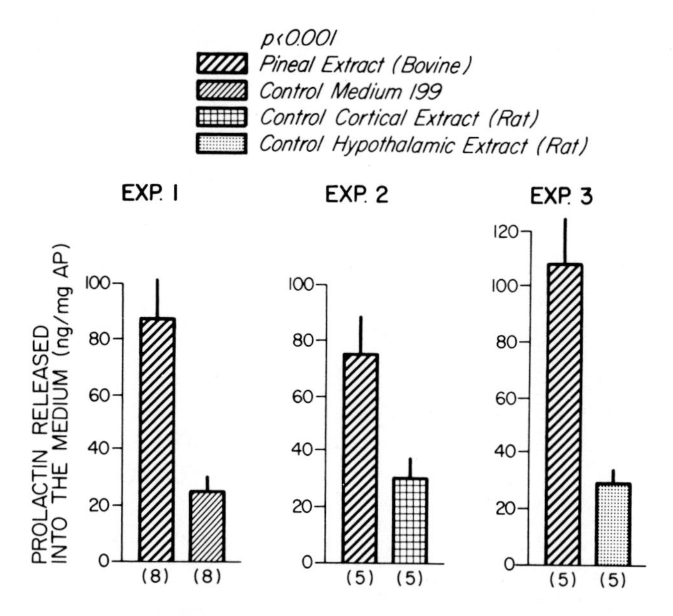

FIGURE 2. Effects of crude bovine pineal extracts on prolactin release from normal male rat hemipituitaries (From Blask, D. E., Vaughan, M. K., Reiter, R. J., Johnson, L. Y., and Vaughan, G. M., *Endocinology,* 199, 152, 1976. With permission.)

presence of high concentrations of radioimmunoassayable gonadotrophin-releasing hormone (GnRH). These results prompted Blask and colleagues[36] to search for the presence of a pineal Prl releasing factor(s) (PPRF) and/or inhibiting factor(s) (PPIF). As a result, pineal extracts from several species including the bovine, rat, and human were shown to stimulate Prl secretion from rat hemipituitaries in vitro (Figure 2). Conversely, partially purified bovine pineal fractions designated A_1 and A_3 were found to inhibit Prl release. Furthermore, bovine PPRF activity may be due to a large molecular weight peptide (MW > 10,000) while PPIF activity may be due to a smaller molecular weight (MW > 500) peptidic substance.[44]

Earlier work revealed that a low molecular weight fraction of sheep pineal tissue possessed FSH-inhibitory activity by virtue of a direct action at the pituitary level. The advent of more sophisticated purification and separation techniques permitted the isolation of low molecular weight sheep pineal fractions that inhibited both LH and FSH release from rat anterior pituitary glands in vitro.[42] Pursuant to the further chemical characterization of these fractions, Ebels et al.[45] isolated a pteridine which in all probability is structurally identical with 6-L-erythrobiopterin. However, it is problematical as to whether this biopterin is responsible for the antigonadotrophic activity of these fractions.

B. Pituitary Responsiveness to Hypothalamic Hormones

In light-deprived hamsters, the response of pituitary LH to exogenously administered GnRH is not altered, suggesting that the pineal gland does not affect pituitary sensitivity to releasing hormones.[46,47] However, GnRH was administered approximately 8 weeks following either blinding or exposure to exposure to short photoperiod, a time when gonadal regression was presumably complete. Therefore, it is conceivable that during the initial few weeks of light deprivation, pituitary responsiveness does decrease but returns to normal by the time gonadal atrophy has been achieved.

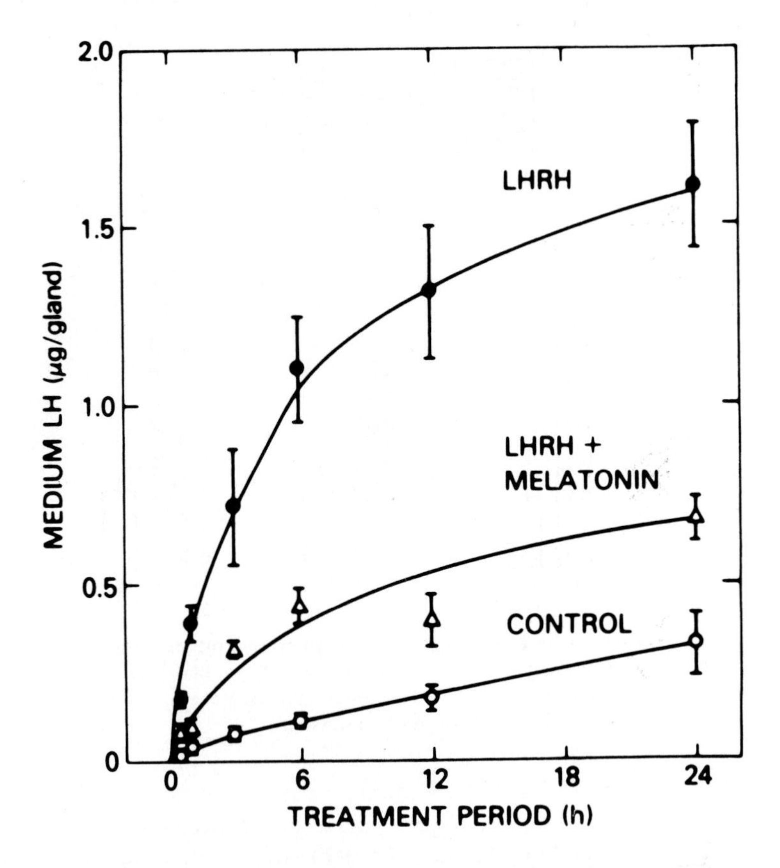

FIGURE 3. Effect of melatonin (10^{-8} *M*) on LH release from neonatal rat pituitary cultures in response to luteinizing hormone-releasing hormone (LHRH) (10^{-9} *M*). (From Martin, J. E., Engel, J. N., and Klein, D. C., *Endocrinology,* 100, 675, 1977.

1. Effects of Indolic Compounds

The important issue of whether pineal indoleamines exert their antigonadotrophic effects by altering the responsiveness of the adenohypophysis to GnRH has been addressed in a number of species. In organ cultures of neonatal rat anterior pituitary glands, the coincubation of GnRH with physiological concentrations of Mel results in as much as an 86% depression in the LH response to GnRH (Figure 3).[37] Moreover, the decrease in pituitary sensitivity to GnRH is an effect not restricted to Mel inasmuch as indoles closely related to Mel, such as serotonin (5-HT) and 5-methoxytryptamine (5-MT) also significantly diminish GnRH-stimulated LH release. Neither the immediate precursor of Mel, *N*-acetyserotonin (NAS), nor the major hepatic metabolite of Mel, 6-hydroxymelatonin (6-HMel) has any measurable effect on either basal or GnRH-stimulated LH secretion in vitro. This latter fact indicates that hydroxylation of Mel at the six position neutralizes its inhibitory effect.[37,48]

Apparently, the ability of Mel to curtail GnRH-stimulated LH secretion is restricted to the neonatal period. As the animal approaches puberty, the anterior pituitary may become relatively refractory to the inhibitory effects of this molecule.[37,49-51] Such a mechanism would not be incompatible with and may actually accompany the processes leading to pubertal development. These include, of course, a decrease in hypothalamic sensitivity to inhibitory steroidal feedback, an event which hails the onset of the pubertal gonadotrophin increase.[52]

Clearly the situation in humans appears to differ considerably from that in lower mammalian species with regard to the effectiveness of Mel in reducing either basal or GnRH-stimulated LH release. Although pharmacological doses of Mel, administered to hyperpigmented Caucasian patients, caused a modest reduction in plasma LH, no effect of FSH was noted.[53] Recent studies in normal young adult men and postmenopausal women indicate that Mel injections have no effect on either basal or GnRH-stimulated release of LH and FSH into the bloodstream.[54,55] These results suggest that if Mel is antigonadotrophic in humans, its site of action is probably not at the pituitary level.

2. Effects of Nonindolic Compounds

Not only does AVT have the ability to stimulate basal LH release from hemipituitaries in vitro but, in contrast to Mel, it enhances the pituitary sensitivity to GnRH in vitro.[27] Attempts to duplicate these results have been largely unsuccessful particularly in dispersed pituitary cell cultures.[41] Another recent report stated that while AVT had no effect on GnRH-stimulated LH release from pituitaries derived from intact or castrated male rats, it significantly inhibited the LH response to GnRH when pituitaries were obtained from castrated estrogen-progesterone (EP)-treated male rats.[56]

Unfortunately, the results of several in vivo experiments have proved to be no less confounding than those obtained with in vitro techniques. An interesting study by Vaughan and associates[57] demonstrated that a single injection of AVT, in combination with GnRH, significantly augmented plasma LH titers in castrated male rats when compared with GnRH-treated control rats. This study corroborated their earlier reports that AVT potentiates the LH-releasing effect of GnRH in vitro.[27] Either opposite or no effects of AVT on GnRH-stimulated LH release have been observed in other species.[58,59]

It appears that both in vivo and in vitro AVT can enhance, diminish, or have no effect on pituitary responsiveness to GnRH depending on a number of factors. The lack of concordance between various studies may be explicable on the basis of any one or combination of these factors which include differences in: (1) the mode of administration of AVT and its temporal relationship to the injection of GnRH, (2) the species, sex, and steroidal milieu of the test animal and finally, (3) whether the animals are anesthetized or unanesthetized.

Demoulin et al.[60] reported that partially purified fractions of ovine pineal glands possess the ability to inhibit GnRH-induced LH release from monolayer cell cultures of rat pituitaries. A similar report revealed that a partially purified pineal extract, obtained from an unspecified species, also inhibited GnRH-induced LH release from monolayer cultures of pituitary cells from 10-day-old rats.[61] The extraction methods employed in these studies indicated that the anti-GnRH activity resides in a small molecular weight substance(s) which is neither Mel nor AVT.

IV. CENTRAL NERVOUS SYSTEM

Of all potential sites of action for pineal hormones studied to date, the brain is perhaps the leading candidate. One or several loci within the brain including the hypothalamus, the mesencephalon, and other extrahypothalamic areas, and perhaps even the pineal gland itself, may represent the primary effector site(s) for pineal antigonadotrophic activity.[1-5]

A. Hypothalamus

Considering the fact that the hypothalamus is the central control and integrative point for the neuroendocrine regulation of anterior pituitary function, it is not surprising that a great deal of attention has been focused on this diencephalic structure as an intermediary between the epiphysis cerebri and the adenohypophysis. With few exceptions, most, if not all of the effects of the pineal gland on reproductive function can be explained, more or less, on the basis of either direct or indirect actions of its hormone(s) on hypothalamic mechanisms governing gonadotrophin release.[1-5]

1. Gonadotrophin Releasing Hormones

For many years it has been proposed that the pineal inhibits pituitary gonadotrophin secretion by altering the synthesis and/or release of hypothalamic releasing and inhibiting hormones,[4,5] yet only recently has this problem been approached head on. Both in vitro and in vivo studies with the blind-anosmic female rat have revealed that dual sensory deprivation increases levels of biologically active GnRH in the medial basal hypothalamus (MBH) (Figure 4). This is a pineal-mediated effect inasmuch as removal or denervation of the gland ameliorates the effects of blinding and anosmia on GnRH activity. We postulated that increased pineal activity results in a decrease in only the release of GnRH, to the extent that it accumulates in the MBH.[62,63] Additionally, hypothalamic levels of PRF activity are altered in blind-anosmic-pinealectomized animals as well.[62,64] The neurosecretory activity in certain hypothalamic nuclei as determined by histochemical methods is also altered by pinealectomy.[65,66]

As in the rat, similar increases in immunoreactive GnRH have been observed in male hamsters blinded for 5 weeks or maintained in short photoperiods for 13 weeks.[67,68] As previously suggested, exposure of hamsters to short photoperiods or complete darkness may inhibit the release and/or stimulate the synthesis of GnRH resulting in its accumulating in the MBH. Unfortunately, the role played by the pineal gland could not be determined from these studies since pinealectomized hamsters were not included.

In marked contrast to these results, we[69] have recently documented a 65% and 80% drop in biologically active GnRH and follicle-stimulating hormone-releasing hormone (FSH-RH) levels respectively (Figure 5) in the MBH(s) of male hamsters blinded for 11 weeks. Whereas pinealectomy totally prevented the effects of blinding on FSH-RH levels, GnRH activity remained low. These results prompted us to postulate that the pineal gland may exert differential effects on hypothalamic GnRH and FSH-RH which may represent two separately controlled releasing hormones for LH and FSH in the hamster. An alternative hypothesis is that the dark-induced decrease in GnRH activity may not be a pineal-mediated event but rather the result of the intervention of some other CNS structure with efferent input to hypothalamic neurosecretory cells. The inconsistencies between the aforementioned investigations and ours may be due to the fact that in our study an in vitro bioassay system was used to assess releasing hormone activity, while in the other studies, the immunoreactive form of GnRH was measured. The possibility arises that the immunoreactive GnRH is not necessarily the molecular equivalent of biologically active GnRH.

2. Biogenic Amines

Changes in the turnover of brain amines, particularly in the hypothalamus, have been correlated with alterations in the output of gonadotrophic hormones. Furthermore, it appears that fluctuations in the availability of biogenic amines to interact, postsynaptically, with releasing and inhibiting hormone neurons is an important mechanism by which gonadotrophin secretion is regulated.[70] Therefore, an alteration in brain amine levels, especially at the hypothalamic level, is a potentially attractive mechanism by which the pineal may influence the activity of the hypothalamo-hypophyseal-gonadal axis. Surprisingly few studies, however, have examined the effects of the pineal or its removal on biogenic amines in discrete brain areas.

Pinealectomy reduces the hypothalamic content of 5-HT and its major metabolite

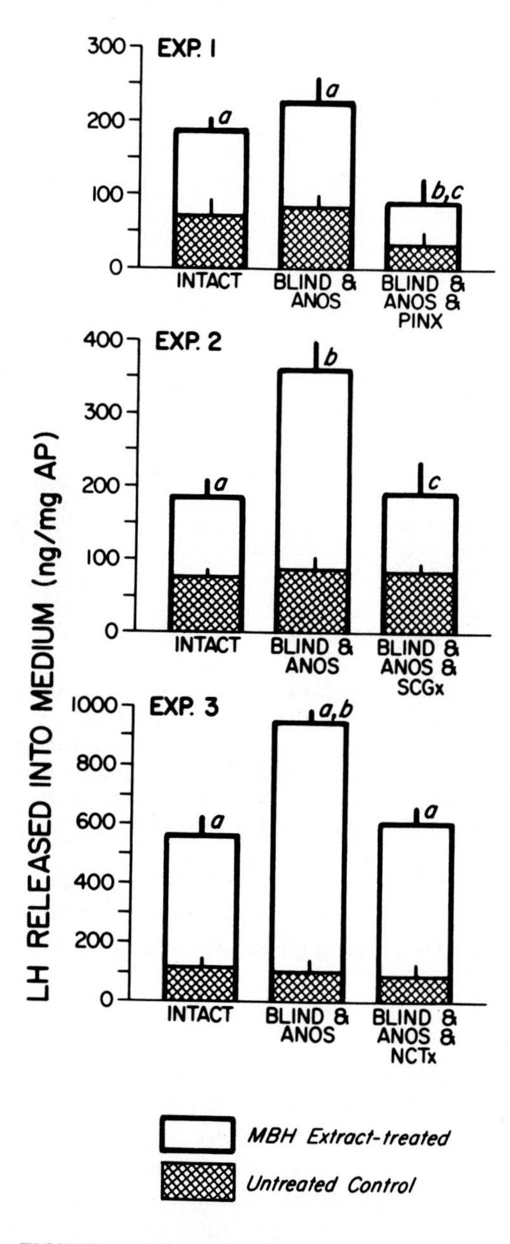

FIGURE 4. Bioreactive GnRH in MBH extracts from either intact, blind-anosmic (ANOS), blind-anosmic-pinealectomized (PINX), blind-anosmic-superior cervical ganglionectomized (SCGx), or blind-anosmic-nervi conarii transected (NCTx) 60-day-old female rats. LH released into the medium by hemipituitaries treated with MBH extracts was used an an index of GnRH activity. (From Blask, D. E., Reiter, R. J., and Johnson, L. Y., *Neurosci. Lett.*, 1, 327, 1975. With permission.)

5-hydroxyindole acetic acid (5-HIAA)[71,72] suggesting a decrease in 5-HT synthesis. Conversely, monoamine oxidase activity increases in pinealectomized animals indicating a concomitant augmentation of 5-HT catabolism.[73] Furthermore, pineal removal or denervation appears to alter the circadian rhythm in hypothalamic 5-HT content as

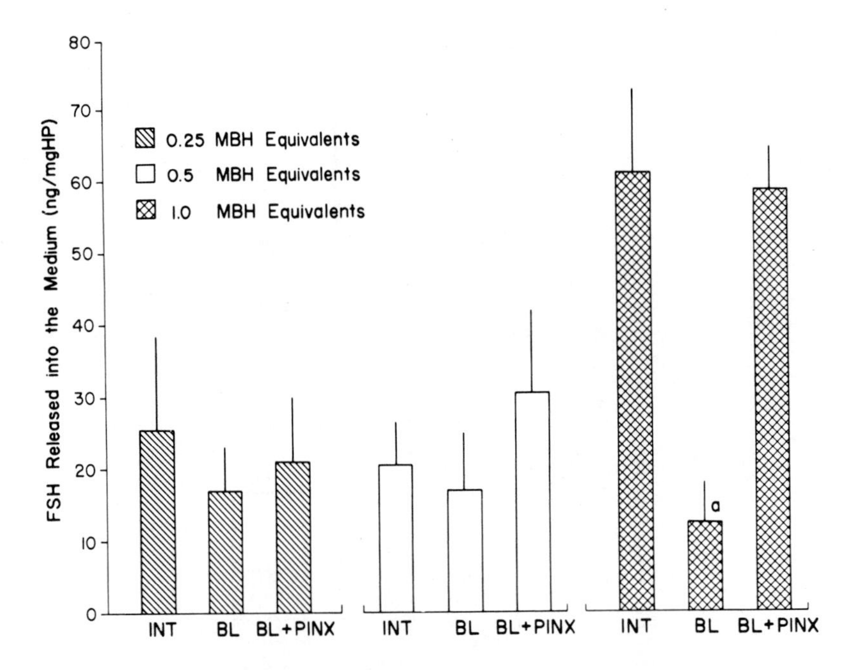

FIGURE 5. Dose-response relationship between the amount of follicle-stimulating hormone-releasing hormone (FSH-RH) activity in MBH extracts from either intact (INT), blind (BL), or blind + pinealectomized (PINX) male hamsters. FSH released into the medium by hemipituitaries treated with MBH extracts was used as an index of FSH-RH activity. (From Blask, D. E., Reiter, R. J., Vaughan, M. K., and Johnson, L. Y., *Neuroendocinology,* 23, 36, 1979. With permission.)

well as presynaptic events such as the uptake of 5-HT by hypothalamic synaptosomes.[74,75] Since 5-HT has been demonstrated to be inhibitory to gonadotrophin release it is conceivable that the pineal gland may stimulate serotoninergic mechanisms to curb gonadotrophin output.

3. Lesion Studies

Surgical interventions into various regions of the hypothalamus have yielded results which provide some important insights into the neural substrates involved in the pineal-mediated inhibition of reproductive physiology. Several studies by Reiter et al.[4] have demonstrated that anterior as well as total hypothalamic deafferentations were effective in completely preventing gonadal atrophy in blinded hamsters. On the other hand, long posterior hypothalamic knife cuts prevented the gonadal response to blinding in only 50% of the animals, whereas short posterior cuts were without effect. Likewise, bilateral electrolytic lesions of the medial forebrain bundle (MFB) were as effective in preventing gonadal regression in blind-anosmic rats, as hypothalamic deafferentations were in blinded hamsters. Thus several surgical lesions of afferent and efferent hypothalamic pathways exactly mimicked the effects of pinealectomy on the reproductive system in animals deprived of light. These intriguing results led to the formulation of several hypotheses: (1) hypothalamic cuts interrupt axons which transmit photic information from the eyes to the pineal gland, (2) the lesions produce a central sympathectomy which functionally inactivates the pineal, (3) the cuts sever inhibitory pathways to the hypothalamus arising from pineal-sensitive neurons in an extrahypothalamic site(s), and (4) the lesions interrupt axons arising from extra-hypothalamic neurons projecting to and sensitizing gonadotrophic releasing hormone cells in the MBH.

Considering recent advances in our knowledge of the neural pathways involved in the control of pineal function,[76] all the effects of hypothalamic deafferentation and MFB lesions on gonadal regression in dark-exposed animals are explicable on the basis of the first postulate proposed by Reiter et al.[4] In this regard, it is interesting to note that simultaneous reports from two independent laboratories showed that bilateral lesions of the suprachiasmatic nuclei (SCN), like hypothalamic deafferentations, stymie gonadal regression in hamsters exposed to short photoperiods or blinding.[77,78] As a result, it was proposed that the SCN under short photoperiodic conditions drive the production and secretion of pineal antigonadotrophic principles.[77]

In rats with SCN lesions or frontal deafferentations, the additional removal of the pineal gland or its denervation by superior cervical ganglionectomy purportedly increases LH levels at proestrus and reverses the anovulatory effect of the hypothalamic lesions.[79-80] These results are difficult to reconcile in view of the fact that hypothalamic lesions of this type interrupt the neural pathways from the retina to the pineal,[76] thus creating a functionally pinealectomized animal.[4] Hence, the mechanism(s) by which an additional surgical pinealectomy promotes ovulation requires further clarification.

4. Steroid Feedback

The inhibitory feedback action of gonadal steroids on the hypothalamic releasing hormone mechanisms controlling gonadotrophin output are well-known.[81] Several well-designed studies have elucidated a potentially important relationship between inhibitory steroidal feedback and photic-induced changes in hypothalamic sensitivity to steroidal feedback. For example, the exposure of male hamsters to short photoperiods makes the hypothalamo-hypophyseal axis more-sensitive to negative feedback effects of testosterone on gonadotrophin release.[82,83] It appears that the pineal gland is responsible for this increase in hypothalamic sensitivity since pinealectomized animals maintained in short photoperiod are not as responsive to negative steroidal feedback as intact hamsters. It has been proposed that the pineal gland, in response to short photoperiod, is secreting a substance which acts on the hypothalamus to increase its sensitivity to steroid feedback.[84]

During the neonatal period, treating rats with large doses of either androgen or estrogen is known to markedly affect the maturation of hypothalamic centers for gonadotrophin control.[85] Interestingly, the treatment of rats with testosterone during the neonatal period greatly amplifies the subsequent gonad-inhibiting effect of blinding.[86] This was later deemed to be a pineal-mediated effect since pinealectomy almost totally negated the reproductive hypotrophy associated with blinding and neonatal androgen treatment. One interpretation of these results is that by the time puberty is reached, the hypothalamic set-point of sensitivity to inhibitory steroidal feedback has increased[4] rather than decreased as it does normally.[52] Under the influence of the pineal gland a new level of steroid uptake by the hypothalamus might be achieved.[87]

From the preceding discussion it is clear that pineal factors must in some way interact with steroids at the hypothalamic level. Perhaps pineal hormones alter the number, as well as the sensitivity of steroid receptors and possibly even the intracellular processing of steroid-receptor complexes.

5. Effects of Indolic Compounds on Hypothalamic Gonadotrophin Control Mechanisms — Indirect Evidence

There is no longer any doubt that Mel has potent antigonadotrophic activity in hamsters[2] and rats[88] when administered at a critical period during a diurnal lighting cycle. A total collapse of the reproductive tract, reminiscent of that which occurs in light-deprived animals, results from daily subcutaneous injections of this indole a few hours prior to lights off. Additionally, significant reductions in gonadotrophin output also occur in Mel-treated animals. Therefore, it is reasonable to assume that Mel exerts its antigonadotrophic effects by acting on the endocrine hypothalamus, or perhaps at another site with efferent input to the hypothalamus. Recall that the demonstration of a direct pituitary site of action of Mel in any adult rodent species has not been forthcoming.[1,2] Furthermore, under certain conditions, Mel as well as a few of its cogeners, also exerts a progonadotrophic or counter-antigonadotrophic effect on reproduction.[2]

Other studies in the rat indirectly implicate the hypothalamus as the primary site of action of Mel and possibly other pineal indoles.[5] For example, the administration of Mel during the neonatal period, either alone or in combination with gonadal steroids, significantly retards the growth and function of the reproductive tract normally associated with the onset of puberty.[89,90] Taking into account the fact that the sexual differentiation of hypothalamic centers controlling gonadotrophin release is virtually complete by the tenth postnatal day,[55] it is not difficult to envisage Mel interfering with the normal development and functioning of hypothalamic-steroidal feedback mechanisms.[89] This may culminate in an alteration of the hormonostat mechanism controlling hypothalamic sensitivity to steroids at puberty as well as the pubertal surge of steroids.[52,89]

6. Effects of Indolic Compounds on Hypothalamic Gonadotrophin Control Mechanisms -- Direct Evidence

In the late 1960s and early 1970s Fraschini et al.[91] reported a series of experiments in which various pineal indoleamines were stereotaxically implanted into specific areas within the CNS. Either Mel or 5-Htol placed into the median eminence resulted in depressed pituitary stores of LH while Mel alone had the additional effect of suppressing plasma LH titers. Two other indoleamines, 5-Metol and 5-HT, had no effect on LH when implanted into the median eminence region. Median eminence implants of pineal fragments produced the same effects as Mel and 5-Htol on pituitary LH. Conversely, median eminence implants of either 5-Metol or 5-HT depressed pituitary FSH stores while 5-Htol and Mel left FSH unaltered. These workers concluded that whereas Mel and 5-Htol are inhibitors of LH synthesis and release, 5-Mtol and 5-HT inhibit FSH production and secretion by virtue of their action on specific receptors in the median eminence.

Testing the possibility that pineal hormones are delivered to the hypothalamus by way of the ventricular CSF, several workers found that intraventricularly injected Mel or 5-HT led to a significant diminution in LH and FSH levels.[34,35,91] Additionally, serum Prl titers were elevated in these animals as well.[35] In an infrequently cited study by Porter et al.[92] the intraventricular administration of 5-HT to intact rats stimulated LH release while neither Mel nor NAS had much of an effect on the levels of this hormone. In contrast to intact animals, castrated rats responded to the intraventricular injection of these indoles with a significant decrease in plasma LH. Interestingly, Porter et al. never offered an explanation for the discrepancy between these findings and those of their initial work[34,35] which are very commonly cited in the pineal literature.[1-5]

Despite these incongruities, the single most consistent observation made in studies of this type is that Mel, as well as some of its precursors and metabolites, given intraventricularly, stimulates the secretion of Prl.[35,91,93] Moreover, neither the intrapituitary nor portal vessel infusion of these substances affects gonadotrophin secretion, a fact frequently construed as indicating a hypothalamic site of action for these indoles.[34,35,91]

The first study to examine the effects of indoleamines directly on the bioassayable GnRH content of the hypothalamus was by Moskowska et al.[94] They documented that the injection of either Mel or 5-HT into the third ventricle of the immature mouse had no effect on hypothalamic GnRH activity. Although identical results were obtained when the indoles were placed into the lateral ventricle, they did observe, however, a significant decrease in hypothalamic GnRH activity if the animals had also received an additional injection of nialamide, a monoamine oxidase inhibitor. Since nialamide blocks the catabolism of 5-HT,[70] it was suggested that Mel and 5-HT suppress GnRH levels by further increasing the hypothalamic content of 5-HT or by acting synergistically with augmented levels of 5-HT. Whether the decrease in GnRH activity due to these indolic substances actually reflected a diminished synthesis and release of GnRH or alternately, an increase in its release in excess of synthesis, could not be determined from this study.

A possible answer to this question can perhaps be found in recent reports by two independent laboratories that Mel stimulates the release of immunoreactive GnRH from rat medial basal hypothalamic tissue in either a continuous flow perifusion system or primary culture.[95,96] Although these data are intriguing, their physiological significance is debatable, since rather pharmacological amounts of Mel (1 m*M*) were required to observe the effect. Hence, the apparent paradox between the ability of Mel to stimulate GnRH release and its capacity to inhibit the action of GnRH may not be easily resolved.

Of potential import is the ostensible existence of an inverse relationship between the diurnal variation in immunoreactive pineal Mel and hypothalamic GnRH in the rat; that is, when pineal Mel is high during the dark phase, hypothalamic GnRH activity is low, the reverse being true with respect to the light phase.[97] It is reasoned that there might be a reciprocal interaction between these two hormones in terms of their synthesis, or that the diurnal rhythms of these substances are under separate but direct photic control. However, these observations and their interpretation are obscured by the fact that exogenously administered Mel actually increases the number of immunoreactive GnRH neurons in the hypothalamus.[98]

That serious consideration should be given to the SCN as potential receptor sites for Mel is emphasized in recent studies in which bilateral lesions of the SCN in hamsters were shown to prevent the testicular regression associated with single daily afternoon injections of Mel. However, if Mel is administered three times per day to SCN-lesioned animals, gonadal atrophy ensues.[99] Interestingly, the authors concluded that the site of action of Mel is not on the SCN since gonadal regression is observed with thrice daily injections in lesioned animals. This conclusion, however, may be erroneous and unwarranted since the lesions did in fact block the action of Mel given once per day. It is my opinion that the SCN are potentially good candidates as sites of action for Mel since these structures are critical in regulating pineal metabolic processes.[76,77] Furthermore, Mel immunofluorescent staining has been observed in the cells of the SCN in the rat.[100]

Certainly, if Mel acts on the hypothalamus then it should be readily taken up by this structure. Apparently, this is in fact the case since the hypothalamus more selectively concentrates exogenously administered ^{3}H-Mel than any other brain region including the cerebral cortex.[101,102] Similarly, the presence of significant levels of endogenous Mel in the rat hypothalamus have been detected by gas chromatographic-mass spectrophotometric techniques.[103] Using a somewhat obscure histochemical technique, one group of investigators claim to have detected large granules of either Mel of 5-HT in presumptive tanycytes lining the optic and infundibular recesses of the third ventricle in humans.[104]

Several structural and biochemical features of the hypothalamus are known to be affected by Mel. The in vitro incubation of either hamster or rat hypothalamic tissue with Mel results in changes in the activity of Δ^4-reductase, an enzyme which converts testosterone to dihydrotestosterone. When incubated with the hamster hypothalamus, micromolar amounts of Mel inhibit the activity of this enzyme whereas nanomolar quantities of Mel significantly stimulate Δ^4-reductase activity in rat hypothalamic tissue.[105] A preliminary hypothesis was advanced that stimulation of this enzyme in the rat hypothalamus by physiological quantities of Mel could result in an increased dihydrotestosterone:testosterone ratio which would then cause an inhibition of gonadotrophin output. The effect of Mel on the activity of hypothalamic monoamine oxidase, has been measured in vitro. The activity of this enzyme was significantly diminished in the presence of Mel, suggesting that it, like Δ^4-reductase, may be a target within the hypothalamus for Mel.[106]

In a study by Cardinali and Freire,[107] Mel was found to interact with microtubular protein in the hypothalamic tissue. These workers detected a 44% reduction in microtubule protein in the arcuate-median eminence region and a smaller decrease in the rest of the hypothalamus. Inasmuch as microtubules may serve an important function in the axonal transport of neurosecretory products to the primary portal vasculature, Mel-induced alterations in microtubular protein may be a mechanism by which this indole modulates the release of gonadotrophin-releasing hormones. Moreover, Mel may also affect the synthesis of hypothalamic proteins as well.[107]

Almost to the exclusion of other brain amines, a substantial amount of evidence has accrued favoring the notion that Mel exerts its antigonadotrophic effects by altering the neurophysiology of hypothalamic 5-HT. For example, the administration of Mel to female rats results in a relatively acute increase in the concentration of 5-HT in the hypothalamus.[108] In related studies, Cardinali and associates[109] made another interesting observation that nanomolar concentrations of Mel, noncompetitively inhibit the uptake of 5-HT, noradrenaline (NA), dopamine (DA), and glutamate by synaptosome-rich homogenates of rat hypothalamus in vitro. In a followup experiment, they tested the effects of the in vivo administration of Mel over 4 days on the uptake of neurotransmitters by hypothalamic synaptosomes in vitro. While Mel effectively blocked the accumulation of 5-HT in hypothalamic nerve terminals, it was ineffective in preventing the uptake of either NA, DA, or glutamate suggesting that of all the hypothalamic neurotransmitters tested, 5-HT was the most susceptible to regulation by Mel. The noncompetitive nature of the inhibition of 5-HT uptake by Mel suggested that Mel interferes with the 5-HT uptake mechanism rather than by competing with 5-HT for its reuptake pump.[75]

Employing a similar in vitro protocol, Meyer et al.[110] examined the effect of several pineal indoles on the uptake of 5-HT by rat hypothalamic homogenates. Contravening the results of Cardinali et al., only 5-Htol inhibited the uptake of 5-HT while Mel had no effect. Nuances in experimental design may account for the differences between these studies with regard to the effects of Mel on 5-HT uptake. Nevertheless, these data do uphold previous interpretations that pineal indoles may be antigonadotrophic by virtue of their capacity to encumber the mechamisms responsible for the reuptake of 5-HT by axon terminals. In this way, more 5-HT could be made available in the synaptic cleft to act postsynaptically on gonadotrophin-releasing hormone neurons and inhibit their activity. Since Mel injections have been shown to alter the circadian rhythms in hypothalamic 5-HT levels in certain species,[74] it is tempting to speculate that it may do so by inducing alterations in 5-HT reuptake by hypothalamic nerve terminals.

7. Effects of Nonindolic Compounds on Hypothalamic Gonadotrophin Control Mechanisms — Indirect Evidence

Partially purified extracts of bovine pineal glands which are devoid of Mel not only inhibit basal LH secretion as well as the postcastration rise in serum LH in male rats but also basal Prl secretion in steroid-primed animals.[42,44] As alluded to previously, a partially purified ovine pineal fraction which in all probability contains a pteridine significantly reduces circulating levels of LH in prepubertal male rats; however, it is doubtful that the LH-inhibitory activity of the fraction is due to the presence of the pteridine.[45] Another presumptive polypeptide, gonadotrophin-inhibiting substance (GIS), presumably of pineal origin,[42] inhibits plasma LH levels and at the same time augments the secretion of Prl in immature female rats treated with PMS.[26] As postulated for the indoleamines, all of the in vivo effects of pineal fractions on gonadotrophins could theoretically be ascribed to an action at the hypothalamic level.

8. Effects of Nonindolic Compounds on Hypothalamic Gonadotrophin Control Mechanisms — Direct Evidence

A number of studies, primarily by European workers, support the supposition that several pineal fractions containing presumptive polypeptides, affect gonadotrophin secretion by a direct action on the hypothalamus. Moskowska et al.[94] were the first to demonstrate that a partially purified ovine pineal fraction designated as F_3, increases the bioassayable activity of GnRH in the hypothalamus both in vitro and when it is injected into the lateral ventricle of the mouse. They concluded that F_3 probably inhibits the release of GnRH, resulting in its accumulation in the hypothalamus. An alternate explanation might be that F_3 stimulates the synthesis of GnRH. As a sequel to these studies, the same group of workers[111] performed experiments in which fraction F_3 was subjected to further ultrafiltration. The two residues obtained from this procedure were referred to as UM-2R and UM-05R. Hypothalamic tissue from mice incubated in the presence of UM-2R (MW $>$ 1000) showed a significant decrease in the secretion of GnRH. Additionally, the hypothalami exhibited a reduced content of GnRH as well. In contrast, UM-05R (MW 500-1000) stimulated hypophysiotrophic activity as evidenced by the increased content of GnRH in both the incubation medium and hypothalamic tissue. Upon reexamination of the effects of the original F_3 fraction they observed that it curtailed the release of GnRH into the medium with a concomitant accumulation of GnRH activity in the hypothalamic tissue, thus confirming their original hypothesis. These data indicate that the F_3 fraction contains a variety of small molecular weight substances (peptides?) which can either inhibit the release, or the release and synthesis, of GnRH.

In addition to GnRH-inhibitory substances, there appears to be a class of molecules which stimulates both the synthesis and release of GnRH. According to Ebels,[112] the GnRH-stimulatory residue, UM-05R, contains a substance(s) with a pteridine-like fluorescence. In a further exploration of this problem, the method of Bensinger et al.[113] was employed to extract sheep pineals with three different organic solvents. The extracts were subsequently subjected to a number of ultrafiltration and purification steps yielding several ultrafiltrates and residues. Virtually without exception, the incubation of rat hypothalami with each of these fractions led to a stimulation of bioassayable GnRH release into the medium. Less consistent, however, was the fact that GnRH activity either increased, decreased, or remained the same in hypothalamic tissue incubated with different fractions.[114] These results indicate that the Bensinger extraction method may uncover several small molecular weight substances with progonadotrophic activity in that they appear to stimulate the synthesis and/or release of GnRH. Moreover, the GnRH-stimulatory or inhibitory effects of pineal extracts might be mediated via changes in the activity of hypothalamic 5-HT neurons, particularly in the ventromedial-arcuate area.[115]

AVT has been shown to possess potent antigonadotrophic activity when administered via either the blood vascular or ventricular system. The gonadotrophin-inhibiting activity of AVT administered by this latter route has led some investigators to postulate that this agent acts on the hypothalamus to inhibit gonadotrophin output.[27] A well executed study by Osland et al.[116] demonstrated that AVT injected into the third ventricle of unanesthetized rats during proestrus inhibits the preovulatory surge of LH and subsequent ovulation. Furthermore, AVT was shown to be ineffective in inhibiting the LH surge caused by the electrochemical stimulation of the medial preoptic area (MPOA) Since this procedure presumably stimulates the release of GnRH from the MBH, it was reasoned that the action of AVT is exerted at the MPOA or a higher neuronal level.

In a related study, Blask et al.[117] tested the effects of subcutaneous (s.c.) injections of AVT on the estrogen-induced LH surge in ovariectomized rats. They found that AVT effectively prevented or delayed the surge of LH which occurred in the late afternoon 2 days following estrogen treatment. Inasmuch as estrogen exerts its positive feedback effect on the MPOA to induce a preovulatory surge of LH,[118] Blask et al. concurred with the conclusions reached by Osland et al. and further suggested that AVT interferes with the positive feedback effects of estrogen on the MPOA or possibly at a higher CNS level. Interestingly, no effects of AVT have been observed on either the preovulatory or estrogen-induced surges of FSH suggesting the possibility that AVT has differential effects on separate hypothalamic releasing mechanisms regulating LH and FSH.[117,119]

As little as 10^{-4} pg of AVT, corresponding to about 6×10^4 molecules, injected into the third ventricle of urethane-anesthetized rats causes a significant decrease in plasma LH titers, no change in plasma FSH, and an increase in hypothalamic 5-HT. Pretreatment of the animals with parachlorophenylalanine (*p*-CPA), an inhibitor of 5-HT synthesis, blocks the inhibitory effect of AVT on LH release, whereas administration of 5-hydroxytryptophan (5-HTP), the precursor of 5-HT, to *p*-CPA-treated rats results in a partial restoration of the ability of AVT to inhibit LH.[120] In a subsequent study, Pavel et al.[121] reported that pinealectomy results in a decrease in hypothalamic 5-HT content and an increase in LH and FSH. As anticipated by these workers AVT completely obviated the effects of pinealectomy on these parameters.[121] Taken together, these results prompted Pavel et al. to emphatically conclude that the pineal, and thus AVT, inhibits GnRH release by increasing the concentration of 5-HT in the hypothalamus.

To date, the only study that has measured the effect of AVT directly on hypothalamic GnRH activity is by Vaughan et al.[122] They found that s.c. injections of AVT every 3 hr for 2 days had no effect on gonadotrophins in intact male rats. However, in acutely orchidectomized rats, this peptide significantly inhibited the postcastration rise in plasma LH and FSH. It is quite possible that AVT blocked the release of GnRH in castrated animals since hypothalamic levels of bioassayable GnRH activity were increased in AVT-treated castrates over castrate controls.

B. Extrahypothalamic Sites

1. Mesencephalon

For a number of reasons, the mesencephalic reticular formation represents another potentially important site of action for pineal colyones. This region not only houses the cell bodies of 5-HT neurons in the nucleus of the raphe, but also the perikarya of NA neurons located in the locus coeruleus. These neurons project, via the MFB to the hypothalamus, and terminate primarily on the SCN (in the case of 5-HT neurons), and the MBH (in the case of NA terminals); of course, some 5-HT neurons also terminate in the MBH while NA axons also project to the SCN.[74] Ostensibly, these nerve terminals make synaptic contacts with releasing and inhibiting hormone cells and regulate their metabolic activity. An additional feature of the mesencephalon is that the ventricular system, i.e., the cerebral aqueduct, serves as a potential route for pineal substances, should they be released into the CSF, to gain access to 5-HT and NA perikarya.

Because of the relationship between the MBH and the mesencephalic reticular formation, Fraschini et al.[91,123] implanted pineal indoles into the latter area in castrated male rats. Whereas mesencephalic implants of either Mel of 5-Htol appeared to specifically reduce pituitary reserves of LH, implants of either 5-Metol or 5-HT diminished only pituitary FSH stores. Control implants of these indoles in the cerebral cortex had no effect on pituitary gonadotrophins. They concluded that specific receptors for these indoles were not only located in the hypothalamus but in the midbrain region as well. They further postulated that the midbrain implants of pineal indoleamines exerted their differential effects on pituitary gonadotrophins by acting via serotoninergic pathways originating in this area. Perhaps, an equally plausible site of action would be the NA neurons in the locus coeruleus.

That pineal indoles can act on extrahypothalamic structure is supported by the fact that ^{3}H-Mel is avidly taken up by the midbrain.[101] In fact, Mel administered intraperitoneally is known for its ability to increase brainstem 5-HT levels, particularly in the midbrain.[108,124] Most evidence indicates that Mel has no effect on catecholamine levels in the brain, notwithstanding, one study reports that Mel significantly augments the whole brain content of catecholamines.[123] Studies such as these emphasize the great need for a comprehensive analysis of the effects of pineal products on the turnover of biogenic amines in discrete brain regions, particularly nuclear groups, since the technology to perform such studies is presently available.

Unfortunately, there exist large gaps in our knowledge concerning the potential of other pineal antigonadotrophic substances to act on extrahypothalamic loci to induce changes in reproductive physiology. However, one group[120] feels that the site of action of AVT within the brain is on the serotoninergic neurons of the nucleus of the raphé, albeit they have no proof of this.

2. Pineal Gland

Possibly the most exciting potential site of action for pineal hormones is the pineal gland itself.[23] Although this is not exactly a new concept,[126] pinealologists have been reluctant to grant it serious consideration until recently.

The first report that a pineal hormone could alter the function of the pineal gland was an outstanding study by Fiske and Huppert.[127] They measured the effects of Mel administered at different times during the lighting cycle on the diurnal changes in pineal 5-HT in the rat. They found that when Mel was given during the midlight period, the diurnal rhythm in 5-HT content was suppressed; however, if Mel was given just prior to lights off, in anticipation of the dark-induced rise in Mel, no changes in the diurnal rhythm of 5-HT were observed. These workers made the very prophetic statement that the effectiveness of Mel may depend on which part of the lighting cycle it is administered. In a similar study, twice daily injections of Mel for 2 weeks induced a stimulation in both pineal NAT and hydroxyindole-*O*-methyltransferase (HIOMT) activity. Unfortunately, the time of day the Mel injections were administered was not specified.[128]

Circumstantial evidence in the hamster points to a possible pineal site of action for Mel in inducing gonadal atrophy. Indeed, Reiter et al.[129] aptly demonstrated that the presence of the pineal gland or its innervation is required, in order for single daily afternoon injections of Mel to cause gonadal atrophy. Related to this issue is a recent study showing that like the pineal, the SCN must also be intact in order for single daily injections of Mel to exert an antigonadotrophic effect.[99] Therefore, Reiter et al. are strongly convinced of the possibility that the exogenous antigonadotrophic effects of Mel are exerted via an action on the pineal gland. Other investigators,[130] however, take issue with this contention inasmuch as they are able to induce gonadal atrophy in pinealectomized, ganglionectomized, or SCN-lesioned[99] hamsters by injecting threefold higher doses of Mel spread over the entire day. As a result, these workers argue that an intact pineal gland is not required for the antigonadotrophic effects of Mel to be manifest. The interpretation of their data is baffling, since much higher doses of Mel are required to induce, in pinealectomized or denervated animals, the same response that occurs in intact hamsters receiving only one injection. Furthermore, it is interesting to note that thrice daily injections of Mel for several weeks have absolutely no effect on the reproductive system in intact hamsters.[99]

Characteristic changes in pineal morphology in response to Mel administration lend further support to the idea that the pineal may be a Mel target. Ultrastructurally, the pinealocytes in Mel-treated rats or hamsters appear to contain an increased number of cellular organelles including cytoplasmic granules, ribosomes, and microtubules. Generally, the pinealocytes appear to be hyperactive as further evidenced by dilated Golgi complexes.[128,131] On the other hand, the ultrastructure of the pineals of dark-exposed hamsters receiving weekly Mel implants, closely resembles the hypoactive features of pineals from hamsters kept in long photoperiods,[132] indicating that perhaps the counter-antigonadotrophic effect of Mel implants in light-deprived hamsters is mediated by the pineal.[2]

If Mel does indeed act at the level of the pineal gland then what is the physiological significance of such an effect? Does this represent some type of ultrashort or autofeedback? Can Mel stimulate its own release and synthesis, or does it stimulate or inhibit the release of other pro- or antigonadotrophic factors from the pineal?

With regard to this last question, it has been proposed that Mel is a releasing factor for other pineal antigonadotrophic substances.[126] In fact, Pavel[133,134] has championed the concept that Mel is the releasing factor for AVT since either intraventricular or intracarotid injections of Mel induce the release of an AVT-like substance into the CSF and plasma of cats. Interestingly, Mel administered during the light period is ineffective in eliciting AVT release, whereas the same dose given during the dark phase results in a marked elevation in AVT in the CSF. This diurnal rhythm in sensitivity of the AVT-release mechanism to Mel is reminiscent of the gonadal sensitivity to Mel in hamsters.[1,2] The specificity as well as the physiological significance of this response is open to question since several synthetic releasing and inhibiting hormones also stimulate the release of AVT.[135]

The demonstration of Mel receptors in the pineal gland would lend further credibility to the concept that Mel plays a physiological role in regulating the gland from whence it originates. Even if Mel is also elaborated in extrapineal loci such as the retina, Harderian gland, or perhaps the GI tract,[1] then teleologically, the presence of Mel receptors in the pineal would make sense, provided of course that the pineal is a target site for extraepiphyseal Mel. Promising work in this area has revealed by fluorescence immunohistochemistry, the presence of a saturable binding material for Mel in the rat pineal gland. Furthermore, the degree of Mel binding apparently increases throughout the light phase and reaches a maximum shortly after lights off.[136]

For all practical purposes there is little evidence to suggest that the pineal gland is a target site for other non-Mel pineal substances. However, a recent report indicated that the basal release of Mel, 5-HT, and 5-HIAA from pineal glands in culture more than triples when the number of glands per incubation well is increased from two to six. This suggests that some secretory product from the pineal exerts a positive autofeedback effect on indole release. This secretory product is apparently not Mel itself since the addition of exogenous Mel to the culture medium failed to have any effect on indoleamine release.[137] That this secretory product might be taurine is suggested by a recent investigation by Wheler et al.[138] demonstrating that in vitro taurine treatment increases both the intraglandular concentration of ^{3}H-Mel and the amount of ^{3}H-Mel in the culture medium of pineal cultures containing the precursor, ^{3}H-tryptophan.

Taurine is an aminosulfonic acid found in the pineal gland in millimolar concentrations in pinealocytes. Furthermore, taurine stimulates pineal *N*-acetyltransferase (NAT) activity ostensibly via an interaction with a beta-adrenergic receptor. A preliminary hypothesis has been advanced proposing that taurine might modulate the transynaptic regulation of pineal function by interacting with either pre- or postsynaptic adrenergic binding sites.

V. PINEAL HORMONE-RECEPTOR INTERACTIONS

A. Melatonin Receptors

The recent demonstration of high affinity binding of Mel in membrane or cytosol preparations of bovine, rat, and hamster brain suggests the presence of Mel receptors in the mammalian CNS.[139,141] That these receptors are concentrated in the MBH is suggested by the fact that of the brain regions examined, Mel binding was maximal in the MBH. Interestingly, in both the rat and hamster, a diurnal variation in the number of Mel receptors in the brain has been demonstrated.[148] The greatest number of receptors is present at a time during the lighting cycle when a Mel injection is effective in inducing gonadal atrophy in both the hamster and rat.[2,88] Conversely, the Mel receptor number is low at a time when a Mel injection is ineffective in inhibiting reproductive physiology.

These data are potentially important in that they may explain why hamsters and rats are sensitive to the antigonadotrophic effects Mel only when it is administered after 6.5 hr into the light phase. As pointed in recent studies the resistance of the reproductive system to inhibition by Mel given early in the light phase implies that Mel receptors are either insensitive,[130] decreased in number,[140] or both. Mel secreted late during the dark phase may act on its receptors, subsequently rendering them transiently insensitive and/or decreased in number; that is, they become "down regulated." Therefore, early in the light phase when the receptors are desensitized they may be unable to respond to additional, exogenously administered Mel. By late afternoon, both the sensitivity and number of receptors might be reconstituted so as to ensure that Mel made available at this time will induce gonadal atrophy.[140,142]

To test this hypothesis, Chen et al.[142] injected hamsters with a large dose of Mel late in the morning to theoretically prolong the down regulated condition of Mel receptors. As the theory goes, if Mel receptors do in fact remain desensitized then daily injections of Mel in the late afternoon should be precluded from causing gonadal involution. As anticipated, the late morning injection of a large dose of Mel totally obviated the ability of single late afternoon injection of Mel to induce gonadal atrophy. Thus, an interaction between endogenous and exogenous Mel at the receptor level might indicate an important role of the pineal gland in synchronizing the responsiveness of target sites to Mel.[140]

B. Interaction of Pineal Hormones with Other Receptors

An increasingly common observation made in this era of modern neuroendocrinology is the fact that many neurohormones and neurotransmitters can interact with and exert biological effects upon receptors other than their own. Also it appears that neurotransmitters can interact with peptidergic receptors while neuropeptides can interact with neurotransmitter receptors.[143] Apparently pineal hormones are no exception to what may become a neuroendocrinological rule. For example, the direct Prl-stimulatory effect of AVT on the anterior pituitary may be mediated via an interaction with a beta-adrenergic receptor, since propranol, a beta blocker, can inhibit the effect of AVT on Prl secretion in vitro.[144]

There may be multiple interactions between pineal constituents and a variety of re-

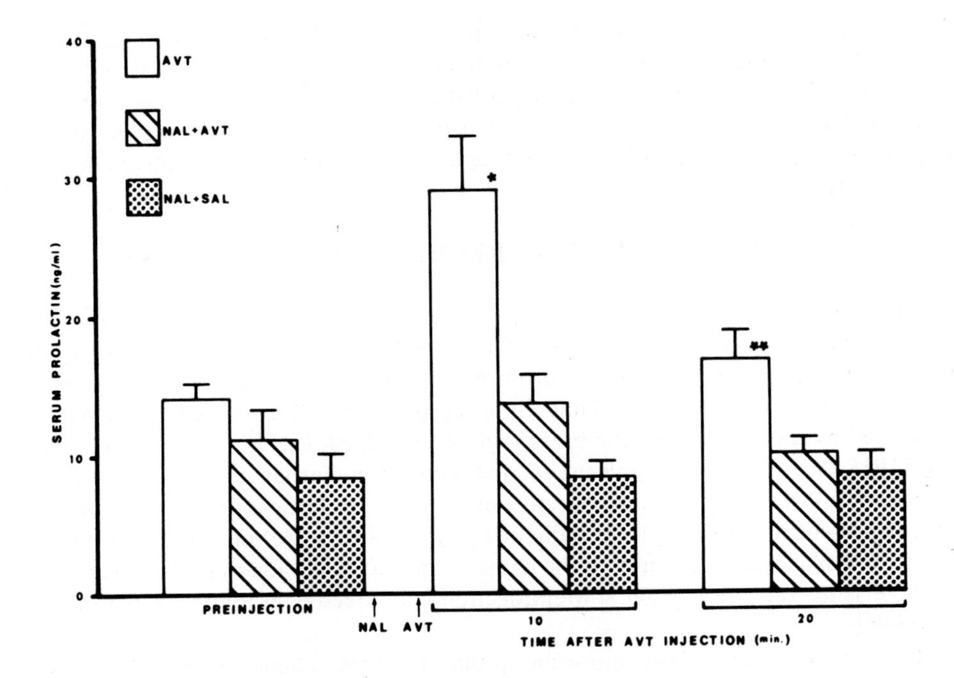

FIGURE 6. Effect of pretreatment with naloxone (NAL) (0.2 mg/kg) on arginine vasotocin (AVT)-induced prolactin release in urethane-anesthetized steroid-treated male rats. Control group received saline (SAL) in lieu of AVT. NAL was administered i.p. 5 minutes prior to AVT or SAL injection i.v. (From Blask, D. E. and Vaughan, M. K., *Neurosci. Lett.*, 18, 181, 1980. With permission.)

ceptors in the CNS. We have recently demonstrated that pretreatment of urethane-anesthetized rats with naloxone, an opiate-receptor antagonist, completely blocks the surge of serum Prl observed 10 minutes after the injection of AVT (Figure 6).[145] Furthermore, naloxone significantly curtails the ability of AVT to inhibit the postcastration rise in serum LH in acutely castrated male rats.[146] These data suggest that AVT may, at least in part, alter pituitary gonadotrophin secretion by an interaction with hypothalamic opiate receptors involved in the control of pituitary function.

In the case of Mel, a similar situation may exist wherein Mel interacts not only with its own receptors but with other receptors as well. An example of this is seen in Figure 7 whereby beeswax implants of pimozide, (dopamine receptor antagonist) completely block the Prl-inhibitory effect of Mel injections in anosmic male rats.[147] This finding indicates that the Prl-inhibitory effect of Mel may require that Mel interact postsynaptically with dopamine receptors located on either PIF neurosecretory cells, dopaminergic neurons, or both.

VI. CONCLUSIONS

From the foregoing discussion, it is clear that virtually no site within the neuroendocrine-reproductive hierachy eludes the influence of pineal hormones (Figure 8). The hormonal envoys secreted from this tiny gland may be capable of diverse and multiple interactions with gonadal steroids, gonadotrophins, hypothalamic hormones, biogenic amines, and both monoaminergic and peptidergic receptors. Though pineal hormones appear to act primarily upon CNS structures, particularly the endocrine hypothalamus, they may also modulate the endocrine activity and responsiveness of secondary effector sites outside of the CNS. Likewise, within the CNS, pineal constituents may not only

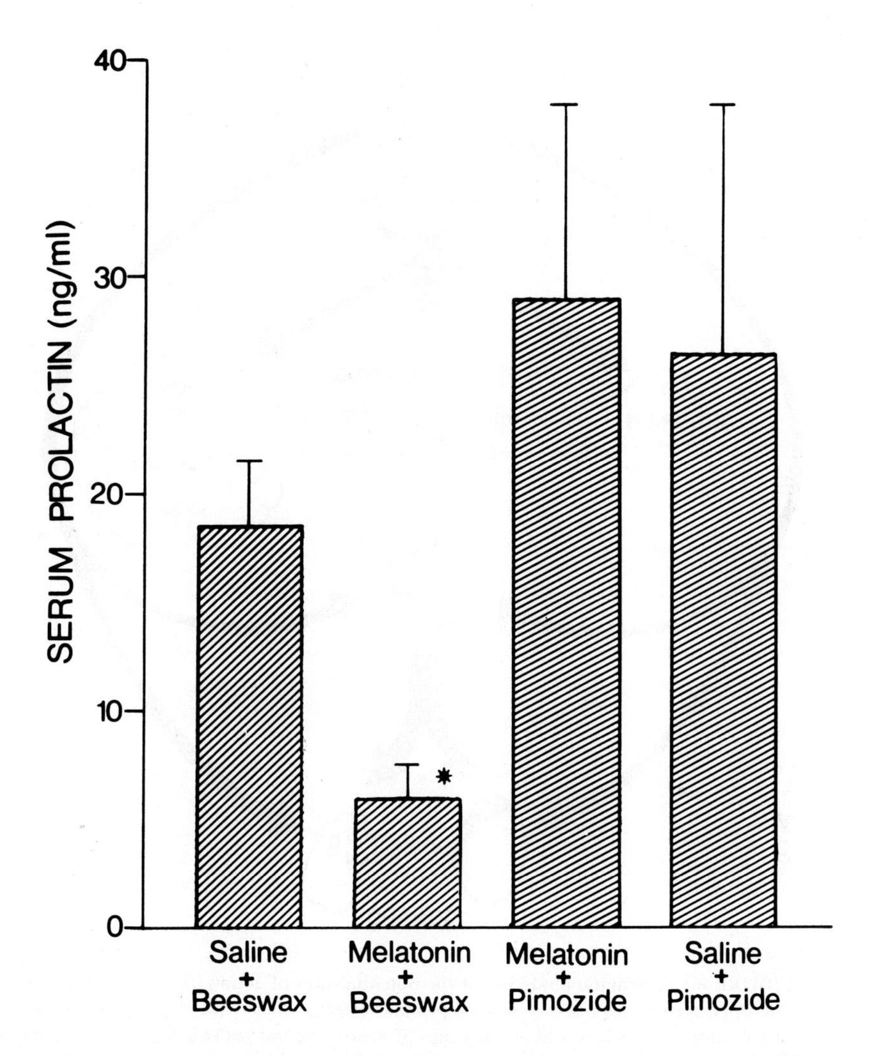

FIGURE 7. Serum prolactin levels in 62-day-old anosmic male rats treated with daily afternoon (1700 to 1800 hr) injections of melatonin (50 μg) and weekly s.c. implants of empty beeswax pellets or pellets containing pimozide (1.2 mg) for 5 weeks. (From Blask, D. E., Nodelman, J. L., and Leadem, C. A., *Experentia,* 36, 1008, 1980. With permission.)

serve a hormonal function but a neuromodulatory role as well by effecting subtle alterations in the neural mechanisms governing gonadotrophin secretion. Through the secretion of a gamut of hormones, the pineal could potentially convey endocrine inhibitory, stimulatory, and modulatory information to diverse loci within the neuroendocrine-reproductive axis. In this way, the pineal gland would be uniquely endowed with a measure of plasticity and versatility in its control of photoperiodically-linked reproductive phenomena.

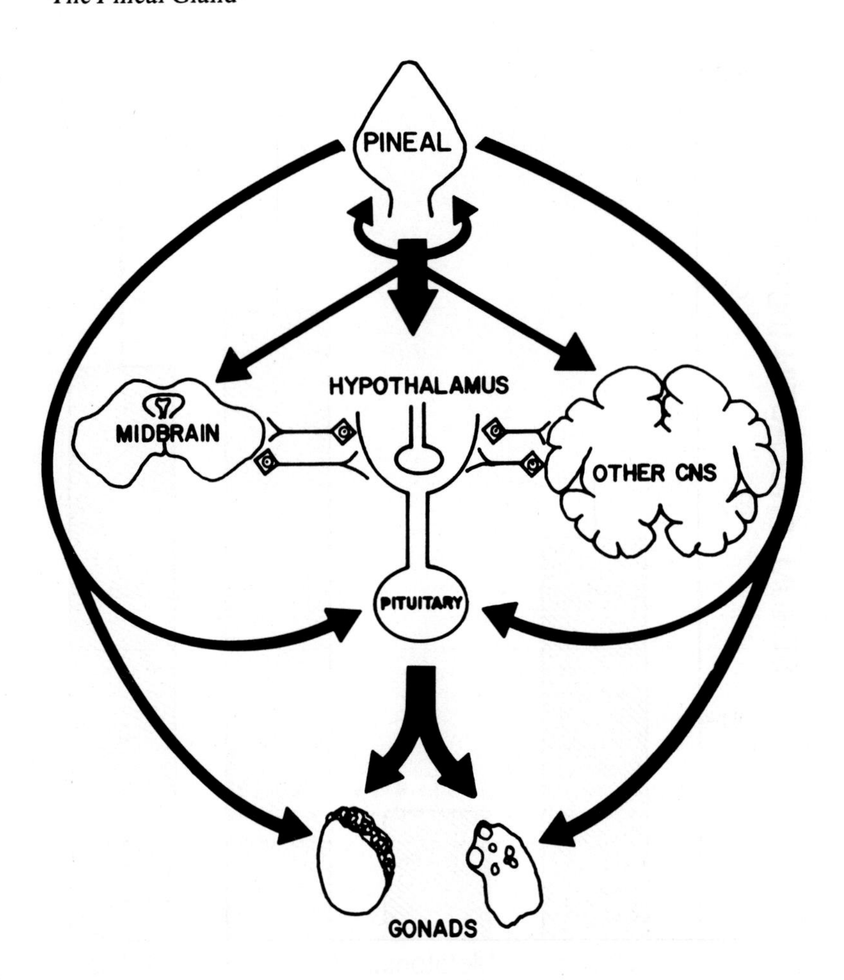

FIGURE 8. Schematic rendering of the potential sites of action of pineal hormones. The thick arrows are directed toward primary effector sites while thin arrows are directed toward secondary and tertiary sites of action. Neural pathways are indicated by ◈—<.

REFERENCES

1. **Reiter, R. J.,** *The Pineal,* Vol. 3, Eden Press, Montreal, 1978, chap. 1.
2. **Reiter, R. J.,** *The Pineal — 1977,* Vol. 2, Eden Press, Montreal, 1977, chap. 9.
3. **Relkin, R.,** *The Pineal,* Vol. 1, Eden Press, Montreal, 1976, chap. 7.
4. **Reiter, R. J.,** Pineal regulation of hypothalamicopituitary axis: gonadotrophins, in *Handbook of Physiology — Endocrinology,* IV, Part 2, Knobil, E. and Sawyer, C., Eds., American Physiological Society, Washington, D.C., 1974, chap. 41.
5. **Reiter, R. J., Vaughan, M. K., Vaughan, G. M., Sorrentino, S., Jr., and Donofrio, R. J.,** The pineal gland as an organ of internal secretion, in *Frontiers of Pineal Physiology,* Altschule, M. D., Ed., MIT Press, Cambridge, 1975, chap. 5.
6. **Simonnet, H. and Thiéblot, L.,** Action de l'épiphyse sur differéntes glandes endocrines, *J. Physiol. Pathol. Gen.,* 42, 726, 1950.
7. **Bindoni, M. and Stanzani, S.,** Effetti dell'asportazione contemporanea della pineale e dell'ipofisi sulla multiplicazione cellulare in alcune organe di ratto, *Boll. Soc. Ital. Biol. Sper.,* 47, 583, 1971.
8. **Hipkin, L. J.,** Effect of 5-methoxytryptophol and melatonin on uterine weight responses to human chorionic gonadotrophin, *J. Endocrinol.,* 48, 287, 1970.
9. **Debeljuk, L., Vilchez, J. A., Schnitman, M. A., Paulucci, O. A., and Feder, V. M.,** Further evidence for a peripheral action of melatonin, *Endocrinology,* 89, 1117, 1971.

10. **Trentini, G. P., Botticelli, A. R., Sannicola, Botticelli, C., and Barbanti Silva, C.,** Decreased ovarian LH incorporation after Mel treatment, *Horm. Metab. Res.*, 8, 234, 1976.
11. **Debeljuk, L., Feder, V. M., and Paulucci, O. A.,** Effects of melatonin on changes induced by castration and testosterone in sexual accessory structures of male rats, *Endocrinology*, 87, 1358, 1970.
12. **Alonso, R., Prieto, L., Hernandez, C., and Mas, M.,** Antiandrogenic effects of the pineal gland and melatonin in castrated and intact prepubertal male rats, *J. Endocrinol.*, 79, 77, 1978.
13. **Fishback, J. B., MacPhee, A. A., and King, J. C.,** Effect of melatonin on reproductive organs and serum gonadotrophins in intact and castrated male hamsters, *Abstr. 60th Ann. Mtg. Endocrine Soc.*, Endocrine Society, Bethesda, Md., 123, 136, 1978.
14. **Peat, F. and Kinson, G. A.,** Testicular steroidogenesis *in vitro* in the rat in response to blinding, pinealectomy and to the addition of melatonin, *Steroids*, 17, 251, 1971.
15. **Ellis, L. C.,** Inhibition of rat testicular androgen synthesis *in vitro* by melatonin and serotonin, *Endocrinology*, 90, 17, 1972.
16. **Cardinali, D. P. and Rosner, J. M.,** Effects of melatonin, serotonin and *N*-acetylserotonin on the production of steroids by duck testicular homogenates, *Steroids*, 18, 25, 1971.
17. **Ellis, L. C. and Buhrley, L. E., Jr.,** Inhibitory effects of melatonin, prostaglandin E_1, cyclic AMP, dibutyryl cyclic AMP and theophylline on rat seminiferous tubular contractility *in vitro, Biol. Reprod.*, 19, 217, 1978.
18. **Kano, T. and Miyachi, Y.,** Direct action of melatonin on testosterone and cyclic GMP production using rat testis tissue *in vitro, Biochem. Biophys. Res. Commun.*, 72, 969, 1976.
19. **Balestreri, R., Foppiani, E., Jacopino, G. E., and Giordano, G.,** Ghiandola pineale, melatonina e steroidogenesi testicolare, *Arch. Maragliano Patol. Clin.*, 25, 119, 1969.
20. **MacPhee, A. A., Cole, F. E., and Rice, B. F.,** The effect of melatonin on steroidogenesis by the human ovary *in vitro, J. Clin. Endocr. Metab.*, 40, 688, 1975.
21. **Young Lai, E. V.,** *In vitro* effects of melatonin on HCG stimulation of steroid accumulation by rabbit ovarian follicles, *J. Steroid Biochem.*, 10, 714, 1979.
22. **Kopin, I. J., Pare, C. M. B., Axelrod, J., and Weissbach, H.,** The fate of melatonin in animals, *J. Biol. Chem.*, 236, 3072, 1979.
23. **Wurtman, R. S., Axelrod, J., and Potter, L. T.,** The uptake of ^{3}H-melatonin in endocrine and nervous tissues and the effects of constant light exposure, *J. Pharmocol. Exp. Ther.*, 143, 314, 1964.
24. **Cohen, M., Roselle, D., Chabner, B., Schmidt, T. J., and Lippman, M.,** Evidence for a cytoplasmic melatonin receptor, *Nature (London)*, 274, 894, 1978.
25. **Norris, J. T.,** The ovary as a potential target organ for pineal principles, *Anat. Rec.*, 169, 389, 1971.
26. **Johnson, L. Y.,** The Effects of Pineal-Related Substances on Preovulatory Hormone and Reproductive Organ Changes in the PMS-treated Immature Rat, Ph. D. thesis, The University of Texas Graduate School of Biomedical Sciences, San Antonio, 1978.
27. **Vaughan, M. K. and Blask, D. E.,** Arginine vasotocin — a search for its function in mammals, in *The Pineal and Reproduction, Progress in Reproductive Biology*, Vol. 4, Reiter, R. J., Ed., S. Karger, Basel, 1978, 90.
28. **Moszkowska, A. and Ebels, I.,** A study of the antigonadotropic action of synthetic arginine vasotocin, *Experientia*, 24, 610, 1968.
29. **Johnson, L. Y., Vaughan, M. K., Reiter, R. S., Blask, D. E., and Rudeen, P. K.,** The effects of arginine vasotocin on pregnant mare's serum-induced ovulation in the immature rat, *Acta Endocrinol.*, 87, 367, 1978.
30. **Vaughan, M. K., Buchanan, J., Blask, D. E., Reiter, R. S., and Sheridan, P. J.,** Diurnal variation in uterine estrogen receptors in immature female rats — inhibition by arginine vasotocin, *Endocr. Res. Commun.*, 6, 191, 1979.
31. **Munsick, R. A., Sawyer, W. H., and van Dyke, H. B.,** Avian neurohypophysial hormones: pharmacological properties and tentative identification, *Endocrinology*, 66, 860, 1960.
32. **Rzasa, J.,** Effects of arginine vasotocin and prostaglandin E_1 on the hen uterus, *Prostaglandins*, 16, 357, 1978.
33. **Fraschini, F., Mess, B., Piva, F., and Martini, L.,** Brain receptors sensitive to indole compounds: function in control of luteinizing hormone secretion, *Science*, 159, 1104, 1968.
34. **Kamberi, I. A., Mical, R. S., and Porter, J. C.,** Effect of anterior pituitary perfusion and intraventricular injection of catecholamines and indoleamines on LH release, *Endocrinology*, 87, 1, 1970.
35. **Kamberi, I. A., Mical, R. S., and Porter, J. C.,** Effects of melatonin and serotonin on the release of FSH and prolactin, *Endocrinology*, 88, 1288, 1971.
36. **Blask, D. E., Vaughan, M. K., Reiter, R. J., Johnson, L. Y., and Vaughan, G. M.,** Prolactin-releasing and release-inhibiting factor activities in the bovine, rat, and human pineal gland: *in vitro* and *in vivo* studies, *Endocrinology*, 99, 152, 1976.
37. **Martin, J. E., Engel, J. N., and Klein, D. C.,** Inhibition of the *in vitro* pituitary response to luteinizing hormone-releasing hormone by melatonin, serotonin and 5-methoxytryptamine, *Endocrinology*, 100, 675, 1977.

38. **Padmanabhan, V., Convey, E. M., and Tucker, E. A.,** Pineal compounds alter prolactin release from bovine pituitary cells, *Proc. Soc. Exp. Biol. Med.,* 160, 340, 1979.
39. **Vaughan, M. K., Blask, D. E., Johnson, L. Y., and Reiter, R. J.,** Prolactin-releasing activity of arginine vasotocin *in vitro, Hormone Res.,* 6, 342, 1975.
40. **Shiino, M., Ishikawa, H., and Rennels, E. G.,** Four subtypes of prolactin producing clonal strains derived from Rathke's pouch epithelium, *Abstr. 59th Ann. Mtg. Endocrine Soc.,* Endocrine Society, Bethesda, Md., 1977, 123.
41. **Demoulin, A., Hudson, B., Franchimont, P., and Legros, J. J.,** Arginine vasotocin does not affect gonadtrophin secretion *in vitro, J. Endocrinol.,* 72, 105, 1977.
42. **Ebels, I. and Benson, B.,** A survey of the evidence that unidentified pineal substances affect the reproductive system in mammals, in *The Pineal and Reproduction, Progress in Reproductive Biology,* Vol. 4, Reiter, R. J., Ed., S. Karger, Basel, 1978, 51.
43. **White, W. F., Hedlund, M. T., Weber, G. F., Rippel, R. H., Johnson, E. S., and Wilber, J. F.,** The pineal gland: a supplemental source of hypothalamic-releasing hormones, *Endocrinology,* 94, 1422, 1974.
44. **Chang, N., Ebels, I., and Benson, B.,** Preliminary characterization of bovine pineal prolactin releasing (PPRF) and release-inhibiting factor (PPIF) activity, *J. Neural Transm.,* 46, 139, 1979.
45. **Ebels, I., de Morée, A., Hus-Citharel, A., and Moszkowska, A.,** A survey of some active sheep pineal fractions and a discussion on the possible significance of pteridines in those fractions in *in vitro* and *in vivo* assays, *J. Neural Transm.,* 44, 97, 1979.
46. **Reiter, R. J. and Johnson, L. Y.,** Elevated pituitary LH and depressed pituitary prolactin levels in female hamsters with pineal-induced gonadal atrophy and the effects of chronic treatment with synthetic LRF, *Neuroendocrinology,* 14, 310, 1974.
47. **Turek, F. W., Alvis, J. D., and Menaker, M.,** Pituitary responsiveness to LRF in castrated male hamsters exposed to different photoperiodic conditions, *Neuroendocrinology,* 24, 140, 1977.
48. **Martin, J. E., Kirk, K. L., and Klein, D. C.,** Effects of 6-hydroxy-6-fluoro-, and 4,6-difluoromelatonin on the *in vitro* pituitary response to luteinizing hormone-releasing hormone, *Endocrinology,* 106, 398, 1980.
49. **Moguilevsky, J. A., Sarchi, P., Deis, R., and Silles, N. O.,** Effect of melatonin on the luteinizing hormone release induced by clomiphene and luteinizing hormone-releasing hormone, *Proc. Soc. Exp. Biol. Med.,* 151, 663, 1976.
50. **Yamashita, K., Mieno, M., Shimizu, T., and Yamashita, E.,** Inhibition by melatonin of the pituitary response to luteinizing hormone releasing hormone *in vivo, J. Endocrinol.,* 76, 487, 1978.
51. **Mieno, M., Yamashita, E., Iimori, M., and Yamashita, K.,** An inhibitory effect of melatonin on the luteinizing hormone releasing activity of luteinizing hormone releasing hormone in immature dogs, *J. Endocrinol.,* 78, 283, 1978.
52. **Ramaley, J. A.,** Development of gonadotrophin regulation in the prepubertal mammal, *Biol. Reprod.,* 20, 1, 1979.
53. **Nordlund, J. J. and Lerner, A. B.,** The effects of oral melatonin on skin color and on the release of pituitary hormones, *J. Clin. Endocrinol. Metab.,* 45, 768, 1977.
54. **Fedeleff, H., Aparicio, N. J., Guitelman, A., Debeljuk, L., Mariani, A., and Cramer, C.,** Effect of melatonin on the basal and stimulated gonadotropin levels in normal men and postmenopausal women, *J. Clin. Endocrinol. Metab.,* 42, 1014, 1976.
55. **Weinberg, V., Weitzman, E. D., Fukushima, D. K., and Rosenfield, R. S.,** Effect of melatonin on the pituitary response to luteinizing hormone releasing hormone (LHRH) in men, *Abstr. 61st Ann. Mtg. Endocrine Soc.,* Endocrine Society, Bethesda, Md., 1979, 241.
56. **Bruot, B. C., Sartin, J. L., and Orts, R. J.,** The effect of arginine vasotocin on the luteinizing hormone releasing hormone stimulated luteinizing hormone release *in vivo* and *in vitro, Physiologist,* 21, 14, 1978.
57. **Vaughan, M. K., Trakulrungsi, C., Petterborg, L. S., Johnson, L. Y., Blask, D. E., Trakulrungsi, W., and Reiter, R. J.,** Interaction of luteinizing hormone-releasing hormone, cyproterone acetate and arginine vasotocin on plasma levels of luteinizing hormone in intact and castrated adult male rats, *Mol. Cell. Endocrinol.,* 14, 59, 1979.
58. **Yamashita, K., Mieno, M., and Yamashita, E.,** Suppression of the luteinizing hormone releasing effect of luteinizing releasing hormone by arginine vasotocin, *J. Endocrinol.,* 81, 103, 1979.
59. **Cheesman, D. W., Osland, R. B., and Forsham, P. H.,** Suppression of the preovulatory surge of luteinizing hormone and subsequent ovulation in the rat by arginine vasotocin, *Endocrinology,* 101, 1194, 1977.
60. **Demoulin, A. and Franchimont, P.,** Influence of sheep pineal extracts on secretion *in vitro* of gonadotrophins and prolactin, *J. Neural Transm.,* Suppl. 13, 360, 1978.
61. **Scott, P. M.,** No effect of arginine vasotocin on LRF-stimulated LH release *in vitro, Anat. Rec.,* 193, 679, 1979.

62. **Blask, D. E. and Reiter, R. J.,** The pineal gland of the blind-anosmic female rat: its influence on medial basal hypothalamic LRH, PIF and/or PRF activity *in vivo, Neuroendocrinology,* 17, 362, 1975.
63. **Blask, D. E., Reiter, R. J., and Johnson, L. Y.,** The influence of pineal activation, removal or denervation on hypothalamic luteinizing hormone-releasing activity: an *in vitro* study, *Neurosci. Lett.,* 1, 327, 1975.
64. **Blask, D. E. and Reiter, R. J.,** Pineal removal or denervation: effects on hypothalamic PRF activity in the rat, *Mol. Cell. Endocrinol.,* 11, 243, 1978.
65. **DeVries, R. A. C. and Kappers, J. A.,** Influence of the pineal gland on neurosecretory activity of the supraoptic hypothalamic nucleus in the male rat, *Neuroendocrinology,* 8, 359, 1971.
66. **DeVries, R. A. C.,** Abolition of the effect of pinealectomy on hypothalamic magnocellular neurosecretory activity in male rats by hypothalamic pineal inplants, *Neuroendocrinology,* 9, 358, 1972.
67. **Knigge, K. M. and Sheridan, M. N.,** Pineal function in hamsters bearing melatonin antibodies, *Life Sci.,* 19, 1235, 1976.
68. **Pickard, G.,** Changes in hypothalamic luteinizing hormone-releasing hormone (LH-RH) in the male golden hamster in response to shortened photoperiods, *Soc. for Neurosci.,* 3, 354, 1977.
69. **Blask, D. E., Reiter, R. J., Vaughan, M. K., and Johnson, L. Y.,** Differential effects of the pineal gland on LHRH and FSH-RH activity in the medial basal hypothalamus of the male golden hamster, *Neuroendocrinology,* 23, 36, 1979.
70. **Kordon, C. and Glowinski, J.,** Role of hypothalamic monoaminergic neurons in the gonadotrophin release-regulating mechanisms, *Neuropharmacology,* 11, 153, 1972.
71. **Moszkowska, A., Kordon, C., and Ebels, I.,** Biochemical fractions and mechanisms involved in the pineal modulation of pituitary gonadotrophin release, in *The Pineal Gland,* Wolstenholme, G. E. W. and Knight, J., Eds., Churchhill Livingstone, London 1971, 241.
72. **Sugden, D. and Morris, R. D.,** Changes in regional brain levels of tryptophan, 5-hydroxytryptamine, 5-hydroxyindoleacetic acid, dopamine and noradrenaline after pinealectomy in the rat, *J. Neurochem.,* 32, 1593, 1979.
73. **Urry, R. L. and Ellis, L. C.,** Monoamine oxidase activity of the hypothalamus and pituitary: alterations after pinealectomy, changes in photoperiod, or additions of melatonin *in vitro, Experientia,* 31, 891, 1972.
74. **Yates, C. A. and Herbert, J.,** Differential circadian rhythms in pineal and hypothalamic 5-HT induced by artificial photoperiods or melatonin, *Nature (London),* 262, 219, 1976.
75. **Cardinali, D. P.,** Changes in hypothalamic neurotransmitter uptake following pinealectomy, superior cervical ganglionectomy or melatonin administration to rats, *Neuroendocrinology,* 19, 91, 1975.
76. **Moore, R. Y.,** Neural control of pineal function in mammals and birds, *J. Neural Transm.,* Suppl. 13, 47, 1978.
77. **Rusak, B. and Morin, L. P.,** Testicular responses to photoperiod are blocked by lesions of the suprachiasmatic nuclei in golden hamsters, *Biol. Reprod.,* 15, 336, 1976.
78. **Stetson, M. H. and Watson-Whitmyre, M.,** Nucleus suprachiasmaticus: the biological clock in the hamster?, *Science,* 191, 197, 1976.
79. **Mess, B., Heizer, A., Tóth, A., and Tima, L.,** Luteinization induced by pinealectomy in the polyfollicular ovaries of rats bearing anterior hypothalamic lesions, in *The Pineal Gland,* Wolstenholme, G. E. W. and Knight, J., Eds., Churchhill Livingstone, London, 1971, 229.
80. **Trentini, G. P., Gaetani, C. F., Martini, L., and Mess, B.,** Effect of pinealectomy and of bilateral cervical ganglionectomy on serum LH levels in constant estrous-anovulatory rats, *Proc. Soc. Exp. Biol. Med.,* 153, 490, 1976.
81. **Schally, A. V., Kastin, A. J., and Arimura, A.,** FSH-releasing hormone and LH-releasing hormone, *Vitam. Horm. (N.Y.),* 30, 83, 1972.
82. **Tamarkin, L., Hutchison, J. S., and Goldman, B. D.,** Regulation of serum gonadotrophins by photoperiod and testicular hormone in the Syrian hamster, *Endocrinology,* 99, 1528, 1976.
83. **Turek, F. W.,** The interaction of the photoperiod and testosterone in regulating serum gonadotrophin levels in castrated male hamsters, *Endocrinology,* 101, 1210, 1977.
84. **Turek, F.,** Role of the pineal gland in photoperiod-induced changes in hypothalamic-pituitary sensitivity to testosterone feedback in castrated male hamsters, *Endocrinology,* 104, 636, 1979.
85. **Gorski, R. A.,** Localization and sexual differentiation of the nervous structures which regulate ovulation, *J. Reprod. Fertil.,* Suppl. 1, 67, 1966.
86. **Kordon, C., and Hoffman, J.,** Mise en évidence d'en effet fortement gonadostimulant de la lumierè chez le rat male prétraité une injection postnatale de testosterone, *C.R. Seances Soc. Biol. Paris,* 161, 1262, 1967.
87. **Gogan, F., Moszkowska, A., Rotsztejn, W. and Slama-Scemama, A.,** Action de l'epiphysectomie sur l'incorporation hypothalamique et hypophysaire d'oestradiol-17-β chez la ratte, *C. R. Acad. Sci. Paris,* 275, 2391, 1972.

88. **Blask, D. E.,** Nutritional status, time of day and pinealectomy: factors influencing the sensitivity of the neuroendocrine-reproductive axis of the rat to melatonin, *Anat. Rec.,* Abstr. 20A, 1980.
89. **Vaughan, M. K.,** Sexual Differentiation of the Rat Hypothalamus: Interaction of Pineal Amines and Steroids, Ph.D. thesis, The University of Texas Medical Branch, Galveston, 1970.
90. **Sorrentino, S., Jr., Reiter, R. J., and Schalch, D. W.,** Hypotrophic reproductive organs and normal growth in male rats treated with melatonin, *J. Endocrinol.,* 51, 213, 1971.
91. **Fraschini, F., Collu, R., and Martini, L.,** Mechanisms of inhibitory action of pineal principles on gonadotropin secretion, in *The Pineal Gland,* Wolstenholme, G. E. W. and Knight, J., Eds., Churchill Livingstone, London, 1971, 259.
92. **Porter, J. C., Mical, R. S., and Cramer, O. M.,** Effect of serotonin and other indoles on the release of LH, FSH and prolactin, *Gynecol. Invest.,* 2, 13, 1972.
93. **Iwasaki, Y., Kato, Y., Ohgo, S., Abe, H., Imura, H., Hiruta, F., Senoh, S., Tokuyama, T., and Hayaishi, D.,** Effects of indoleamines and their newly identified and metabolites on prolactin release in rats, *Endocrinology,* 103, 254, 1978.
94. **Moszkowska, A., Slama-Scemama, A., Lombard, M. N., and Hery, M.,** Experimental modulation of hypothalamic content of the gonadotropic releasing factors by pineal factors in the rat, *J. Neural Transm.,* 34, 11, 1973.
95. **Kao, L. W. L. and Weis, J.,** Release of gonadotrophin-releasing hormone (Gn-RH) from isolated, perifused medial-basal hypothalamus by melatonin, *Endocrinology,* 100, 1723, 1977.
96. **Hollander, C. S., Prasad, R., Richardson, S., Hirooka, Y., and Suzuki, S.,** Melatonin modulates hormonal release from organ cultures of rat hypothalamus, *J. Neural Transm.,* Suppl. 13, 369, 1978.
97. **Brown, G. M., Pang, S. F., Friend, W., Seggie, J., and Grota, L. J.,** Inverse relationship of hypothalamic LHRH to pineal melatonin and *N*-acetylserotonin, *J. Neural Transm.,* Suppl. 13, 356, 1978.
98. **Leonardelli, J., Tramu, G., and Hermand, E.,** Mélantonine et cellules à gonadolibérine (LH-RH) de l'hypothalamus du rat, *C.R. Seances Soc. Biol. Paris,* 172, 481, 1978.
99. **Bittman, E. L., Goldman, B. D., and Zucker, I.,** Testicular responses to melatonin are altered by lesions of the suprachiasmatic nuclei in golden hamsters, *Biol. Reprod.,* 21, 647, 1979.
100. **Bubenik, G. A., Brown, G. M., and Grota, L. S.,** Differential localization of *N*-acetylated indolealkylamines in CNS and the Harderian gland using immunohistology, *Brain Res.,* 118, 417, 1976.
101. **Anton-Tay, F. and Wurtman, R. J.,** Regional uptake of ^{3}H-melatonin from blood or cerebrospinal fluid by rat brain, *Nature (London),* 221, 474, 1969.
102. **Cardinali, D. P., Hyyppa, M. T., and Wurtman, R. J.,** Fate of intracisternally injected melatonin in the rat brain, *Neuroendocrinology,* 12, 30, 1973.
103. **Green, A. R., Koslow, S. H., and Costa, E.,** Identification and quantitation of a new indolealkylamine in rat hypothalamus, *Brain Res.,* 51, 371, 1973.
104. **Symington, R. B., Marks, S. M., and Ryan, P. M.,** Secretory cells of the human hypothalamus, *S. Afr. Med. J.,* 46, 1484, 1972.
105. **Frehn, J. L., Urry, R. L., and Ellis, L. C.,** Effect of melatonin and short photoperiod on Δ^4-reductase activity in liver and hypothalamus of the hamster and rat, *J. Endocrinol.,* 60, 507, 1974.
106. **Urry, R. L. and Ellis, L. C.,** Monoamine oxidase activity of the hypothalamus and pituitary: alterations after pinealectomy, changes in photoperiod, or additions of melatonin *in vitro, Experientia,* 31, 891, 1975.
107. **Cardinali, D. P. and Freire, F.,** Melatonin effects on brain. Interaction with microtubule protein, inhibition of fast axoplasmic flow and induction of crystaloid and tubular formation in the hypothalamus, *Mol. Cell. Endocrinol.,* 2, 317, 1975.
108. **Anton-Tay, F., Chov, C., Anton, S., and Wurtman, R. J.,** Brain serotonin concentration: elevation following intraperitoneal administration of melatonin, *Science,* 162, 277, 1968.
109. **Cardinali, D. P., Nagle, C. A., Freire, F., and Rosner, J. M.,** Effects of melatonin on neurotransmitter uptake and release by synaptosome-rich homogenates of the rat hypothalamus, *Neuroendocrinology,* 18, 72, 1975.
110. **Meyer, D. C., Quay, W. B., and Ma, Y.-H.,** Comparative inhibition of hypothalamic uptake of 5-hydroxytryptamine and norepinephrine by 5-hydroxy- and 5-methoxyindole derivatives, *Gen. Pharmacol.,* 6, 285, 1975.
111. **Citharel, A., Ebels, I., L'Héritier, and Moszkowska, A.,** Epiphyseal-hypothalamic interaction. An *in vitro* study with some sheep pineal fractions, *Experientia,* 29, 718, 1973.
112. **Ebels, I.,** A chemical study of some biologically active pineal fractions, *Prog. Brain Res.,* 52, 309, 1979.
113. **Bensinger, R., Vaughan, M., and Klein, D. C.,** Isolation of a non-melatonin lipophilic antigonadotrophic factor from the bovine pineal gland, *Fed. Proc. Fed. Am. Soc. Exp. Biol.,* 32, 225, 1973.
114. **Slama-Scemama, A., L'Hértier, A., Moszkowska, A., van der Horst, C. J. G., Noteborn, H. P. J. M., de Morée, A., and Ebels, I.,** Effects of sheep pineal fractions on the activity of male rat hypothalami *in vitro, J. Neural Transm.,* 46, 47, 1979.

115. **Smith, A. R. and Kappers, J. A.**, Effect of pinealectomy, gonadectomy, *p*-CPA and pineal extracts on the rat parvocellular neurosecretory hypothalamic systems; a fluorescence histochemical investigation, *Brain Res.*, 86, 353, 1975.
116. **Osland, K. B., Cheesman, D. W., and Forsham, P. H.**, Studies on the mechanism of the suppression of the preovulatory surge of luteinizing hormone in the rat by arginine vasotocin, *Endocrinology*, 101, 1203, 1977.
117. **Blask, D. E., Vaughan, M. K., Reiter, R. J., and Johnson, L. Y.**, Influence of arginine vasotocin on the estrogen-induced surge of LH and FSH in adult ovariectomized rats, *Life Sci.*, 23, 1035, 1978.
118. **Goodman, R. L.**, The site of the positive feedback action of estradiol in the rat, *Endocrinology*, 102, 151, 1978.
119. **Cheesman, D. W., Osland, R. B., and Forshan, P. H.**, Effects of 8-arginine vasotocin on plasma prolactin and follicle-stimulating hormone surges in the proestrous rat, *Proc. Soc. Exp. Biol. Med.*, 156, 369, 1977.
120. **Pavel, S., Luca, N., Calb, M., and Goldstein, R.**, Inhibition of release of luteinizing hormone in the male rat by extremely small amounts of arginine vasotocin: further evidence for the involvement of 5-hydroxytryptamine-containing neurons in the mechanism of action of arginine vasotocin, *Endocrinology*, 104, 517, 1979.
121. **Pavel, S., Luca, N., Calb, M., and Goldstein, R.**, Reversal by arginine vasotocin of the effects of pinealectomy on the amount of 5-hydroxytryptamine in the hypothalamus and the concentration of luteinizing hormone and follicle-stimulating hormone in the plasma of immature male rats, *J. Endocrinol.*, 84, 159, 1980.
122. **Vaughan, M. K., Blask, D. E., Johnson, L. Y., and Reiter, R. J.**, The effect of subcutaneous injections of melatonin, arginine vasotocin, and related peptides on pituitary and plasma levels of luteinizing hormone, follicle-stimulating hormone, and prolactin in castrated adult male rats, *Endocrinology*, 104, 212, 1979.
123. **Fraschini, F.**, The pineal gland and the control of LH and FSH secretion, in *Progress in Endocrinology*, Gual, C., Ed., Excerpta Medica, Amsterdam, 1969, 637.
124. **Cotzias, G. C., Tang, L. C., Miller, S. I., and Ginos, J. Z.**, Melatonin and abnormal movements induced by L-dopa in mice, *Science*, 173, 450, 1971.
125. **Wendel, O. T., Waterbury, L. D., and Pearce, L. A.**, Increase in monoamine concentration in rat brain following melatonin administration, *Experientia*, 30, 1167, 1974.
126. **Quay, W. B., Indoleamines, in** *Pineal Chemistry In Cellular and Physiological Mechanisms*, Charles C Thomas, Springfield, Ill., 1974, chap. 7.
127. **Fiske, V. M. and Huppert, L. C.**, Melatonin action on pineal varies with photoperiod, *Science*, 162, 279, 1968.
128. **Freire, F. and Cardinali, D. P.**, Effects of melatonin treatment and environmental lighting on the ultrastructural appearance, melatonin synthesis, norepinephrine turnover and microtubule protein content of the rat pineal gland, *J. Neural Transm.*, 37, 237, 1975.
129. **Reiter, R. J., Blask, D. E., Johnson, L. Y., Rudeen, P. K., Vaughan, M. K., and Waring, P. J.**, Melatonin inhibition of reproduction in the male hamster: its dependency on time of day of administration and on an intact and sympathetically innervated pineal gland, *Neuroendocrinology*, 22, 107, 1976.
130. **Goldman, B. D., Hall, V., Hollister, C., Roychoudhary, P., Tamarkin, L., and Westrom, W.**, Effects of melatonin on the reproductive system in intact and pinealectomized male hamsters maintained under various photoperiods, *Endocrinology*, 104, 82, 1979.
131. **El-Domeiri, A. A. and Das Gupta, T. K.**, The influence of pineal ablation and administration of melatonin on growth and spread of hamster tumors, *J. Surg. Oncol.*, 8, 197, 1976.
132. **Barrat, G. F., Nadakavukaren, M. J., and Frehn, J. L.**, Effect of melatonin implants on gonadal weights and pineal gland fine structure of the golden hamster, *Tissue Cell*, 9, 335, 1977.
133. **Pavel, S.**, Arginine vasotocin release into cerebrospinal fluid of cats induced by melatonin, *Nature (London)*, 246, 183, 1973.
134. **Pavel, S. and Goldstein, R.**, Further evidence that melatonin represents the releasing hormone for pineal vasotocin, *J. Endocrinol.*, 82, 1, 1979.
135. **Goldstein, R. and Pavel, S.**, Vasotocin release into the cerebrospinal fluid of cats induced by luteinizing hormone releasing hormone, thyrotrophin releasing hormone and growth hormone release-inhibiting hormone, *J. Endocrinol.*, 75, 175, 1977.
136. **Holloway, W. R., Grota, L. J., and Brown, G. M.**, Immunocytochemical identification of melatonin binding material in the pineal gland of the rat, *Soc. Neurosci.*, 5(1515), 484, 1979.
137. **Miele, B. D.**, *In vitro* studies on ultra-short feedback control of pineal indole secretion: lack of effect of melatonin, *Anat. Rec.*, 193, 622, 1979.
138. **Wheler, G. H. T., Weller, J. L., and Klein, D. C.**, Taurine stimulation of pineal *N*-acetyltransferase activity and melatontin production via beta-adrenergic mechanism, *Brain Res.*, 166, 65, 1979.

139. **Cardinali, D. P., Vacas, M. I., and Boyer, E. E.,** Specific binding of melatonin in bovine brain, *Endocrinology,* 105, 437, 1979.
140. **Vacas, M. I. and Cardinali, D. P.,** Diurnal changes in melatonin binding sites of hamster and rat brains, correlation with neuroendocrine responsiveness to melatonin, *Neurosci. Lett.,* 15, 259, 1979.
141. **Nilea, L. P., Wong, Y.-W., Mishra, R. K., and Brown, G. M.,** Melatonin receptors in brain, *Eur. J. Pharmacol.,* 55, 219, 1979.
142. **Chen, H. J., Brainard, G. C., III, and Reiter, R. J.,** Melatonin given in the morning prevents the suppressive action on the reproductive system of melatonin given in late afternoon, *Neuroendocrinology,* 31, 129, 1980.
143. **Renaud, L. P., Blume, H. W., and Pittman, Q. S.,** Neurophysiology and neuropharmacology of the hypothalamic tuberoinfundibular system, in *Frontiers in Neuroendocrinology,* Vol. 5, Ganong, W. F. and Martini, L., Eds., Raven Press, New York, 1978, chap. 5.
144. **Blask, D. E., Vaughan, M. K., and Reiter, R. J.,** Modification by adrenergic blocking agents of arginine vasotocin-induced prolactin secretion *in vitro, Neurosci. Lett.,* 6, 91, 1977.
145. **Blask, D. E. and Vaughan, M. K.,** Naloxone inhibits arginine vasotocin-(AVT) induced prolactin release in urethane-anesthetized male rats *in vivo, Neurosci. Lett.,* 18, 181, 1980.
146. **Vaughan, M. K., Johnson, L. Y., Blask, D. E., and Reiter, R. J.,** Reversal by naloxone of the arginine vasotocin (AVT)-induced depression of plasma luteinizing hormone in castrated adult male rats, *Abstr. 6th Int. Cong. Endocrine Soc.,* Endocrine Society, Bethesda, Md., 277, 348, 1980.
147. **Blask, D. E., Nodelman, J. L., and Leadem, C. A.,** Preliminary evidence that a dopamine receptor antagonist blocks the prolactin-inhibitory effects of melatonin in anosmic male rats, *Experentia,* 36, 1008, 1980.

INDEX

A

B

C

D

E

F

G

I

J

K

L

M

R

S

T

U

V

W

X

Z